Phenomenology, Logic, and the Philosophy of Mathematics

This book is a collection of fifteen essays that deal with issues at the intersection of phenomenology, logic, and the philosophy of mathematics. The first of the three parts, "Reason, Science, and Mathematics," contains a general essay on Husserl's conception of science and logic, an essay on mathematics and transcendental phenomenology, and an essay on phenomenology and modern pure geometry. Part II is focused on Kurt Gödel's interest in phenomenology. It explores Gödel's ideas and also some work of Quine, Penelope Maddy, and Roger Penrose. Part III deals with elementary, constructive areas of mathematics – areas of mathematics that are closer to their origins in simple cognitive activities and in everyday experience. This part of the book contains essays on intuitionism, Hermann Weyl, the notion of constructive proof, Poincaré, and Frege.

Richard Tieszen is Professor of Philosophy at San Jose State University.

To my parents,
James D. and Beverly J. Tieszen

Phenomenology, Logic, and the Philosophy of Mathematics

RICHARD TIESZEN

San Jose State University

CAMBRIDGE
UNIVERSITY PRESS

CAMBRIDGE UNIVERSITY PRESS
Cambridge, New York, Melbourne, Madrid, Cape Town, Singapore, São Paulo, Delhi

Cambridge University Press
32 Avenue of the Americas, New York, NY 10013-2473, USA

www.cambridge.org
Information on this title: www.cambridge.org/9780521119986

First published 2005
This digitally printed version 2009

A catalog record for this publication is available from the British Library

Library of Congress Cataloging in Publication data
Tieszen, Richard L.
Phenomenology, logic, and the philosophy of mathematics / Richard Tieszen.
p. cm.
Includes bibliographical references and index.
ISBN 0-521-83782-0
1. Mathematics – Philosophy. 2. Phenomenology. 3. Logic, Symbolic and
mathematical. 4. Constructive mathematics. 5. Intuitionistic mathematics. I. Title.
QA8.4.T534 2005
510′.1 – dc22

2004062888

ISBN 978-0-521-83782-8 hardback
ISBN 978-0-521-11998-6 paperback

Contents

Acknowledgments

The essays collected here were written over a period of fifteen years. In preparing them for publication in this volume I have modified them in a few places, mostly for clarity and for continuity with other chapters in the collection. I have also cut some material from a few of the essays. Some overlap or repetition remains here and there, but the trade-off is that such overlapping allows the essays to be read independently of one another. In any case, I think that a little repetition is not onerous. It may even be helpful to some readers. The Bibliography has been standardized, and in the case of Husserl's work in particular I have provided references to the original publications, the relevant *Husserliana* editions, and the English translations. Since Husserl's works are usually composed in short sections I have adopted the convention in the essays of referring to quotations from Husserl's works by providing the title of the work and the section number. This method puts the reader in touch with the relevant texts but allows for choice in consulting the different editions and languages. In the case of Gödel's writings I have followed the citation style used in *Kurt Gödel: Collected Works*. The Bibliography for the present volume includes very few works that were not cited in the original essays. I have included a few new references where there has been some clear line of development of an argument or point in the essays.

Chapter 1 was written for the volume *Continental Philosophy of Science*, edited by Gary Gutting (Oxford: Blackwell, 2004). It appears here with the permission of Blackwell Publishing.

Chapter 2 originally appeared under the title "Mathematics" in *The Cambridge Companion to Husserl*, B. Smith and D. Smith (eds.) (Cambridge:

Cambridge University Press, 1995), pp. 438–462. It is reprinted here with the permission of Cambridge University Press.

Chapter 3, "Free Variation and the Intuition of Geometric Essences: Some Reflections on Phenomenology and Modern Geometry," will appear in *Philosophy and Phenomenological Research*. It is reprinted here with the permission of the editors.

Chapter 4 appeared as "Kurt Gödel and Phenomenology," *Philosophy of Science* 59, 2 (1992), pp. 176–194. It is reprinted here with the permission of the Philosophy of Science Association.

Chapter 5 was originally published as "Gödel's Philosophical Remarks on Logic and Mathematics: Critical Notice of *Kurt Gödel: Collected Works*, Vols. I, II, III," *Mind* 107 (1998), pp. 219–232. It is reprinted by permission of Oxford University Press.

Chapter 6 appeared as "Gödel's Path from the Incompleteness Theorems (1931) to Phenomenology (1961)," *Bulletin of Symbolic Logic* 4, 2 (1998), pp. 181–203, © copyright 1998 Association for Symbolic Logic. It is reprinted here with permission of the Association for Symbolic Logic.

Chapter 7 appeared as "Gödel and the Intuition of Concepts," *Synthese* 133, 3 (2002), pp. 363–391, © copyright 2002 Kluwer Academic Publishers. It appears here with kind permission of Kluwer Academic Publishers.

Chapter 8 appeared as "Gödel and Quine on Meaning and Mathematics," in *Between Logic and Intuition: Essays in Honor of Charles Parsons*, R. Tieszen and G. Sher (eds.) (Cambridge University Press, 2000), pp. 232–254. It is reprinted with the permission of Cambridge University Press.

Chapter 9 was originally published as "Review of *Mathematical Realism*, by Penelope Maddy," *Philosophia Mathematica* 3, 2 (1994), pp. 69–81. It is reprinted here with the permission of the editor, Robert S. D. Thomas.

Chapter 10 was originally published as "Review of *Shadows of the Mind: A Search for the Missing Science of Consciousness*, by Roger Penrose," *Philosophia Mathematica* 4, 3 (1996), pp. 281–290. It appears here with the permission of the editor, Robert S. D. Thomas.

Chapter 11 appeared as "Intuitionism, Meaning Theory and Cognition," *History and Philosophy of Logic* 21, 3 (2001), pp. 179–194. It is reprinted here with the permission of Taylor & Francis Limited (http://www.tandf.co.uk).

Chapter 12 was published as "The Philosophical Background of Weyl's Mathematical Constructivism," *Philosophia Mathematica* 3, 8 (2000), pp. 274–301. It is reprinted with the permission of the editor, Robert S. D. Thomas.

Chapter 13 was originally published under the title "What Is a Proof?" in *Proof, Logic and Formalization*, M. Detlefsen (ed.) (London: Routledge, 1992), pp. 57–76. It appears here with the permission of Routledge.

Chapter 14 was originally published under the title "Logicism, Impredicativity, Formalism," in *Henri Poincaré: Science and Philosophy*, J. L. Greffe, G. Heinzmann, and K. Lorenz (eds.) (Berlin: Akademie Verlag and Paris: Albert Blanchard, 1996), pp. 399–415. It is reprinted with the permission of Gerhard Heinzmann and Akademie Verlag.

Chapter 15 appeared as "The Philosophy of Arithmetic: Frege and Husserl," in *Mind, Meaning and Mathematics*, L. Haaparanta (ed.) (Dordrecht: Kluwer Academic Publishers, 1994), pp. 85–112, © copyright 1994 Kluwer Academic Publishers. It is reprinted here with kind permission of Kluwer Academic Publishers.

Over the years I have discussed ideas about phenomenology, logic, and mathematics with many friends and colleagues. I have included the original acknowledgments in the chapters themselves, but there have been many other people who also deserve to be thanked for discussion, comments, and suggestions.

My interest in the relationship of phenomenology to the exact sciences goes back to my days in graduate school and college. During this period I benefited most from interactions with Charles Parsons, Wilfried Sieg, Howard Stein, Shaughan Levine, and Isaac Levi at Columbia University; Robert Tragesser at Barnard; and J. N. Mohanty and Izchak Miller at the Graduate Faculty of the New School for Social Research. Soon after I arrived at Columbia, Charles Parsons gave a seminar on Husserl's *Logical Investigations*. This allowed me to deepen the study of Husserl's works that I had already begun at the New School and in college. At the New School, J. N. Mohanty gave a year-long seminar on Husserl's logical works. It was this kind of in-depth seminar that made the Graduate Faculty unique. As an undergraduate I benefited from studies of phenomenology with Robert Welsh Jordan and of modal logic and other systems of logic with Fred Johnson. It was during my time at Columbia that I met and began the first of many discussions with Hao Wang about Gödel's philosophical views. Wang was one of the people I knew who most encouraged thinking about mathematics from a phenomenological standpoint. I also met Dagfinn Føllesdal while I was a graduate student, and although he was never formally my teacher I have been discussing Husserl's work with him since that time. Several other people were especially helpful and encouraging early on: Michael Resnik, Gian-Carlo Rota, Bill Tait, and Bill McKenna come to mind. I am indebted to Mike Resnik in particular.

As I began to focus even more on constructive mathematics I benefited from discussions with Dirk van Dalen, Per Martin-Löf, Anne Troelstra, Dag Prawitz, and Göran Sundholm. Van Dalen has been especially helpful. After I arrived in California I began to attend many of the logic events at Stanford. These were often organized by Sol Feferman, and I have profited from many exchanges with him over the years.

In addition to all of those mentioned, I would like to thank many other people with whom I have discussed my work: Mark van Atten, Michael Friedman, David Smith, Barry Smith, Ed Zalta, John Corcoran, Albert Visser, Karl Schuhmann, Robin Rollinger, Paolo Mancosu, Thomas Ryckman, Karl Ameriks, Guglielmo Tamburrini, Ernan McMullin, Gary Gutting, Charles Chihara, Jerrold Katz, Paul Cortois, Grisha Mints, Dieter Lohmar, Claire Hill, Jairo da Silva, Peter Hadreas, and Kai Hauser. (It is possible that as I write this I have not remembered everyone who deserves to be mentioned.)

Finally, I thank my wife, Nancy, for her patience and her meditative ways. *Nam Mô A Di Đà Phật.* I dedicate this book to my parents, James D. and Beverly J. Tieszen.

Introduction

Themes and Issues

This volume is a collection of fifteen of my thematically related papers on phenomenology, logic, and the philosophy of mathematics. All of the essays are concerned with the interpretation, analysis, and development of ideas in Husserlian phenomenology in connection with recent and historical issues in the philosophy of mathematics and philosophy of logic.

Many of the interesting questions about phenomenology and the exact sciences that engaged such thinkers as Frege, Carnap, Schlick, and Weyl with the early phenomenologists have unfortunately been neglected in more recent times. One could speculate on why this has happened. On the one hand, it no doubt resulted from the development of certain trends in what has since been called 'analytic' philosophy. On the other hand, it resulted from the particular trajectory of Continental philosophy after Husserl. Husserl's emphasis on science and the analysis of essence gave way almost immediately on the Continent to various philosophies of human existence under the influence of Husserl's student Heidegger. In addition, there were long delays in the English translation of many of Husserl's writings, and to complicate matters further, philosophy curricula at many universities came to be organized around the division between analytic and Continental philosophy.

In my view, this general division between the analytic and Continental traditions has not always been good for philosophy. Least of all should it be maintained in the case of philosophers such as Brentano, Husserl, and some others in the early phases of the phenomenological movement, for here there is still direct engagement with major figures in Anglo-American philosophy. Many of the essays in this book are concerned with issues and ideas that are common to both the Continental and the analytic

1

traditions of philosophical thinking about logic and mathematics. I hope
the book can be seen as adding to the growing literature that encourages
communication between the traditions and puts some of these artificial
divisions behind us.[1] The methods and ideas of analytic and Continental
philosophy, where they are in fact different, can inform and enrich each
other in many ways. There are of course significant disagreements on
some of the issues, but I expect that by examining them we will only
reach a deeper understanding.

§ 1

In order to appreciate the approach I take to phenomenology in this
book it is necessary to realize that Husserl's own thinking about logic and
mathematics went through several transformations. It will be useful to
situate my work in a general way with respect to three main stages that
can be discerned in Husserl's writings on these subjects. Roughly speaking, there is the early work of the *Philosophy of Arithmetic (PA)* and related
writings (1891–1900), the middle period of the *Logical Investigations (LI)*
(1900–1907), and the later period starting with the *Ideas Pertaining to a
Pure Phenomenology and to a Phenomenological Philosophy (Ideas I)* (1907–
1938). (For much more detail see Tieszen 2004.) Some of the central
changes that divide these periods from one another and that are relevant
to my work in this volume are as follows: the *Philosophy of Arithmetic* consists of descriptive psychological and 'logical' investigations of arithmetic.
Husserl's ontology at this point includes physical entities and processes
and mental entities and processes. The 'ideal' or abstract objects that are
clearly part of Husserl's ontology from 1900 onward are not to be found
(in any obvious way) in *PA*. Husserl analyzed the origins of the natural
number concept in *PA* by focusing on mental processes of abstraction,
collection, and so on. As a result, some of the critics of the book, most
notably Frege, thought they detected a form of psychologism in *PA*. Psychologism is the view that mathematics and logic are concerned with
mental entities and processes, and that these sciences are in some sense
branches of empirical psychology. Psychologism was a popular form of
naturalism about logic and mathematics in the late nineteenth century.

---

[1] See, for example, Michael Dummett's *Origins of Analytic Philosophy*, Michael Friedman's *A
Parting of the Ways: Carnap, Cassirer, and Heidegger* and *Reconsidering Logical Positivism*, and
many of the writings of Dagfinn Føllesdal and J. N. Mohanty.

Another important point about *PA* is that the view of 'intuitive' or 'authentic' knowledge in arithmetic in the book is very limited. Almost all of our arithmetical knowledge is held to be 'symbolic' and in Part II of *PA* Husserl investigates this symbolic or 'logical' component of arithmetic. In this second half of *PA* Husserl develops a kind of formalism about arithmetic and some other parts of mathematics. It appears that parts of mathematics in which we cannot have 'authentic presentations' of objects are to be understood in terms of the inauthentic, merely symbolic representation of the objects. It is held that we cannot have authentic presentations of most of the natural numbers and that the same is true, for example, for the objects of classical real analysis and the objects of n-dimensional geometries with n > 3. In these cases, all we can do is to try to show, for example, that the formal systems that we take to be about such objects are consistent. Aspects of this kind of formalism are retained in the later writings but against a wider background that includes ideal meanings, essences, a retooled notion of intuition, and the like.

In the *LI* Husserl introduces ideal objects into his ontology and this in fact underwrites his own extended critique of psychologism in the "Prolegomena to Pure Logic." Husserl now recognizes physical entities and processes, psychical entities and processes, and ideal objects of different kinds (e.g., meanings, universals, natural numbers, sets). There are many claims in the *LI* about intentionality and the commitment of logic to objects such as ideal meanings and universals. It seems to me that much of what Husserl says about logic in the book would be best modeled in various higher-order intensional logics. (None of the presently existing systems of higher-order intensional logic, however, seems quite right.) The conception of intuition in the *LI* also changes in several ways. Now there is intuition not only of 'real' objects (physical or mental entities) but also of ideal objects. There is, in other words, ordinary sensory intuition or perception but also 'categorial' intuition of ideal objects. In relation to *PA*, the account of intuition is refined and extended in several ways. Intuition is now explicitly characterized in terms of the 'fulfillment' of intentions. The intentionality of logical and mathematical cognition comes to the fore in the *LI*. Husserl now holds that there can be static or dynamic fulfillment of intentions in either sensory intuition or categorial intuition. In dynamic fulfillment the intuition of an object is developed through sequences of acts in time. This raises the question of what could possibly be intuited through such sequences of acts, either in sensory or in categorial intuition. One can ask, for example, about the extent of the possibilities of dynamic intuition concerning particular mathematical objects such as natural numbers.

By the time of *Ideas I* there are other important changes. Phenomenology is now characterized as a transcendental idealism and is said to be an eidetic science, albeit a 'material' (as opposed to 'formal') eidetic science. The language of essences is now prominent and Husserl spells out some of the differences between material and formal eidetic sciences. We are still supposed to investigate mental activities of abstraction, collection, reflection, and the like, in connection with logic and mathematics, but this is now presented as an eidetic, descriptive, and epistemological (not psychological) undertaking. It is part of a transcendental philosophy. Although Husserl continues to recognize ideal objects such as meanings, essences, and natural numbers, along with ideal truths about these objects, the seemingly more robust realism about these objects and truths in the *LI* undergoes a shift. It is now taken up into Husserl's phenomenological idealism. The exact nature of this transition is still a subject of much discussion and debate among Husserl scholars. Husserl, however, now clearly speaks about how the 'ideality' or 'irreality' of such objects as meanings and essences is itself constituted, nonarbitrarily, in consciousness. It is not that ideal meanings or essences are mental objects. Indeed, they are constituted in consciousness as nonmental and as transcending consciousness. A logical truth such as the principle of noncontradiction, for example, is constituted as being true even if there were no conscious constituting beings. It is as if one takes a form of realism or platonism about logic and mathematics as a whole and places it inside a transcendental framework that is now meant to do justice to sensory experience and mathematical experience.

Husserl's mature views on formal and regional ontologies are also now in place. There is still an important place for purely formal logic and formal mathematics in Husserl's philosophy of the exact sciences and Husserl continues to speak about what he calls definite, formal axiom systems and their ontological correlates, definite manifolds. Roughly speaking, a 'definite' formal axiom system appears to be a consistent and complete axiom system, and a definite manifold is the system of formal objects, relations, and so on, to which a definite axiom system refers.

In connection with Husserl's views on formal systems, one can distinguish the meaning-intentions or ideal meanings that can be associated with signs in a formal system from the mere 'games meaning' that one can attach to signs in formal systems solely on the basis of manipulating sign configurations according to sets of rules (see *LI*, Investigation I,

§ 20). It is this 'games meaning' that takes center stage in strict formalist views of mathematics. In order to illustrate the distinction, let us create a miniature formal system right now. Suppose this formal system has only two rules of inference:

Rule 1: If there are sign configurations of the form $\Phi \sim \Psi$ and Φ then derive the sign configuration Ψ.

Rule 2: If there is a sign configuration of the form $\Phi \oplus \Psi$ then derive the sign configuration Φ.

Suppose the alphabet of the formal system consists of the symbols P . . . Z, P' . . . Z', P'' . . . Z'', the constants $\sim$, $\oplus$, and the signs (,), and that we have an appropriate definition of well-formed expressions.

Query: If we have the sign configurations $P \sim (Q \oplus R)$ and $P \oplus S$, then is the sign configuration Q derivable? Yes. Apply Rule 2 to $P \oplus S$ to obtain P. Apply Rule 1 to P and $P \sim (Q \oplus R)$ to obtain $Q \oplus R$. An application of Rule 2 to $Q \oplus R$ will give us Q as output.

We could try many other queries and play with this little formal system for a while. What, however, is this formal system *about?* Who knows, I just concocted it. It need not be about anything. In order to derive Q from the sign configurations that are given it is not necessary or even useful to know the meaning of the expression, if any, for which Q has been chosen. It is as though we are playing by the rules of a particular game and we need not be concerned with what the signs are about.

One can create indefinitely many formal systems at will. In the space of all possible formal systems, however, we find that some formal systems are correlated with existing parts of mathematics, are meaningful to us, interesting, or useful. If mathematics consists only of formal systems, then how could we single out those systems that actually correlate with existing parts of mathematics, that are meaningful, interesting, or useful?

Consider another example. Suppose I construct an axiomatic formal theory in the language of first-order logic with identity and function symbols. One can use standard rules of inference that are associated with such a theory. The theory has three axioms:

Axiom 1: $(\forall x)(\forall y)(\forall z)\, g(x, g(y, z)) = g(g(x, y), z)$
Axiom 2: $(\forall x)\, g(x, a) = x$
Axiom 3: $(\forall x)\, g(x, fx) = a$

Now would a sign configuration such as

$$(\forall x)(\forall y)(\forall z)(g(x, z) = g(y, z) \rightarrow x = y)$$

be derivable from these axioms on the basis of the standard rules of inference for quantification theory with '=' and function symbols? It turns out that it can be derived. One could again play with this system. What, however, is the point? What does it mean?

Suppose we interpret the system as follows: let the domain consist of the integers, 'g' be + (so that $g(x, y)$ is just $x + y$), 'f' be negation (i.e., $\neg$), and 'a' name 0. It is now possible to see this theory as an axiomatization of group theory, and under this particular interpretation we are looking at the additive group of the integers. In deriving

$$(\forall x)\, (\forall y)\, (\forall z)\, (g(x, z) = g(y, z) \rightarrow x = y)$$

we are thus deriving a theorem of group theory. Notice how much our thinking about the formal system changes once we see it as an axiomatization of *group theory*. There is a change in the directedness of consciousness. Instead of being directed toward the signs and how they can be manipulated according to rules, we are, against this interpretive background, now directed in our thinking toward a rich mathematical theory that happens to have many applications.

On Husserl's view, it is by virtue of ideal meanings that we are referred to or directed toward certain kinds of objects or states of affairs. In mathematics and logic, in particular, we can be directed toward certain kinds of objects or states of affairs. In our mathematical thinking we may be directed, for example, to sets, natural numbers, groups, or spaces, where these are not to be understood as just more concrete signs. There is a kind of meaningfulness and directedness toward objects in many parts of mathematics that goes beyond the mere manipulation of sign configurations on the basis of rules. The signs are often interpreted in particular ways, and we are thereby directed in our thinking in a manner that would be absent short of such interpretation. I take this to be an important theme in the latter two stages of Husserl's thinking about logic and mathematics. It is a theme that attracted Gödel to Husserl's work even though critics have argued that Gödel's incompleteness theorems are incompatible with some of Husserl's claims about, or at least hopes for, definite axiom systems and definite manifolds. What Gödel was interested in, however, was the prospect offered by phenomenology for the (nonreductionistic) clarification of the meaning of basic terms in mathematics. I will have more to say about this later in this Introduction.

In the essays included in this volume it should be clear that I favor the middle and late stages of Husserl's phenomenology. If pressed, I would choose the third stage over the work at the other stages. Of course, there are also some themes (e.g., the concern for the 'origins' of concepts of logic and mathematics) that remain more or less constant throughout the stages. In the essays that follow I am always writing from the point of view of the later two stages in Husserl's work, even when I write about Husserl's earliest work. It will not be possible to understand what I am doing in the chapters unless this is kept in mind. Husserl himself did not have the opportunity to rewrite such works as the *Philosophy of Arithmetic* from the perspective of his more mature philosophy. It is left to his readers to try to rethink the early work on the philosophy of arithmetic and the philosophy of geometry in terms of the later ideas and methods.

My earlier book, *Mathematical Intuition: Phenomenology and Mathematical Knowledge*, was also written from the perspective of ideas in transcendental phenomenology. The book was focused on mathematical intuition in the case of natural numbers and finite sets, and the idea was to see what kind of account one could develop in this case, on the basis of some of the recent literature on mathematical intuition but also on Husserl's later ideas on intentionality, meaning, ideal objects, possibilities of dynamic fulfillment of mathematical intentions, the analysis of the origins of mathematical concepts in everyday experience, and related views about acts of abstraction, reflection, formalization, and so on. Some of the essays in the present volume, especially Chapters 11 through 15, need to be understood from this perspective. I am concerned with the matter of how far we can push the idea of the fulfillment or fulfillability of intentions that are directed toward particular mathematical objects. What is the best way, for example, to understand talk about the (potential) presence or absence to human consciousness of particular natural numbers? Are there limits on the intuition of particular mathematical objects? This can be seen as an effort to understand the relationship between the more *intuitive* parts of mathematics that have their origins in everyday experience (for example, arithmetic and elementary geometry) and the more *conceptual* and rarefied parts of mathematics (such as higher set theory) where it appears that we cannot have complete or fully determinate intuitions of particular mathematical objects (even if we can engage in 'objective', meaningful, eidetic thinking that appears to be directed toward such objects).

Because I have compared some of Husserl's ideas in the elementary parts of mathematics that have their origins in everyday experience with ideas in intuitionism (see Part III) one might form the impression that I think Husserl himself was an intuitionist. As should be obvious in many of the following chapters, this would amount to a misunderstanding of what I am doing. Husserl was not an intuitionist in the style of Brouwer or

Heyting. There are many differences between his views and those of the traditional intuitionists. One could list the differences. One of the central differences, for example, is that from 1900 onward Husserl has ideal objects such as meanings in his ontology and he speaks about these objects in a way that is more platonistic than would be pleasing to Brouwer or Heyting. Mental entities and processes are required for knowing about natural numbers, for example, but the objects known about – natural numbers – are not themselves mental entities. Rather, they are ideal objects. There are also some important differences concerning solipsism, the explicit recognition of the intentionality of human consciousness, meaning theory, the place of formalization, the views of what can be intuited, and the like. Husserl holds, for example, that there is intuition of meanings or of essences, and of the relations of essences to one another, but it is not clear that traditional intuitionism would have a place for this kind of intuition.

At the same time, it is very interesting to compare Husserl's explorations of logic and mathematics with those of the intuitionists. Husserl never tired of arguing that in order to do justice to logic and mathematics we must investigate the subjective side as well as the objective side of these sciences. We need to consider not only the ideal objects and truths of logic and mathematics but also the subjective acts and processes by virtue of which we come to know about objects and truths. Such acts and processes include carrying out sequences of acts in time, abstracting, collecting, reflecting, and various forms of memory and imagination. No one in recent times has done more to investigate the subjective side of logic and mathematics than Husserl and people such as Brouwer and Heyting. There are bound to be interesting and important points of contact. Some of these points of contact were in fact already being explored in Husserl's time by such individuals as Oskar Becker (who was one of Husserl's research assistants), Hermann Weyl (who was influenced in some ways by Husserl), Felix Kaufmann (also influenced by Husserl), and Heyting himself. Weyl, for example, comes close to identifying the phenomenology of mathematics with intuitionism in some of his comments. As I have indicated, I think this identification goes too far. It does seem to me, however, that there are ideas about subjectivity, intersubjectivity, meaning, intuition, internal time, and other topics in transcendental phenomenology that can be used to support and develop some ideas in intuitionism. Indeed, this is the tack I take in some of the chapters in Part III. In any case, I think it is not a good idea to act as if this period in the development of phenomenological ideas about mathematics did not exist.

There are also interesting points of contact with some of Hilbert's philosophical ideas about mathematics, especially in Husserl's early work. Some of the connections were already being discussed by people around Husserl, for instance, Dietrich Mahnke and Oskar Becker. Just as Husserl was no Brouwerian intuitionist, however, he was also not what we might nowadays think of as a Hilbertian formalist or finitist. One could again list some significant differences in their general philosophical views, especially if one considers Husserl's work from 1900 onward. The kind of 'games meaning' that one can associate with signs in formal systems plays a central role in Hilbert's formalism, and Husserl certainly favors the axiomatization and formalization of mathematical theories. One should try to show that formal axiom systems are consistent and complete, and that they possess other desirable properties. This is all a legitimate and important part of formal logic and formal mathematics. The meaningfulness of mathematics, however, is not everywhere exhausted by the games meaning associated with symbols. There are also ideal meanings that can be expressed by symbols, and it is by virtue of these meanings that there is reference to ideal objects. The meaning theory is different from what we find in Hilbert. It is a bit more like what we find in Frege, although here too there are differences. There is also intuition not only of finite sign configurations (as tokens or as types) but of natural numbers, universals, and meanings themselves. There is an explicit concern with the intentionality of mathematical thinking and with the 'origins' of mathematical concepts of a sort that is not to be found in Hilbert. There are also other interesting points of comparison.

In a similar manner, one could compare and contrast Husserl's views on logic and mathematics with those of many of his contemporaries. It is useful and interesting to look at Husserl's views in connection with those of Cantor, Frege, Carnap, Russell, and others. I think that all of this needs to be explored. Husserl's views, however, cannot be straightforwardly identified with any of these other positions. One needs to sort through the details.

One way in which Husserl's general approach stands out from that of many other philosophers of mathematics is that he combines an investigation of the central feature of human consciousness – intentionality – with an investigation of fundamental issues in the philosophy of mathematics and logic. In the philosophy of mind it is usually taken as a basic fact that human consciousness, especially in scientific modes of thinking, exhibits intentionality. If we recognize this, then we will not be able to ignore the fact that the objectivity of such sciences as logic and mathematics must always be correlated somehow with the subjectivity (and intersubjectivity) of the scientists engaged in acquiring scientific knowledge. Husserl's work is especially interesting for its claim that there are subjective and objective sides to mathematics and logic and for the way it investigates both sides of these sciences in an effort to mesh their epistemologies with their ontological claims.

§ 2

It will be useful to include some specific comments on each of the chapters in this collection in order to put them into proper perspective and to indicate some connections between them. The essays are grouped into three categories. Part I of the volume contains three essays on reason, science, and mathematics from the viewpoint of phenomenology. The first chapter provides an overview of Husserl's claims about reason and about logic as a theory of science. It takes up the idea of logic as the science of all possible sciences (as *mathesis universalis*), including Husserl's mature view (in *Formal and Transcendental Logic*) of the three levels of what he calls 'objective formal logic', along with some of his ideas on manifold theory and formal ontology. The essay follows Husserl's thinking about science through his late work on the crisis of the modern sciences. That crisis stems from misapplications of various forms of naturalism and objectivism. The scientism that comes in for criticism in Husserl's late work is a view that has abandoned philosophical rationalism. Just as there can be science within reason, there can also be science without reason. The second chapter in Part I focuses on some problems in the philosophy of mathematics in particular and on ways they might be approached from the perspective of transcendental phenomenology. It should be read as a general overview that is filled out in various ways in subsequent chapters. The third essay focuses in more detail on geometry. It discusses Husserlian views about 'ideation' or the intuition of essences in connection with modern pure geometry, the idea that there are different formal systems of geometry and different spatial ontologies, and some Husserlian themes about the origins of geometry. It considers the idea of creating new variants of geometry based on the formalization of Euclidean geometry. The essay also discusses the concern for the consistency of the resulting systems, especially when we have left behind the more familiar and intuitive domains of two and three dimensions.

Part II of the volume is centered on a series of essays on Kurt Gödel's interest in Husserl's phenomenology. Gödel had turned his attention to philosophy already by 1943, and by 1959 he was enamored of many ideas in Husserl's work. His interest in Husserl continued until his death in 1978. In Part II the focus shifts primarily to Gödel's work, and the chapters are intended to contribute as much to Gödel studies as to the phenomenological study of mathematics. The essays in this part treat a number of central issues in the philosophy of mathematics and logic: issues about intentionality in mathematics, rational intuition, meaning clarification and concept analysis, the intuition of essences, platonism, minds and machines, and the place of formal systems in mathematical thinking. Gödel claims that Husserlian transcendental phenomenology offers a better perspective on mathematics than positivist and naturalistic views, and he links his claims directly to many of his well-known results in mathematical logic. All of this is explored in some detail. The final three chapters in this part of the volume discuss the views of Quine, Penelope Maddy, and Roger Penrose from the perspective of the ideas discussed in the earlier essays on Gödel. Gödel's views on phenomenology and meaning clarification in mathematics are contrasted with Quine's view of mathematics, and Maddy's efforts to develop a 'compromise platonism' (compromising Quine and Gödel) and a defensible account of intuition are critically examined. Penrose's arguments on minds, machines, and Gödel are discussed in the final chapter of this part.

A few additional remarks might be helpful to frame the chapters on Gödel properly. Readers who are acquainted with Husserl's work will see that I do not discuss Husserl's ideas on definite axiom systems and definite manifolds in these chapters. The reason I do not discuss these ideas is straightforward: Gödel does not discuss them. It has occasionally been claimed that Husserl's views about definite axiom systems and definite manifolds are refuted by Gödel's incompleteness theorems. This is perhaps true. There may be some specific remarks about definite axiom systems that the incompleteness theorems refute. There may have been a kind of naiveté in some of Husserl's writing about the prospects for showing formalized mathematical theories to be consistent and complete. It seems to me, however, that the matter is complicated by a number of considerations. One general problem is that the notion of 'definiteness' at work in these discussions is not nearly as clear as one would like it to be in order to resolve the issues once and for all. Is it, for example, a notion that applies only to purely extensional systems? Does it include the notions of consistency, completeness, *and* categoricity? Does it include the idea of finite axiomatizability? Does it incorporate the sharp distinction between syntax and semantics that we now routinely presuppose in metamathematical considerations? There is also language in some of Husserl's writings, for example, that suggests that he thought of definite axiom systems and definite manifolds as representing an 'ideal' in mathematics, an ideal we should aim for even if we cannot always achieve it. Even Gödel speaks about overcoming particular cases of incompleteness by finding new evident axioms that would allow us to decide previously undecidable propositions. Much more could be said about all of this.

In any case, if one focuses on Husserl's ideas on definite axiom systems and definite manifolds one might wonder how Gödel could have been interested in Husserl's work at all. Gödel himself gives us the answer. He tells us that it is precisely because of his incompleteness theorems that he is interested in ideas in phenomenology about clarification of the meaning of mathematical expressions. How would such meaning clarification be possible if mathematics just consisted of the rule-governed manipulation of meaningless or uninterpreted symbols in purely formal systems? Gödel is pointing to a different or an additional dimension of Husserl's conception of logic and mathematics. It is not Husserl's contribution to the modern conception of what a formal system is that is primarily of interest to him. The notion of a formal system of the sort relevant to Gödel's incompleteness theorems reached a level of precision in the work of Hilbert, Gödel, and other proof theorists that cannot be found in Husserl's writings. Of course, Gödel liked Husserl's Leibnizian idea of logic as *mathesis universalis* and related ideas about the place of axiomatic formal systems in unified exact science. The promise he found in Husserl's work, however, was in the prospect for meaning clarification and for achievement of a more refined intuition of essences. This is what I explore in many of the essays in Part II.

A second point concerns this idea of meaning clarification. In the short paper from 1961 in which Gödel explicitly discusses Husserl the emphasis is on the prospects for the clarification of meaning of basic expressions of mathematics offered by phenomenology. In discussions with Hao Wang he also speaks about clarification of the intuition of essences. These are of course central themes in Husserl's later phenomenology, but to put a somewhat finer point on it, it might be useful to note that in Husserl's work there is a general distinction between meaning and essence of the following sort: there are no essences that correspond to expressions that are meaningful but (formally or materially) contradictory. An ideal meaning is expressed, for example, by an expression such as "round rectangle," but there is no essence corresponding to this expression. In order to judge that there is no round rectangle it is presupposed that the expression "round rectangle" has a meaning. There is, however, a "dissonant unity" of the intended essences "round" and "rectangle" here. These essences are incompatible. One can of course also reflect upon and try to clarify the meaning of instances of expressions of the form "P is essential to x." It might be helpful to keep these points in mind if one wants to read Gödel's ideas on phenomenology in connection with Husserl's own views.

Finally, I would like to note that although there are many interesting connections between Gödel and Husserl, it might be the case that Gödel's platonism differs in some respects from what we find in Husserl. Gödel evidently favored Husserl's transcendental phenomenology, and it seems to be possible to make fairly robust platonistic claims from this perspective. We can evidently say, for example, that mathematical objects are mind independent and unchanging, but now we always add that they are constituted in consciousness in this manner, or that they are constituted by consciousness as having this sense (meaning). One can evidently say that it is not necessary for the existence of mathematical objects or truths that anyone ever be aware of them or that there ever be expressions for them, but we now add that they are constituted as having this sense. They are constituted in consciousness, nonarbitrarily, in such a way that it is unnecessary to their existence that there be expressions for them or that there ever be awareness of them. Gödel would perhaps be happy with such a view, but I can imagine objections to it as a reading of either Gödel or Husserl. The matter certainly deserves further investigation. Another issue is that modern, transfinite set theory of the sort that Gödel was concerned with in some of his work raises problems that were not directly addressed by Husserl. Should we really accept the existence of the huge, impredicatively specified transfinite sets that we speak about in higher set theory? Can we legitimately speak about the constitution of the being of such sets? What about the sets postulated by some of the axioms that have been offered in order to extend Zermelo-Fraenkel set theory? Some of Gödel's comments indicate that he was prepared to accept the full ontology of Zermelo-Fraenkel set theory (and more), but it is possible that Husserl would not go so far. In any case, I think that according to Husserl's meaning theory these parts of mathematics would indeed count as meaningful. It is just that they demand a more penetrating clarification of meaning than some other parts of mathematics. Gödel evidently thought that there was a sharp distinction between Husserl's meaning theory and other theories of meaning that were on offer but that were antithetical to his purposes.

Part III of the volume is called "Constructivism, Fulfillable Intentions, and Origins" because here I focus on elementary parts of mathematics whose origins are closer to everyday experience and practice. The kinds of cognitive activities involved at these levels are relatively simple. We can carry out sequences of acts in time, see ordinary objects in small groups, collect objects, make some simple abstractions on this basis, and so on. Constructive mathematics has in particular always been concerned, in Husserlian language, with those mathematical intentions that we humans can actually or potentially fulfill, especially when we are directed toward particular mathematical objects, such as particular natural numbers. The idea of tracing the origins of mathematical and logical concepts back to everyday experience is a theme that runs through all of Husserl's thinking about these subjects. A number of the essays in Part III discuss the idea that there is a '*founding*' level of experience, and that various kinds of cognitive activities, such as abstracting, collecting, reflecting, and comparing collections, are *founded* on this basic experience and make the sciences of mathematics and logic possible. It is on the basis of this founding level that we ascend to the higher abstractions, greater generalizations, and formalizations that we find at the more rarefied theoretical levels of mathematics. (This idea of founding and founded levels is also discussed earlier in the book, in Chapter 3, in connection with geometry.)

The essays in Part III center around themes already discussed in earlier times by Husserl, Oskar Becker, Hermann Weyl, Arend Heyting, and others. The first essay in this part uses ideas from phenomenology and the philosophy of mind to support some claims in intuitionism. It contrasts Michael's Dummett's thinking about intuitionism with the view (discussed by Becker, Weyl, and Heyting) that constructions can be viewed as fulfilled (or fulfillable) mathematical intentions. The second essay in the section explores Weyl's ideas on constructivism, influenced as they were by some of Husserl's views. The third essay in the section sets out some possibilities for thinking about proofs, either as processes or as products, as fulfillable mathematical intentions. The focus here is on the parts of mathematics and logic that comply with the idea of intuiting particular mathematical objects. A lot more could be said about the ideas in this essay. The final two essays continue to focus on the intuition of mathematical objects and the origins of the most elementary parts of mathematics in connection with ideas of Poincaré and Frege.

One of Husserl's claims in *PA* concerning Frege in particular is that the concept of number is logically simple or primitive and that the analyzed sense (meaning) in Frege's definition of number is not the same as the original sense. What is needed, according to Husserl, is an analysis of the origin of the concept. Let me say a little bit more about this here.

Part of the Fregean analysis, in modern terms, would go something like this: we use higher-order logic to analyze number expressions. If we want to say, for example, that there are exactly two cars in the parking lot (Cx: x is a car in the parking lot), we say that $(\exists x)(\exists y)(x \neq y \wedge (\forall z)(Cz \leftrightarrow (z = x \vee z = y)))$. Generalizing on the predicate place, we can say that exactly two things have the property X:

$$(\exists x)(\exists y)(x \neq y \wedge (\forall z)(Xz \leftrightarrow (z = x \vee z = y))).$$

To say explicitly that exactly two things have the property X we adopt the following characterization:

$$(\forall X)(2X \leftrightarrow (\exists x)(\exists y)(x \neq y \wedge (\forall z)(Xz \leftrightarrow (z = x \vee z = y)))).$$

Similarly,

$$(\forall X)(0X \leftrightarrow \neg(\exists y)Xy),$$
$$(\forall X)(1X \leftrightarrow (\exists x)(\forall y)(Xy \leftrightarrow y = x)),$$
$$(\forall X)(3X \leftrightarrow (\exists x)(\exists y)(\exists z)((x \neq y \wedge y \neq z \wedge x \neq z) \wedge$$
$$(\forall w)(Xw \leftrightarrow (w = x \vee w = y \vee w = z)))),$$

and so on.

Now consider a statement such as $2 + 2 = 4$. On the kind of analysis that Frege gives this goes over into a statement of higher-order logic of the form

$$(\forall X)(\forall Y)((\neg(\exists z)(Xz \wedge Yz) \wedge (2X \wedge 2Y)) \rightarrow 4(X \vee Y)).$$

To see the point of Husserl's claim (and compare this with Poincaré's remark in § 3 of Chapter 14) we can use a substitution test. Suppose a person P knows that $2 + 2 = 4$. Does it follow with necessity that

P knows that $(\forall X)(\forall Y)((\neg(\exists z)(Xz \wedge Yz) \wedge (2X \wedge 2Y)) \rightarrow 4(X \vee Y))$?

Does '$2 + 2 = 4$' have the *same meaning* as

'$(\forall X)(\forall Y)((\neg(\exists z)(Xz \wedge Yz) \wedge (2X \wedge 2Y)) \rightarrow 4(X \vee Y))$'?

If so, we would expect the substitution to hold. It seems, however, that a person might very well know that $2 + 2 = 4$ without knowing that

$$(\forall X)(\forall Y)((\neg(\exists z)(Xz \wedge Yz) \wedge (2X \wedge 2Y)) \rightarrow 4(X \vee Y)).$$

Consider, after all, what the domain of quantification must consist of in the latter statement, whether impredicativity is involved, and so on. Generally speaking, is a natural number really just an infinite equivalence class? Our ancestors knew that $2 + 2 = 4$, our children know it, and it seems that many people know it without knowing what goes into making a statement such as

$$(\forall X)(\forall Y)((\neg(\exists z)(Xz \wedge Yz) \wedge (2X \wedge 2Y)) \rightarrow 4(X \vee Y))$$

true. How could this be? One can see Husserl as trying to provide an answer to this latter question by going back to the origins of the concept of natural number in everyday experience and in simple cognitive activities such as running through a group of objects or experiencing objects in groups, counting, and so on. Frege's analysis of such statements as '$2 + 2 = 4$' is interesting and valuable, but the analyzed sense of number statements that it presents is not the same as the original sense. Frege tells us that he is not really interested in a philosophical analysis of the origins of the concept of natural number anyway. As I point out in some of the chapters in Part III, there are several ways in which Frege and Husserl were working at cross purposes in their investigations of number.

The book is thus organized to open with ideas about the very general conception of logic in Husserl's work and the subsumption of mathematics and the regional sciences under this conception. After focusing in a little more detail on mathematics in the second essay and geometry in particular in the third essay, it takes up issues about methods of meaning clarification and essence analysis in connection with Gödel. Gödel appears to think that these methods have as much of a role to play in the most abstract and rarefied parts of mathematics (for example, higher, impredicative set theory) as they have in the most elementary parts of mathematics. The final part of the book then focuses on the most elementary parts of mathematics and logic, the parts that are closer in their genesis to ordinary sensory experience. It is in these parts of mathematics in particular that one can make the best case for the claim that intentions directed toward particular mathematical objects are in principle fulfillable. It is argued that the fulfillability of intentions directed toward particular mathematical objects, however, is independent of the method of clarification of the meaning of mathematical concepts as this might be used in higher mathematics. Pure mathematics and pure logic are eidetic sciences, but it does not follow that we can have determinate and complete intuitions of all the particular objects toward which we might be directed in mathematical thinking.

I believe that presenting these essays in one volume will allow the reader to see many connections that would be missed if only a few of the essays were read or if they were read in isolation. The essays can be read independently, but they are probably most profitably read in connection with one another. It is my hope that the book will help to carry forward ideas in the phenomenological tradition about logic and mathematics. It contains new perspectives on and new ideas about a variety of topics in the philosophy of logic and mathematics: the role of intentionality in mathematics, mathematical intuition, platonism, constructivism, impredicativity, logicism, varieties of formalism, nominalism, minds and machines, and the like.

PART I

REASON, SCIENCE, AND MATHEMATICS

1

Science as a Triumph of the Human Spirit and Science in Crisis

Husserl and the Fortunes of Reason

> The reason for the failure of rational culture, as we said, lies not in the essence of rationalism itself but solely in its being rendered superficial, in its entanglement in 'naturalism' and 'objectivism'.
>
> (Husserl, "The Vienna Lecture," 1935)

Husserl's later philosophy contains an extensive critique of the modern sciences. This critique of what Husserl calls the 'positive', 'naive', or 'objective' sciences has been very influential in Continental philosophy and, in particular, in the retreat from holding up science and technology as models for philosophy. Philosophers such as Heidegger, Merleau-Ponty, Ricoeur, Habermas, and Derrida, along with many others on the Continent, have been influenced by this part of Husserl's thought. There is also in Husserl's work, however, a very grand view of the value and possibilities of science, provided that science is understood appropriately and in a broad sense as a theory of the many forms of reason and evidence. It cannot be adequately understood as a narrow, technical, naturalistic, and 'one-sided' specialization of one sort or another. Indeed, in its broadest sense, science would coincide with the rigorous exercise of philosophical reason. In Husserl's writings we find (1) reflections on the nature of science as a whole; (2) studies of particular sciences, especially geometry, arithmetic, logic, and natural science; but also (3) a far-reaching analysis

In preparing the final version of this chapter I benefited from the comments of the many participants in the Science and Continental Philosophy conference organized by Gary Gutting and held at the University of Notre Dame in September 2002. I especially thank Ernan McMullin, Gary Gutting, Karl Ameriks, Michael Friedman, Philip Quinn, Hans-Jörg Rheinberger, Terry Pinkard, and Simon Critchley.

of how modern scientific culture has fallen away from its higher calling. I will discuss each of these aspects of Husserl's philosophical thinking. (For some earlier studies of these topics see, e.g., Gurwitsch 1974; Heelan 1989; Ströker 1988, 1987a, 1979.)

The critique of the modern sciences in Husserl's later work did not spring forth ex nihilo. There are seeds of it in his earliest work. The *Philosophie der Arithmetik (PA)*, for example, is already premised on the view that only the philosopher can provide the kind of deeper reflection on arithmetic that the mathematical technician either cannot or will not provide. In this first book we are presented with an analysis of how this science is founded on the everyday experience of groups of sensory objects, how abstraction from this basis and a kind of formalization are required to get the science of arithmetic off the ground. Arithmetic is built up from 'higher' cognitive activities that are founded on ordinary perceptual acts. The idea of analyzing 'origins' of concepts, which Husserl inherited from some of his teachers (especially Brentano), is already at work here, and Husserl will continue to insist throughout his career on the value of this kind of analysis. Ideas of this type will figure into his later charge that the modern sciences have forgotten their origins in the everyday practices of the lifeworld.

By the time of the *Logical Investigations (LI)* Husserl tells us that philosophers have the right, indeed the duty, to examine critically the foundations of the sciences. Philosophy and, in particular, phenomenology are needed to supplement the sciences. After 1907 or so phenomenology is portrayed in ever more detail as the transcendental and a priori 'science' that investigates the essence of the sciences (as well as other domains of human experience). Scientists themselves are technicians who build up theories and methods without insight into the essence of these theories and methods or into the conditions for their possibility. Scientists are concerned more with practical results and mastery than with essential insight. For just this reason, the sciences are in need of continual epistemological reflection and critique of a sort that only the philosopher can provide. Scientists are oriented toward their objects but not toward the scientific thinking itself in which the objects are given. It is the phenomenologist who will study the essential features of this thinking. Husserl pictures the work of the philosopher and the scientist as mutually complementary (Husserl *LI*, "Prolegomena to Pure Logic," § 71). The philosopher does what the scientist cannot and will not do if she is to practice her science, and the scientist does what the philosopher cannot do *qua* philosopher.

Phenomenological philosophy, it is thus argued, has the right to investigate the sciences and subject them to critical scrutiny. With the 'transcendental' and 'eidetic' turn of phenomenology we see, in particular, more and more development in Husserl's critique of naturalism, empiricism, naïve 'objectivism', positivism, and related positions. Psychologism and a kind of evolutionary biologism about logic, as forms of empiricism, are already subjected to criticism in the "Prolegomena to Pure Logic" of 1900. All of these views entail a kind of relativism about science and logic that Husserl rejects. Toward the end of his career he rejects historicism for similar reasons.

Although many of the ideas that figure into Husserl's later critique of the sciences can be found in his earlier writings, the language of the 'crisis' and the 'danger' of the sciences does not emerge in his publications until the nineteen thirties. If Husserl had his worries about various aspects of the sciences earlier on, he evidently did not yet see the sciences as being at a point of crisis. From the early thirties on, however, the idea of a crisis moves from the margins to the center of his thinking. There has been, Husserl declares, a 'superficialization' of reason in the modern sciences. The problems with the modern sciences are being unjustly imputed to reason itself. Reason is under attack. The real problem, however, is that the modern sciences themselves can be practiced and are being practiced, in a manner of speaking, without reason. In the earlier writings the 'superficialization' of reason in the positive sciences had not yet impressed Husserl so deeply. The abstractions, idealizations, formalizations, quantifications, technization, mechanization, specialization, and other activities required by the modern sciences had not yet been plumbed so deeply with respect to their potentially negative consequences for human existence. In developing his work in this direction Husserl wanted to speak to the more general crises of human existence in European culture that were all around him at the time. He was no doubt responding to some extent to other philosophers with whose work he was engaged, such as that of his own student Martin Heidegger. Unlike Heidegger, however, he saw himself as trying to preserve the value and possibilities of human reason in dark times. Irrationalism and feelings of meaninglessness were breaking out everywhere and science, supposedly the very embodiment of reason, seemed powerless to stop it. Husserl thought that there was a fundamental difference, however, between the existing positive sciences and 'universal, responsible science'.

I will open my treatment of Husserl's views on science in this chapter with a consideration of his pre-*Crisis* writings on the subject. In these writings we find general reflections on science, but also specific analyses of arithmetic, geometry, logic, and natural science. These analyses are of lasting interest in their own right, quite independently of his later worries about the dangers to civilization that arise when the positive sciences are not seen in their proper perspective. I will then make a transition to the specific arguments on and analyses of the crisis of the modern sciences.

§ 1　Arithmetic, Geometry, Logic, and the Science of All Possible Sciences

The grand view of the sciences mentioned a moment ago appears for the first time in Husserl's publications in the *LI* of 1900. By this time Husserl had already worked in mathematics, completing a doctoral thesis on the calculus of variations. He had published his *PA* and had continued to work on material related to his philosophy of arithmetic. He had also been working on the foundations of geometry. In the *LI*, however, the scope broadens considerably. The focus is now on logic, but on logic in the very broad sense of a theory of reason, a *Wissenschaftslehre*. Leibniz, Bolzano, and Lotze in particular are cited as the philosophers who have seen most deeply into logic as the theory of science or as the 'science of all possible sciences'. Husserl was to continue to develop these ideas on logic and science in all of his later publications, from the *LI* to the *Ideas Pertaining to Pure Phenomenology and a Phenomenological Philosophy* (in three books) up through the *Formal and Transcendental Logic* and other writings. In what follows I will concentrate on the later, more mature views on the science of all possible sciences. Arithmetic, geometry, real analysis, physics, and all of the other sciences are subsumed under this conception. Indeed, what Husserl has in mind, at least as an ideal, is a unified and systematic conception of the sciences of the sort found in the work of many of his rationalist predecessors.

Much of Husserl's work on science, like that of other rationalists, is focused on the exact sciences. Pure logic occupies a central role because it studies the most fundamental ideas that underlie all of the sciences. Logic, as the study of reason in a very broad sense, is a condition for the possibility of any science. All testing, invention, and discovery rest on regularities of form, and it is the science of logic that focuses on form. All of the sciences will require logic, but they will then add to the basic forms and structures of logic, filling them out in various ways. As we will see, Husserl elaborates on these ideas quite extensively.

Husserl tells us that he is interested in the old and venerable idea of pure logic as *mathesis universalis*. One already finds this idea in such philosophers as Descartes and Leibniz. *Mathesis universalis* includes the idea of a mathematics of judgments. Mathematicians had always been focused on their own objects in different domains of mathematical thought, but they had not focused on judgments themselves. Logicians, on the other hand, had focused on judgments, and their logical properties and relations to one another. *Mathesis universalis* should, among other things, unite the two. After all, the objects of mathematics (and logic) are referred to and given through judgments. Indeed, all sciences are composed of sets of judgments, and it is by virtue of these judgments that they refer to their own objects and states of affairs.

Now it is 'pure' logic that we are to think of as the science of all possible sciences. Husserl quotes Kant's words on logic with approval: we do not augment but rather subvert the sciences if we allow their boundaries to run together (Husserl *LI*, "Prolegomena to Pure Logic," § 2). Logic is its own autonomous, a priori subject. In particular, it ought to be kept distinct from psychology, anthropology, biology, and other empirical sciences. The view that logic was concerned with mental processes and entities, as these would be studied in empirical psychology, was especially prevalent at the time. Frege railed against this psychologism about logic, and by 1900 Husserl also subjected it to extensive criticism. Logic is a formal, deductive, and a priori discipline, and as such it is distinct from all of the empirical sciences. Husserl's critique of psychologism in particular broadens to include any effort to found logic on an empirical science. To found logic thus would be to involve it in a relativism that is in fact foreign to it.

Pure logic is not about 'real' mental processes or mental entities. Rather, it is concerned with 'ideal' meanings. Husserl's critique of psychologistic and other empiricist views of logic is underwritten by an ontology that recognizes both real and ideal objects. Real objects, in the first instance, are objects to which temporal predicates apply, in the sense that the objects come into being and pass away in time. They have a temporal extension. Some real objects also have spatial extension. Spatial predicates apply to them. Thus, an ordinary physical object is one to which both spatial and temporal predicates apply. By way of contrast, a thought process has a temporal but not a spatial extension. Ideal objects and truths have neither temporal nor spatial extension. In his later writings Husserl shifts from saying that they are atemporal or nontemporal to saying that they are 'omnitemporal'. One does not somehow locate them outside time altogether but holds that they exist at all possible times. As one would expect, ideal objects are also acausal. Thus, we seem to be presented with a rather platonistic view of ideal objects, but this 'platonism' becomes rather nuanced after Husserl takes his turn into transcendental idealism around 1908. In these later writings Husserl speaks of how the sense of every existent is constituted in the subjectivity of consciousness. This means that the sense of ideal objects and truths as transcendent, nonmental (and, hence, in a sense, as mind independent), partially given, omnitemporal, and acausal is itself constituted in the subjectivity of consciousness (see, e.g., Husserl's *Formal and Transcendental Logic (FTL)*, § 94).

Pure logic is concerned with ideal meanings, and it is judgments that express these ideal meanings. Judgments refer to objects or states of affairs by way of their meanings. The meanings expressed by judgments are ideal and the objects to which we are referred by judgments may themselves be either real or ideal. The objects of pure logic and pure mathematics are ideal. Judgments have a form and a 'matter' (or content). Two judgments with different 'matters', for example, may have the same form: 'This house is red' and 'This table is blue' both have the form 'This S is P' (this is only a partial formalization). Among other things, pure logic will therefore need to track features of judgments and other types of expressions, the ideal meanings expressed by judgments, the objects referred to, and the form and matter in each case. Husserl's view of all of this, inaugurated in the *LI* and developed up through *FTL*, is mapped out in his stratification of 'objective formal logic' into three levels. I will give a brief overview of the conditions for the possibility of science that are included in this threefold stratification.

A fundamental condition for the possibility of any science is that it operate with judgments that not only are formally well formed but express unified or coherent meanings. Thus, at the bottom or first level of any possible science we have a priori or 'universal' grammar. This is not to be a psychologistic or anthropological science. Husserl thinks there will be a priori rules of grammar for both the form and the matter of expressions. The basic idea at this level is to lay out the rules and methods for determining whether or not a string of signs (words) is meaningful. This will require purely formal grammar, in order to distinguish formally well-formed strings of signs from strings that are not well formed. Here Husserl seems to have in mind what we now think of in laying out the grammar of purely formal languages. Using one of his own examples, we could say that 'This S is P' meets the relevant formal conditions, whereas a string such as 'This is or' does not. Husserl distinguishes simple from complex meanings and he wants to know the rules for forming judgments as meaningful wholes from meaningful parts. Already in Investigation IV of his *Logical Investigations* he had applied the theory of parts and wholes from Investigation III to this question and concluded that there must be a priori laws of grammar. What is needed, in addition to a purely formal grammar, is a theory of semantic categories to determine which substitutions of matter in the forms will give us something meaningful and which will give us mere nonsense (*Unsinn*). Consider, for example, the following two substitutions for 'This S is P': 'This tree is green' and 'This careless is green'. The former is a judgment. It expresses a unified meaning. The latter, however, is simply nonsense. Each part of it is meaningful but the whole formed from the parts is not. An expression from the wrong category has been substituted in place of 'S'. Roughly speaking, a categorial grammar would allow nominal material to be replaced by nominal material, adjectival material by adjectival material, relational material by relational material, and so on. Adjectival material could not be freely replaced by nominal material, and so on. Husserl allows that false, foolish, or silly judgments may result from the substitutions permitted but they would still be meaningful judgments. We might, for example, obtain a judgment such as 'This blue raven is green'. This is not meaningless (nonsense), but it is false. Indeed, it is an a priori material absurdity or inconsistency, on the assumption that nothing that is blue all over can be green. On the other hand, "This careless is green" is nonsense.

Level two of 'objective formal logic' is the 'logic of noncontradiction' or 'consistency logic'. The strings of words admitted at level one express unified meanings. The next question we can ask about such (sets of) judgments is whether or not they are consistent. Here we also need to track formal and material versions of this question. A form such as 'There is an S and there is no S' is formally contradictory. It is a purely formal a priori absurdity. There may also be judgments that are not formally contradictory but that are instead materially inconsistent, for example, 'This blue raven is green'. The latter expression is meaningful but Husserl says it is countersensical (*Widersinn*). It is a material a priori absurdity. Thus, there are both formal and material a priori absurdities. A judgment such as 'This raven is orange', on the other hand, is meaningful and is not a material a priori absurdity. It is a synthetic a posteriori judgment, the kind of judgment whose truth or falsity requires sense experience.

Husserl in fact distinguishes synthetic a priori from synthetic a posteriori judgments along these lines (see Husserl *LI*, Investigation III, and

Husserl *Ideas I*, §§ 10 and 16). Both require material concepts but the former depend on purely a priori relations among such concepts. For example, if an object x is blue all over, then it cannot be green. If an object x is red, then x is spatially extended, but the converse does not hold. If an object x is red, then it is colored, but again the converse does not hold. Sense experience could not tell against such truths. What Husserl calls 'regional ontologies' (discussed later) are made up of synthetic a priori truths for different domains of cognition. Analytic a priori truths, on the other hand, are purely formal. They depend on purely formal a priori laws. Husserl thus speaks of purely formal logic as 'apophantic analytics'. (*Apophansis* is the Greek term for judgment or proposition.) The ontological correlate of purely formal logic is formal ontology. In this connection, Husserl also distinguishes formalization from generalization or, if you like, formal from material abstraction. Abstraction of the form from a material proposition is different from the kind of abstraction involved in moving from species to genus. The relationship of the form of a proposition to its material instances does not require concepts with content in the same way that these are required in setting out genus/species relationships. For example, a concept lower in a species/genus hierarchy (e.g., being red) implies all of the concepts (e.g., being extended) above it (but not the converse). It is for reasons of this type that there can be synthetic a priori judgments and regional ontologies.

At the second level of logic we should aim to distinguish judgments that are countersensical from those that are not. If a judgment is countersensical, it is meaningful, but it is not possible that there are objects corresponding to it. There could be no state of affairs that corresponded to it. On the other hand, if it is consistent, then it can have corresponding objects or states of affairs.

Husserl develops at this level the idea of purely formal 'apophantic' logic(s). Apophantic logic would just be the logic of forms of judgments, broadly conceived. It would presumably include forms of any kinds of meaningful judgments. One might speak of different logics here, but Husserl seems to have in mind some unified conception of these in one overarching logic. Consistent apophantic logic(s) would stand as a condition for the possibility of any science since we cannot have sciences that contain formally contradictory judgments.

Apophantic logic is conceived of in terms of axiomatic formal systems. Husserl sometimes refers to this level of consistency logic as 'consequence logic'. We view logic as an axiomatic formal system in which one derives consequences of the axioms on the basis of formal rules of inference.

The idea that judgments express meanings by virtue of which they refer to objects or states of affairs is mirrored at this second level of logic, in the following sense: if a set of judgments (including a formal system) is consistent, then objects or states of affairs corresponding to the set of judgments are possible. Husserl calls the 'ontological correlate' of a consistent formal axiomatic system a manifold. Since it is the correlate of the mere forms of judgments it constitutes a purely formal ontology. In apophantic logic we are focused on the forms of judgments and the features of judgments as expressions of meanings, whereas in formal ontology we are focused on the possible objects or states of affairs referred to by such judgments, solely with respect to their form. In formal ontology we are concerned with the most general and formal notions of object, state of affairs, property, relation, whole, part, number, set, and so on. In the particular sciences these notions would be 'materialized' or specified in different ways. Geologists, for example, would speak about particular kinds of objects (e.g., rocks, mountains) and particular kinds of parts, properties, and relations of these objects. Physicists would specify their objects and properties differently than geologists or biologists but would still be using the notions of object, part, property, and relation. In the different sciences we would have different regional ontologies. (For studies of the natural sciences in particular see Hardy and Embree 1993; Heelan 1983; Kockelmans 1993; and Kockelmans and Kisiel 1970.)

We can ask not only whether formal systems are consistent but also whether they are complete. Husserl introduced the notion of a 'definite' formal system early on in this thinking, and it seems to be the notion of a formal system that is both consistent and complete. He frequently mentions how he and Hilbert arrived at such a notion independently of one another (see, e.g., Husserl *Ideas I*, § 72). Husserl also introduces the notion of a definite manifold as the ontological correlate of a definite formal system. It is the definite purely formal 'world' that corresponds to a definite axiomatic formal system.

At the third level we have what Husserl calls 'truth logic' (*Wahrheitslogic*). At this level we are concerned with the truth or falsity of judgments. If judgments are consistent, then it is possible that there are objects corresponding to the judgments and it is possible that the judgments are true. For judgments to be considered true, however, it appears that we need more than mere consistency. Truth or falsity is determined by intuition, or by what Husserl calls 'meaning-fulfillment'. Intuition takes place in sequences of acts carried out through time, and it provides evidence that there are objects corresponding to judgments, or, on the other hand, it can show us how the intentions expressed by our judgments are frustrated. There are different degrees and types of evidence: clear and distinct, adequate, apodictic. If we have evidence for a state of affairs, then we can return again and again to precisely the same state of affairs. Moreover, the idea of intersubjective confirmation is built into Husserl's conception of evidence.

Thus, Husserl pictures level two as being concerned with the senses or meanings of judgments apart from the 'truth' or 'falsity' of the judgments (see Husserl *FTL*, Chp. 5). It is concerned with mere meaning-intentions and their forms. But when we are concerned with the truth or falsity of judgments, with meaning-fulfillment, we are at the level of truth-logic. There is, evidently, a notion of 'truth' (and 'existence') operative at level two, but, relative to level three, it is merely a notion of possible truth (and possible being). At one point, Husserl says it is a notion of truth that simply means 'derivable from the axioms', given whatever set of axioms we start with. At level three, however, we need to include structures of evidence and meaning-fulfillment.

It thus appears that at level three we need not only judgments and their consistency but also some kind of fulfillment procedure for making the object or state of affairs referred to by the judgment present. We need a judgment plus an intuitive ('verification' or 'construction') process for finding the object or state of affairs. Consistency alone does not guarantee this. Husserl seems to hold that consistency does not by itself imply existence but only possible existence. This notion of possible existence may suffice, in particular, in purely formal mathematics. It does not, however, imply that there is an intuition of objects or states of affairs referred to by mathematical judgments (see later discussion). Something more is needed. In previous work I have discussed how we might understand this idea of judgment + fulfillment procedure in the case of natural numbers and finite sets (see Tieszen 1989).

If we now consider the relationships between the levels in this broad conception of logic as the science of all possible sciences, we see that each level, starting from the lowest, is a condition for the possibility of the next level. In this account of the possibility of science the higher levels presuppose the lower levels. We cannot have a true scientific theory unless we have a consistent scientific theory and consistency itself presupposes that we have meaningful expressions, but meaningfulness does not itself guarantee consistency and consistency does not itself guarantee (intuition of) truth.

The relationship, in fact, is somewhat more nuanced. It appears that level two, with its emphasis on consistency, can serve where intuition fails. As long as a set of judgments is consistent we can feel secure in operating with the judgments and they can have an important role to play in science, even if we can have no intuitions corresponding to the judgments. Husserl emphasizes this in connection with theories of various kinds of numbers (e.g., negative, real, complex) and with n-dimensional Euclidean and non-Euclidean geometries where n > 3. He in fact cites Riemann, Grassmann, and others as central influences on his own conception of manifold theory. Thus, consider Euclidean geometry of two or three dimensions. This geometry existed for many years as a 'material' eidetic (a priori) science before it was formalized. The objects of Euclidean geometry are taken to be exact and ideal objects. They are idealizations of objects given to us in everyday sensory perception. The objects of everyday sensory intuition are inexact. Husserl says that the essences of such sensory objects are imprecise or 'morphological' (see Husserl *Ideas I*, § 74). An idealization of the shapes given to us in everyday perception takes place through the conception of greater and greater perfectings of the shapes until we form the conceptions of ideal circles, triangles, and so on. Euclidean geometry can thus be viewed as an idealization of structures found in everyday sensory perception. Once Euclidean geometry is *formalized* we obtain for the first time a Euclidean manifold of two or three dimensions. The foundation is then in place for generalizing and constructing n-dimensional Euclidean or even non-Euclidean manifolds. These latter constructions arise out of free mathematical imagination by playing with and making variations on the Euclidean manifold. In this manner Husserl wants to account for the origin and constitution of different 'spaces' and 'geometries', starting from the space of everyday perceptual intuition. He wants to allow complete freedom to devise new such constructions and to regard the resulting mathematics as perfectly legitimate, subject only to maintaining consistency (see Chapter 3).

Now as soon as we go beyond three-dimensional geometry, our ordinary spatial intuition begins to fail. But what if we could show that the resulting geometries were indeed consistent? Then there would be nothing illogical or incoherent about our conceptions (intentions) in these cases. The existence of relative consistency proofs of different geometries by way of various kinds of models seems to lie behind Husserl's conception of manifolds and formal theories. If we generalize this perspective to any part of mathematics or logic, or to science as a whole, then we seem to have some conception of what is supposed to be covered by level two of the theory of science.

Once we obtain new manifolds and formal systems through imagination we can reinterpret them any way we like. We are free to reinstantiate them with 'matter' in a variety of ways. Although Husserl does not discuss the topic, one can say that this is just what happened in the case of the application of non-Euclidean manifolds in relativistic physics, in which we see a surprising application of what had previously appeared to be merely conceptual or symbolic mathematics (see Becker 1923 and Weyl 1918b).

Of course we do not always know that a given set of judgments is consistent. In some cases it may be a very long time before we know, if we ever do. We also know, on the basis of the work of Gödel, that there are certain limitations on providing consistency and completeness proofs for mathematical theories. It seems that Husserl already held that level three considerations can help us in science even if we do not get an answer to our question at level two whether a set of judgments is consistent. That is, we have an intuitive basis for some parts of mathematics that can provide some security or reliability. We can sometimes appeal to (possible) intuition of mathematical objects for our evidence. Husserl's studies of arithmetic and Euclidean geometry show that he thought these sciences had a foundation in intuition. In effect, we start with shapes given to us in everyday perception or with our awareness of groups of everyday perceptual objects and trace out the cognitive acts of abstraction, idealization, reflection, and formalization that would lead to the Euclidean manifold or to the natural number manifold.

In his earliest studies of arithmetic Husserl held that most of our arithmetic knowledge was purely symbolic. Only a small part of it was actually intuitive. In fact, we cannot make number determinations much beyond 5 in single acts of intuition. Everything else amounts to purely symbolic thinking that does duty for the objects and laws that we cannot make present to ourselves in intuition. At this point in his career Husserl placed great emphasis on the nature and use of formal symbol systems in mathematics. These early studies abound in investigations of purely algorithmic methods, formal systems, and manifold theory (see, e.g., Husserl 1979, 1983). By the time of the *LI* Husserl's conception of intuition broadened and was explicated in terms of the distinction between mere meaning-intention and meaning-fulfillment. He also now distinguished the 'static' from the 'dynamic' fulfillment of a meaning-intention. On this view there could be at least partial fulfillment of intentions to larger numbers, and one could begin to speak about what is in principle fulfillable and what is not. Natural numbers themselves were now clearly taken to be ideal objects. In any case, the distinction between the intuitive and the merely symbolic was always present in Husserl's philosophy of science. Some parts of mathematics, logic, and even natural science would evidently always have to be considered merely symbolic or conceptual. (We might keep the *symbolic* and the *conceptual* separate here, reserving the former term for the merely algorithmic 'games-meaning' of signs and the latter term for the full meaning-intention of signs.)

Within this architectonic Husserl, as we said, distinguishes formal ontology from various regional ontologies. Regional ontologies are a priori ontologies of particular regions of being. When forms are filled in with 'material' or content, we are considering specific kinds of objects, properties, relations, parts, wholes, sets, and so on. All of the different specific scientific theories would have different regional ontologies. Transcendental phenomenology itself is a regional science. It is the science of consciousness with all of its various structures and characteristics. Each science has its own objects, with their own properties, and so forth. Husserl is generally an antireductionist about the specific sciences. Part of the reason for this derives from his view that judgments in the various sciences express their own meaning-intentions, and it is by virtue of these meaning-intentions that we are directed toward objects. Reductionist schemes may very well fail to respect differences in meaning-intention, along with the different implications and purposes reflected in these. For certain scientific purposes they might, if feasible, be useful. Eliminative reductionism in particular might diminish the many aspects or perspectives under which the world could be viewed and investigated. Eliminativism could hinder scientific work and perhaps even be dangerous, blinding us to important phenomena.

One very broad distinction among the sciences, as has been suggested, is that some are eidetic or a priori and some are empirical. Formal and regional ontologies are a priori, dealing, respectively, with the formal (analytic) and material (synthetic) a priori. Husserl also says that among the eidetic sciences some are exact and some are inexact. Mathematics and logic are exact. They trade in exact essences and exact objects. Among the inexact eidetic sciences Husserl counts transcendental phenomenology. It is, Husserl says, a 'descriptive' science that deals with inexact or 'morphological' essences, for example, the essence 'consciousness'. Unlike empirical psychology, however, it is not a causal-explanatory science. It does not seek causal generalizations. Underneath this rationalistic superstructure of the eidetic sciences we have the empirical sciences, the sciences that depend essentially on the evidence of sense experience and induction. Husserl's rationalism comes across quite clearly in many of his writings. The empirical sciences contain various imprecise, vague, or inexact components. They deal in probabilities and contingencies, not necessities. Whereas mathematics and logic set the standard for what is clear, distinct, and precise, the empirical sciences deal with indistinct or vague typifications of or generalizations from sense experience. The empirical sciences depend on the inexact essences associated with sensory objects. Even though the empirical sciences may be vague in this way and trade on various contingencies, they will presuppose various kinds of essences and essential truths. It is this latter 'material' a priori domain that will form in each case the subject matter of a regional ontology (which then has lying behind it the purely formal level). To work at developing a regional ontology is not to engage in the work of the empirical sciences. Rather, it is to explore and map out the essences and essential laws pertaining to a given domain. This, in effect, gives us the a priori conditions for the possibility of the empirical science of the domain. The phenomenologist, for example, is to do this for the domain of human consciousness. As such, she is not simply doing empirical psychology. Husserl returns again and again in his work to the topic of how phenomenology is distinct from empirical psychology.

§ 2   Phenomenological Philosophy and the Foundation of the Sciences

The science of all possible sciences focuses on the expression of meanings in judgments, and the reference to objects or states-of-affairs by way of these meanings. At the different levels and in different domains of inquiry we are to distinguish judgments from nonjudgments, consistent judgments from inconsistent judgments, and fulfilled from frustrated or unfulfilled judgments. The scientist may deal with her judgments and objects, but she *abstracts* from the role of the human subject in the sciences. She does not consider the scientific thinking itself in which her objects are given. We have the following picture:

scientific judgment J, expresses a meaning

↓   by which it refers to

[object or state of affairs]

This is really only part of a larger whole, however, that includes human subjects as those who are doing the judging, thinking, remembering, and so on. The whole picture is more like this:

Subject S thinks that J, where J expresses a meaning
↓   by which the subject refers to
[object or state of affairs]

The positive or 'objective' sciences deal with only part of a larger whole, and a dependent part at that. Husserl calls the broad conception of logic or of science in which we do not omit the role of the human subject a 'transcendental logic'. As in Kant's conception, it is a 'logic' or 'science' in which we do not abstract from the possible experience of human beings. The expression *transcendental logic* is in fact often used as another expression for transcendental phenomenology itself. Transcendental phenomenology is to be the science of the subjective and intersubjective side of experience, of consciousness and its object-directedness in any domain of conscious experience. It is to be the science of the essential features and structures of consciousness that provides the philosophical foundation of the sciences. If we are really interested in the conditions for the possibility of science, we cannot forget that, at bottom, it is the human subject who makes science possible. It is human subjects who bring about or constitute the sciences over time and who hand down the sciences from generation to generation.

On Husserl's analysis there are different types and levels of consciousness. Science is built up from the lifeworld experience of human subjects on the basis of acts of abstraction, idealization, reflection, formalization, and so on. The most basic, founding experiences are the everyday lifeworld activities, practices, and perceptions of people. The lifeworld, Husserl says, is "the intuitive surrounding world, pregiven as existing for all in common" (Husserl *The Crisis of the European Sciences and Transcendental Phenomenology (Crisis)*, §§ 33–34). All of our activities, including our loftiest sciences, presuppose the everyday practical and situational truths of the lifeworld. Our praxis and our prescientifc knowledge in the lifeworld play a constant role in all of our activities. The lifeworld was there for us before science, and even now human beings do not always have scientific interests. We have the intuited, everyday world that is prior to theory and then the various theories that are built up from this basis. Science thus presupposes the lifeworld as its starting point and cannot therefore replace this world or substitute something else for it. Human subjects are the meaning-givers and interpreters who produce the sciences from this basis and who can choose to be responsible about them or not.

Various fields of 'objective' science must therefore be correlated with the subjective if we are to do justice to the sciences. To see things whole we must deal with both objectivity and subjectivity. Otherwise science is naïve and one-sided. We would have only the positive, objective sciences in which the human subject is not remembered in the scientist's work. The human subject we are talking about here, moreover, should not be the reduced, dessicated subject that is presented to us by the *objective* sciences. It should not be the subject as interpreted through these sciences, which are themselves already founded on more basic forms of human experience. As we said, the positive 'objective' sciences are not foundational but are themselves founded on our lifeworld experience. In this manner, Husserl turns the table on those who think of the sciences and technology as providing the fundamental ways of knowing, understanding, and being, or who would value the positive sciences and technology above all else. It is this turn of thought, this critique of scientism, that has resonated with so many Continental philosophers for so long, from existentialists up through various postmodern philosophers.

Husserl was arguing long before such philosophers as Thomas Nagel that the positive sciences cannot in principle understand or do justice to the human subject (see, e.g., "The Vienna Lecture" included with Husserl's *Crisis*). They are 'objective' sciences that make of the human subject a kind of scientific object, an objectified subject. Husserl's view thus has many implications for psychology and the social or human sciences, as can be seen in the secondary literature on the subject (see, e.g., Natanson 1973).

In order to obtain the objective sciences one *abstracts* away from many features of experience. To abstract and to idealize are automatically to simplify, leaving behind some of the complexity and richness of concrete experience. Of course in the sciences such simplifications and idealizations help us to get a grip on things and make the work manageable. The simplification serves a purpose and might sometimes be put to good use, but one should not forget that it does not give us the complete picture. In the objective sciences one abstracts, for example, from the qualitative features of experience to focus on quantitative features. Calculation and calculational techniques come to the fore. Primary qualities of phenomena are highlighted and secondary qualities are marginalized.

As part of this abstraction one tends to focus on formal or structural features of experience, as distinct from contentual features. This shift to form or structure also involves a shift away from the full meaning of experience. Empirical, third-person observation is prized above all else while eidetic, consciousness-oriented observation is ignored or dogmatically held to be impossible. Add to this the specialized character of each of the positive sciences and we can see that, at best, they can treat of only part of the human subject, the outer shell, as it were. The subject with his or her 'lived body' is completely overlooked (see, e.g., Husserl *Ideas II* on the 'lived body'). Indeed, much of what is so important about human subjectivity and intersubjectivity will in principle be missed by the methods of the positive sciences. It is in the very nature of the methods of the objective sciences that they cannot get at human subjectivity with all of its first-person qualitative, contentual, and meaning-giving features.

For Husserl it does not follow that there cannot be a 'science' of human subjectivity. Transcendental, eidetic phenomenology is supposed to be (or to become) just such a science. It seeks essential, not accidental or contingent, features of human subjectivity. It does not itself seek empirical facts about human subjectivity but is concerned with necessary conditions for the possibility of such facts. It seems undeniable, for example, that human consciousness exhibits intentionality. How could we deny that a belief is always a belief about something, that it is always object-directed, unless we were already in the grip of an objective (founded) science that tells us there can be no such thing as intentionality (or even as belief)? One of the dangers of the positive sciences is just that they can blind us to phenomena that would otherwise be obvious. They can create prejudices of various types. One unconsciously accepts various background assumptions that are passed along in the positive sciences without subjecting them to critical scrutiny. It is phenomenology that should uncover such assumptions and critically analyze them.

If we can say that human consciousness exhibits intentionality, then we can also say that the human conscious subject is directed toward certain objects or states of affairs. We can distinguish different modes of consciousness (e.g., believing, knowing, remembering, desiring) and investigate and compare these modes with respect to their essential features. In further exploring features of consciousness we can see that acts of consciousness involve a kind of meaning-giving, that they are always perspectival, that they will have inner and outer horizons and certain Gestalt qualities, that there is always a time structure to consciousness (including a retention-protention structure and a secondary memory structure), and that the human subject is an embodied subject with particular forms of bodily intentionality. We see that perceptual observation is underdetermined by sensation (or 'hyletic data'), that there is a kind of hylomorphism in perception, and so on. Phenomenology is to investigate all of this material in detail. In connection with the positive sciences in particular it will investigate the types of consciousness in which science has its origins. In such a genetic, constitutional phenomenology one investigates the various founding and founded acts of consciousness that make science possible.

At the most fundamental level of experience transcendental phenomenology is to investigate the lifeworld and its structures (Husserl *Crisis*, § 36). We can say that the world is, prescientifically, already a spatiotemporal world in which there are bodies with particular shapes and material qualities (e.g., color, warmth, hardness). There is no question here of ideal mathematical points, or 'pure' straight lines, of the exactness belonging to geometry, and the like. It is a world in which certain shapes would have stood out in connection with practical needs. It would have been a world in which various relations between objects would have been noticed, and in which there would have been some awareness of causality. Primitive forms of measurement would have emerged, and so on. In works such as the *Crisis* Husserl therefore begins to plumb the formal and most general features of the lifeworld. This project was already under way in earlier manuscripts in which Husserl investigated the constitution of material nature and the constitution of animal nature (see Husserl *Ideas II*).

The investigation of the meaning and the manner of being of the lifeworld can itself be 'scientific' in a broad sense. This will be a notion of 'science' that does not systematically exclude our subjectivity as human beings situated in history. There is a difference, Husserl says, between 'objective' science and science in general (Husserl *Crisis*, § 34). Philosophy itself, as rational and critical investigation, may have as its goal the ideal of becoming scientific (Husserl, "Philosophy as Rigorous Science"). This ideal is very much alive even now except that it has been transmuted into the naturalism that dominates our age. Phenomenological philosophy, however, should aim to be a science that encompasses both objectivity and subjectivity and their relation to one another. Given this view of phenomenological philosophy as science, along with its relation to the other sciences we see that, in the end, it is transcendental phenomenology itself that is to be the science of all possible sciences.

§ 3 The Crisis of the Modern Sciences

The 'crisis' of the modern sciences has resulted from the attempt of these sciences to be 'merely factual' and to place increasing value on technization, mathematization, formalization, and specialization at the expense of other aspects of experience. There has been, Husserl argues, a positivistic reduction of science to mere factual science. The crisis of science, in this sense, is its loss of meaning for life. What is the meaning of science for human existence? The modern sciences seem to exclude precisely these kinds of questions. The physical sciences abstract from everything subjective and have nothing to say in response to such questions. The 'human sciences' on the other hand are busy trying to exclude all valuative positions and are attempting to be merely factual. They also fail to speak to such questions. It is for just these reasons that there is hostility in some quarters toward the modern sciences. Such human questions, however, were not always banned from the realm of science, especially if we think of the earlier aspirations of philosophical reason to become science. Positivism, however, 'decapitates philosophy' (Husserl *Crisis*, § 3). Various forms of skepticism about reason have set in. Skepticism insists on the validity of the factually experienced world and finds in the world nothing of reason or its ideas. Reason itself becomes more and more enigmatic under these conditions. In attempting to combat such trends one need not resort to naive and even absurd forms of rationalism. Rather, one needs to find the genuine sense of rationalism.

The modern sciences are made possible by mathematics and, in particular, by formal mathematics of the type that we see in algebra, analytic geometry, and so on. In the *Crisis* and related writings Husserl focuses on Galileo's mathematization of nature and the role of 'pure geometry' in making modern natural science possible. With Galileo nature becomes a mathematical manifold. What is involved in this mathematization of nature? Husserl discusses this at some length. It depends on the rise of pure geometry with its idealizations and its exactness. Galileo inherits pure geometry and with it he begins to mathematize nature. In interpreting nature through pure geometry we begin to view it in a different manner. It is no longer the 'nature' of prescientific, lifeworld experience. Various idealizations of nature are involved in seeing nature in terms of pure geometry. In the process we leave out or abstract from some of the aspects of the lifeworld experience of nature. The 'plenum' of original intuitive experience, however, is not fully mathematizable. Husserl thus says that the mathematization of nature is an achievement that is 'decisive for life' (Husserl *Crisis*, § 9f).

With mathematics one has at hand the various formulae that are used in scientific method. It is understandable, Husserl says, that some people were misled into taking these formulae and their 'formula-meaning' for the true being of nature itself. This 'formula-meaning', however, constitutes a kind of superficialization of meaning that unavoidably accompanies the technical development and practice of method. With the arithmetization of geometry there is already a kind of emptying of its meaning. The spatiotemporal idealities of geometric thinking are transformed into numerical configurations or algebraic structures. In algebraic calculation one lets the geometric signification recede into the background. Indeed, one drops it altogether. One calculates, remembering only at the end of the calculation that the numerals signify multitudes. This process of method transformation eventually leads to completely universal 'formalization' and to the kind of formalism we see applied in so many places in the modern sciences.

Various misunderstandings arise from the lack of clarity about the meaning of mathematization (Husserl *Crisis*, § 9i). For example, one holds to the merely subjective character of specific sense qualities. All concrete phenomena of sensibly intuited nature come to be viewed as 'merely subjective'. If the intuited world of our life is merely subjective, then all the truths of pre- and extrascientifc life that have nothing to do with its factual being are deprived of value. Nature, in its 'true' being, is taken to be mathematical. The obscurity is strengthened and transformed all the more with the development and constant application of pure formal mathematics. 'Space' and the purely formally defined 'Euclidean manifold' are confused. The true axiom is confused with the 'inauthentic' axiom (of manifold theory). In the theory of manifolds, however, the term *axiom* does not signify judgments or propositions but forms of propositions, where these forms are to be combined without contradiction.

In the same vein, Husserl speaks of emptying the meaning of mathematical natural science through 'technization'. Through calculating techniques we can become involved in the mere art of achieving results the genuine sense and truth of which can be attained only by concrete intuitive thinking actually directed at the subject matter itself (Husserl *Crisis*, § 9g). Only the modes of thinking and the type of clarity that are indispensable for technique as such are in action in calculation. One operates with symbols according to the rules of the game (as in the 'games-meaning' Husserl had discussed in *LI*, Investigation I, § 20). Here the original thinking that genuinely gives meaning to this technical process and gives truth to the correct results is excluded. In this manner it is also excluded in the formal theory of manifolds itself. The process whereby material mathematics is put into such formal logical form is perfectly legitimate. Indeed, it is necessary. Technization is also necessary, even though it sometimes completely loses itself in merely technical thinking. All of this must, however, be a method that is practiced in a fully conscious way. Care must be taken to prevent dangerous shifts of meaning by keeping in mind always the original bestowal of meaning upon the method, through which it has the sense of achieving knowledge about the world. Even more, it must be freed of the character of an unquestioned tradition whose meaning has been obscured in certain ways. This technization in which one operates with symbolic concepts often admits of mechanization. All of this leads to a transformation of our experience and thought. Natural science undergoes a far-reaching transformation and there is a covering over of its meaning.

As an example of some of Husserl's ideas here suppose I give you the formula

$$P(A_1/B) = \frac{P(A_1) \times P(B/A_1)}{(P(A_1) \times P(B/A_1)) + (P(A_2) \times P(B/A_2))}$$

and ask you to compute the values of $P(A_1/B)$ in case after case in which the values of the terms on the right-hand side of the equation are supplied and they always fall between 1 and 0. It is obviously possible to carry out this kind of input-computation-output procedure without knowing anything about what the formula is or means, what the numbers represent, what the origins of the formula are, what the purpose of this task is, and so on. Of course this might be only a small subroutine of a much more extensive routine and there is a very definite sense in which one need not know what any of it is about. Indeed, the entire process could be computerized. There is no need to reflect on or to understand the meaning of the formula. One might be carrying out computations about the space shuttle, or nuclear weapons, or economic aspects of homelessness. One could be quite blind to all of this and still carry out all of the calculations. Indeed, the goal may now simply be to solve such computational problems quite independently of what the computations are about. All of one's energies may go into this end and it is possible to become quite submerged in this kind of technical work, as happens in the case of many engineering problems. There is of course conscious directedness in all of this, but it is quite different from being directed toward the objects to which the formula is being applied. The objects are now the signs that are being manipulated in the calculation. In this manner there can be a complete displacement of concern (see Tieszen 1997b).

The skills and abilities associated with calculation and other forms of technical work will obviously be valued all the more in this kind of environment. This in turn fosters specialization in and professionalization of domains of scientific work. Technical knowledge and pragmatic success take center stage while other forms of knowledge and understanding tend to be marginalized. A pragmatic instrumentalism is the natural outcome of this shift in values and goals. This might be a pragmatic instrumentalism in which efficiency and the control and domination of nature are first principles, an instrumentalism that tends to view nature and everything in it as resources or 'input' for scientific/technological processes.

There is clearly potential here for a kind of alienation from reality, a distancing from the basic foundations of knowledge and understanding. This is an alienation made even worse by forgetting about these origins. If we think of Husserl's slogan "Back to the things themselves," then it is an alienation from 'the things themselves'. One might say that the possibility of such alienation attends the possibility of inauthentic ways of thinking and being, as opposed to more authentic ways. All of this arises with the abstractions and idealizations that make science possible. It arises from taking something as an independent whole that is really only a dependent part of a larger whole. It is to recede from a more holistic perspective.

In the modern natural sciences there is, Husserl thus says, a surreptitious substitution of the mathematically structured world of idealities for the only real world, the one that is actually given through perception: the everyday lifeworld. This substitution already occured as early as Galileo, and it was subsequently passed down through the generations (Husserl *Crisis*, § 9h). What has happened is that the lifeworld, which is the foundation of the meaning of natural science, has been forgotten. A type of naiveté has developed. Galileo is at once both a discovering and a concealing genius.

Although the techniques and methods of the sciences are handed down through the generations their true meanings, as we have said, are not necessarily handed down with them. It is the business of the philosopher and phenomenologist to inquire back into the original meanings through an eidetic analysis of the sedimentation involved. There is a 'historical meaning' associated with the formations of the sciences, but, as mentioned earlier, Husserl is not interested primarily in empirical history.

He is interested in finding the a priori unity that runs through all of the different phases of the historical becoming and the teleology of philosophy and the sciences (Husserl *Crisis*, § 15). Husserl says that it is a ruling dogma that there is a separation in principle between epistemological elucidation and historical explanation, or between epistemological and genetic origin. Epistemology cannot, however, be separated in this way from genetic analysis. To know something is to be aware of its historicity, if only implicitly. Every effort at explication and clarification is nothing other than a kind of historical disclosure. The whole of the cultural present implies the whole of the cultural past in a contentually undetermined but structurally determined generality. It implies a continuity of pasts that imply one another. This whole continuity is a unity of traditionalization up to the present. Here Husserl speaks about unity across difference on a global scale, not just in the case of the unities through difference that arise for us in our own personal cognitive life. Of course in the latter case too there is a temporality and a 'history'. Anything historical has an inner structure of meaning. There is an immense structural a priori to history.

Historicism is the view that there could be no such historical a priori, no supertemporal validity. It claims that every people has its own world. Every people has its own logic. Husserl responds by pointing out some of the background asumptions that are necessary for factual historical investigation to occur at all (see Husserl *Crisis*, Appendix VI). These are what we must know or assume before we can even get started with any factual historical investigation. In spite of all the indeterminacy in the horizon of 'history', it is through this concept or intention that we make our historical investigations. This is a presupposition of all determinability. But what kind of method can we use to make apparent to ourselves the universal and a priori features? We need to use the method of free variation in imagination in which we run through the conceivable possibilities for the lifeworld. In this way we remove all bonds to the factually valid historical world. We determine what is necessary and invariant through all of the contingencies and variations.

In the case of Euclidean geometry, for example, Husserl says that only if the necessary and most general content (invariant through all conceivable variation) of the spatiotemporal sphere of shapes is taken into account can an ideal construction arise that can be understood for all future time and thus be capable of being handed down and reproduced with an identical intersubjective meaning. Were the thinking scientist to introduce something time bound into her thinking, something bound to what is merely factual about her present, her construction would likewise have a merely time-bound validity or meaning. This meaning would be understandable only to those who shared the same merely factual presuppositions of understanding. Geometry as we know it would therefore not be possible.

Husserl thus argues against a relativism that would deny the ideal, omnitemporal character of sciences such as geometry, arithmetic, and logic, but he also argues against any absolutism that would cut off the truths of such sciences from their relation to human subjects. The objective and the subjective can only be properly understood in relation to one another, each conditioned by the other. One can say the same thing about the real and the ideal.

The things taken for granted in the positive sciences should, according to phenomenological philosophy, be viewed as 'prejudices'. As ideas and methods become sedimented over the years they become, quite literally, part of a prejudged mass of conditions and assumptions. In this manner various obscurities arise out of a sedimentation of tradition. One should subject such prejudgments again and again to critical, rational judgment. The genetic investigation of phenomenology is thus meant to allow us to become aware of such prejudices and to enable us to free ourselves of various presuppositions. It is therefore supposed to be the deepest kind of self-reflection aimed at self-understanding. Husserl sets up such presuppositionlessness as an infinite, regulative idea that we may never reach but that we should nonetheless never abandon. Husserlian hermeneutics would thus be rather different in some ways from that, for example, of Heidegger or Gadamer.

§ 4 Conclusion

These kinds of reflections lead to a host of issues that have distinguished Continental thinking about science from the analytic philosophy of science. Much of the analytic philosophy of science has tended to focus on questions that are more internal to the scientific enterprise. There have been periods during which some of the important questions raised by Husserl's critique of the sciences would simply never have occurred to those working in the analytic philosophy of science. Consider those circumstances, for example, in which the historical dimensions of science have been overlooked or devalued or in which the hypotheses and explanatory schemes of the positive sciences have been viewed as nonhermeneutical or value free. Husserl's work, by way of contrast, opens onto many issues involving the broader social, political, historical, and ethical dimensions of science and technology.

Husserl can be read as providing a corrective to scientism, to the view that it is only the positive sciences that supply knowledge, understanding, and truth. He is not arguing that the positive sciences do not have their place or do not have any value. Rather, we need to see them in their proper perspective. We need the correct balance. If scientism is extreme in one direction, then a view that completely rejects a place for the positive sciences is too extreme in the other direction. Scientism can be a kind of irrationalism and it can in its own way be blind, but there can also be an antiscientific kind of irrationalism and blindness. It seems to me that neither form of irrationalism would be acceptable to Husserl. It is Husserl's emphasis on reason that distinguishes him both from many analytic philosophers and from many Continental philosophers.

The positive, objective sciences both reveal and conceal. Phenomeno-logical philosophy, aiming toward science in a broader sense, tells us to retain what is revealed by the positive sciences (subject to critical scrutiny, responsibility, and broader values) and to reveal what is concealed by the positive sciences. This would be rational enlightenment at its best, for we would then be casting the light of reason as widely as possible.

2

Mathematics and Transcendental Phenomenology

Husserl began to publish on problems in the philosophy of mathematics and logic soon after he received his doctoral degree in mathematics in 1883 and he continued to publish on them throughout his lifetime. Although much has happened in the foundations of mathematics since the beginning of the twentieth century, many of Husserl's ideas are still relevant to recent issues in the philosophy of mathematics. In this chapter I argue that a number of the views on mathematics that are part of Husserl's transcendental phenomenology are more compelling than current alternative views in the philosophy of mathematics. In particular, I provide an overview of how Husserl's ideas can be used to solve some basic problems in the philosophy of mathematics that arise for (naive) platonism, nominalism, fictionalism, Hilbertian formalism, pragmatism, and conventionalism.

§ 1 A Précis of Problems in the Philosophy of Mathematics

Many of the basic problems in the philosophy of mathematics center around the positions just mentioned. It will not be possible to discuss these problems in any detail here, but at least some general indications can be given.

A major difficulty for platonism has been to explain how it is possible to have knowledge of immutable, acausal, abstract entities such as numbers, sets, and functions. Once it is argued that these entities are abstract and mind independent there seems to be no way to establish an

I thank Barry Smith for some helpful comments and suggestions.

epistemic link with them that does not involve mysticism. The apparent insurmountability of this problem might persuade one to abandon platonism altogether in favor of some form of nominalism. The nominalist will at least not have the problem of explaining how knowledge of abstract entities or universals is possible because on this view there simply are no abstract entities or universals. There are only concrete spatiotemporal particulars, and it is argued that however one ends up construing these there will be no great mystery about how we could come to know about them. One could work quite naturally, for example, with a causal account of knowledge.

The problems for nominalism lie elsewhere. Nominalism just does not appear to do justice to existing parts of mathematics, and especially to set theory. Mathematical statements do not appear, prima facie, to be about concrete spatiotemporal particulars, even though they may in some cases be applied to such entities. The language of mathematics does not itself mention such objects. So one of the first problems for the nominalist is to explain how and why mathematics is really so much different from the way it appears to be, and from the way it is taken to be by practicing mathematicians. For example, it is a fundamental assumption of different mathematicians at different times and places that they are discussing the *same number* (e.g., the number π) in their research. Mathematical practice suggests that *there is* an identity through difference here, but how could this be possible on a nominalist view?

We are also supposed to believe that we are systematically misled by the language of mathematics but not, for some reason, by language that refers to physical objects, or to concrete spatiotemporal particulars. Nominalist enterprises are reductive, for they propose schemes for reducing the language of mathematics to the language of concrete spatiotemporal particulars. The reductive schemes proposed even for elementary number theory, however, have turned out either to employ notions that resist nominalistic treatment or to be rather far-fetched. A major barrier to a nominalistic treatment of mathematics lies in the fact that many mathematical propositions are about infinite sets of objects, such as the set of natural numbers, but seeing how such propositions could be reduced to a language in which only spatiotemporal particulars are mentioned is difficult. The assumption that there is an infinite number of spatiotemporal particulars goes beyond what is needed in physical theory and may in fact be false. On the other hand, one might try to introduce modal notions into the reductive scheme to provide for at least a potential infinity of natural numbers. Here too there are problems. A coherent nominalist account of modality must be provided, but no such account has been forthcoming.

Should the problems of nominalism be enough to make one rebound into platonism? Not according to fictionalists. Fictionalism, roughly speaking, is the view that assimilates the language of mathematics to the language of fiction. Arguments for fictionalism have recently been put back into circulation in some of the work of Hartry Field.[1] As does nominalism, it denies that there are mind-independent, abstract objects. Mathematical objects are fictions, albeit sometimes convenient fictions. Unlike nominalism, it need not attempt to reduce the language of mathematics to a language of concrete spatiotemporal particulars. A fictionalist can argue that we understand mathematics independently of any such reductive scheme, just as we understand fiction independently of such a scheme. There are many problems with fictionalism and I shall return to some of them later.

Hilbertian formalism, as an attempt to secure the foundations of mathematics via finitist consistency proofs, is also beset with many problems. In particular, Gödel's incompleteness theorems show that we will not be able to have consistency proofs for interesting parts of mathematics, even for elementary number theory, if these proofs are to be based, as Hilbert wished, on only the immediate intuition of concrete, 'meaningless' finite sign-configurations. Consistency proofs for interesting mathematical theories will evidently have to involve reflection on or analysis of the *meanings* of the sign configurations or to introduce more 'abstract' elements (e.g., objects that cannot be encoded in the natural numbers) into our thinking, such as the elements we find in the Gödel and Gentzen consistency proofs for number theory. Traditional Hilbertian formalism has no place for such meaning-theoretic considerations or for such abstract elements. On the other hand, modifications of Hilbert's program that allow for such considerations, or for such abstract objects as primitive recursive functionals, take a decisive step away from strict formalism. The reliability or security of mathematics that is supposed to be based on metamathematical consistency proofs must now rest on insights of a different character.[2]

Hilbert's emphasis on the role of 'meaningless' syntax is also related to other problems of formalism, for example, the problem of why

---

[1] See Hartry Field 1989.

[2] Gödel links these claims directly to Husserlian phenomenology in "The Modern Development of the Foundations of Mathematics in the Light of Philosophy." This is a translation of a shorthand draft of a lecture that was never given. See Gödel *1961/?.

mathematicians are so interested in some systems of sign-configurations but not in others. This difference must have something to do with the meaning or reference of the sign-configurations. Moreover, how is the strict formalist to account for the understanding we have of the applications of mathematics?

Another perspective on mathematics is provided by pragmatism. In some of its formulations pragmatism appears to be simply antimathematical. For example, if we are not to accept any distinction which does not make a difference to practice it would appear that we ought not to accept large parts of 'pure' or theoretical mathematics. Pragmatists do not recognize any form of evidence intrinsic to mathematics that is supposed to support our mathematical beliefs. Our mathematical beliefs are to be judged solely on the basis of their fruitfulness, not on their faithfulness to some intuitive or informal mathematical concept that we believe we are developing. On Quine's view, for example, we are justified in believing axioms of mathematical theories only insofar as they form part of our best-confirmed scientific theories. But then what of unapplied parts of mathematics? One might wonder why mathematics should be beholden in this way to the natural sciences.

Another problem for the Quinean view has been pointed out by Charles Parsons.[3] Quine's view avoids the difficulties of earlier, cruder forms of empiricism about mathematics by assimilating mathematics to the most theoretical part of natural science, the part furthest removed from observation. But then how can it account for the obviousness of elementary mathematics? Are we really to believe that '$7 + 5 = 12$' is a highly theoretical assertion, one which is even more rarefied than the highly theoretical assertions of the physical sciences?

Finally, a consideration of how mathematics is actually done, of mathematical practice, shows that pragmatism simply does not respect the facts of mathematical experience. Fruitfulness does not always figure into which definitions, rules, or axioms are accepted by mathematicians. There are many examples in mathematics of efforts to be faithful to some intuitive or informal mathematical concept that is under investigation, examples of what Georg Kreisel has called 'informal rigor'.[4] One of the best recent examples of such informal but rigorous concept analysis can be found in Zermelo's description of an iterative concept of set, which has since been extended by set theorists.[5] Efforts to settle open problems

---

[3] See Parsons' "Mathematical Intuition" 1980, p. 151.
[4] See Georg Kreisel "Informal Rigour and Completeness Proofs" 1967.
[5] Zermelo's earliest description is in Zermelo 1930.

in set theory (e.g., the continuum hypothesis) by adopting new axioms have not been focused only on the question of which axioms are most fruitful. Another example of informal rigor in constructive mathematics can be found in the descriptions of the concept of lawless sequences that Troelstra and van Dalen use to obtain axioms for the intuitionistic theory of these objects.[6]

Conventionalism about definitions, rules, and axioms is subject to similar objections about informal rigor. It too fails to recognize any form of evidence unique to mathematics. It fails to do justice to the way open problems are approached and solved in practice. For conventionalism there is no question of trying to be faithful to some intuitive mathematical concept about which we are developing an understanding because in no particular case is there such a concept. Coventionalism is in fact unable to recognize any intrinsic constraints on mathematical concepts, as though concepts can be changed at will and solutions to problems adopted by decree. There are, in addition, many other objections that can be raised to conventionalism.

Husserl's position in the philosophy of mathematics, or at least the part of it that is worth saving, cuts across the various positions we have been discussing. It comprises an effort to avoid the kinds of problems described earlier and to arrive at a more refined view, and at the same time to do justice to mathematics as it is actually given and practiced.

§ 2 The Background of Husserl's View: Objectivity and Subjectivity in Mathematics

Husserl, having been trained in mathematics late in the nineteenth century, was witness to advances in formalization, generalization, and abstraction that were unprecedented in the history of mathematics.[7] He was evidently deeply impressed through his own work in the field with the 'objective' nature of mathematics. Early on, in the *Philosophy of Arithmetic*, we find him attempting to reconcile the 'psychological' or subjective aspects of our mathematical experience with the 'logical' or objective

---

[6] See A. Troelstra and D. van Dalen 1988, pp. 645–657.

[7] For some of the background of Husserl's mathematical training see especially J. P. Miller 1982; also, D. Lohmar 1989. Other works on or related to Husserl's philosophy of mathematics are Rosado-Haddock 1987; G.-C. Rota "Husserl and the Reform of Logic" and "Husserl" in M. Kac, G-C. Rota, and J. Schwartz (eds.) 1986; R. Schmit 1981; T. Seebohm, D. Føllesdal, and J. N. Mohanty (eds.) 1991; R. Tieszen 1989; R. Tragesser 1977 and 1984; and D. Willard 1984.

aspects. The relationship between the subjective and objective aspects of our experience in mathematics and logic was to become a central theme in his work, from the earliest to the latest stages of his career.[8] In § 2 of the Introduction to the second volume of the *Logical Investigations* (*LI*), for example, we find him asking:

How can that which is intrinsically objective become a presentation, and thus, so to speak, something subjective? What does it mean to say that an object exists both 'in itself' and 'given' in knowledge? How can the ideal nature of what is universal, [for example] a concept or a law, enter the stream of real mental events and become an item of knowledge for a thinking person?

Husserl's questions here bear a striking parallel to questions that had been discussed earlier by Bolzano and that were also being confronted by Husserl's great contemporary, Frege. In "The Thought" we find Frege trying to answer exactly the same questions about timeless and immutable 'thoughts'.[9] Recently, we find a similar concern in Gödel's effort to explain how we could have knowledge of transfinite sets which "clearly do not belong to the physical world."[10] The tenor of Husserl's questions suggests the perspective of a platonist, although, as we shall see, he is in some respects not a classical metaphysical platonist and he wishes to avoid the plight of the platonist mentioned earlier. Although it is as important for Frege and Gödel to answer these kinds of questions as it is for Husserl, we find that Husserl devoted far more effort to the project and obtained far better results. Husserl's answer to the questions show that he is also not a nominalist, fictionalist, or strict Hilbertian formalist, and that he disagrees with pragmatism and conventionalism about mathematics. At the center of his approach lies the concept of intentionality.

§ 3   Intentionality

It takes no great insight to notice that our mathematical beliefs are always *about* something. They are about certain objects, such as numbers, sets, functions, or groups, or they are directed to states of affairs concerning such objects. This 'aboutness' or 'directedness' of mathematical beliefs

---

[8] See, for example, Husserl's *Philosophy of Arithmetic*; "Philosophy as Rigorous Science" 1965, p. 172; *Formal and Transcendental Logic*, § 100; *The Crisis of European Sciences and Transcendental Phenomenology*, Appendix VI. We refer to this latter work hereafter as the *Crisis*. Also, see *Experience and Judgment*, §§ 3–4.

[9] See Gottlob Frege "The Thought" 1918.

[10] Kurt Gödel "What Is Cantor's Continuum Problem?" 1964, p. 267. See also Chapters 4, 6, 7.

is referred to as the 'intentionality' of our beliefs. That is, intentionality is just the characteristic of 'aboutness' or 'directedness' possessed by various kinds of cognitive acts, for instance, acts of believing, knowing, remembering, imagining, willing, desiring, and so on.

It is curious that although the intentionality of cognition is widely recognized and discussed in other areas of philosophy, there has been a veritable blind spot about it in the literature in the philosophy of mathematics. As a consequence, it has been easier for philosophers of mathematics to ignore what appear to be insurmountable problems in the efforts to naturalize intentionality.

A standard (if simplified) way to analyze the concept of intentionality is to say that an act of cognition is directed toward, or refers to, an object (or state of affairs) by way of the 'content' of the act, where the object (or state of affairs) the act is about may or may not exist. Historically, a variety of terms have been used in place of the term *content*, for example, *ideas*, *concepts*, *intentions*, and in Husserl's later technical terminology, *noematic nucleus*.[11] We can picture the general structure of the intentionality of our acts in the following way,

$$\text{Act(Content)} \longrightarrow [\text{object}],$$

where we 'bracket' the object in the sense that we do not assume that the object of an act always exists. Husserl is famous for suggesting that we bracket the object, and that we then focus our attention on the act and act-content, where we think of an act as directed toward a particular object by way of its content.

The contents of acts can be determined by considering 'that'-clauses in attributions of beliefs and other cognitive states to persons. Consider, for example, the following expressions:

Mathematician M believes that $7 + 5 = 12$,
M knows that there is no largest prime number,
M believes that every even integer >2 is the sum of two primes (Goldbach conjecture).

Or consider an example from Zermelo-Fraenkel (ZF) set theory,

M remembers that $(\forall x)(x \neq \varphi \rightarrow (\exists y)(y \, \varepsilon \, x \wedge y \cap x = \varphi))$ (the axiom of foundation).

---

[11] See especially Edmund Husserl *Ideas I*, §§ 36–37, 84, 87–102, 128–135. Also, *LI*, Investigations I and V.

Or a belief about extending ZF,

> M believes that a supercompact cardinal exists.

In these examples, the contents of a mathematicians' acts are expressed by the propositions following the word *that.* Different act-characters are also indicated: believing, knowing, remembering. Thus, a mathematician might have different types of acts with the same content, or acts of the same act-character with different contents. Note also the various differences in the contents expressed here. We have, respectively, a simple singular proposition about natural numbers, a theorem, a conjecture, an axiom, and a proposed axiom. Also, the propositions may be expressed in a formal language or not, and may be, for example, singular or universal, and so forth.

Different sets of considerations would be brought to bear in the phenomenological analysis of these propositions, depending on the features we have noted, but on Husserl's view what they have in common, as expressions of intentions, is that they are *about* numbers or sets, and properties of or relations between these objects. They are not directed toward other kinds of objects, such as physical aggregates or strings of signs. Husserl's view comports well with the ordinary view that the grammar and logic of mathematical language are not somehow deceptive and radically different from other parts of language. In fact, we can avoid from the outset the nominalist's problem, and all of the related problems, of having to explain how and why mathematics is really so much different from the way it appears to be and from the way it is actually taken to be in practice.[12] On Husserl's view we are to take mathematical language at face value, and to take mathematical theorems as true.

It should be noticed that we are of course taking these expressions of content to have a meaning. On Husserl's view language is in fact meaningful only insofar as it expresses intentions.[13] We can think of the content of an act as the meaning of the act by virtue of which we refer to an object or state of affairs. As Husserl puts it: "Under content we understand the 'meaning' of which we say that in it or through it consciousness refers to an object as its own" (*Ideas I*, § 129). Intentionality and meaning theory thus go hand-in-hand. Husserl in fact calls for just the kind of reflection

---

[12] For Husserl's specific criticisms of nominalism see especially *LI*, Investigation II; §§ 2, 7–8, 14, 26 speak to the issue of taking language about mathematical objects and universals at face value. See also *Ideas I*, §§ 1–26.

[13] See the *LI*, Investigations I and V; and *Ideas I*, §§ 87–91 and 128–133.

on or analysis of the meaning of sign-configurations for which there is no official place in Hilbertian formalism but which Gödel's incompleteness theorems suggest we cannot avoid. Husserl thought it possible to do a good deal of phenomenology on the basis of reflecting on the content of our acts without being concerned about whether objects or referents of the acts existed or not. For example, we could seek to clarify descriptively the content (meaning) of our acts through genetic or 'origins' analysis or through a procedure that Husserl called 'free variation in imagination'.[14] I shall have more to say about Husserl's views on the descriptive clarification of meaning later.

Husserl also has an account of the reliability of mathematics that is untouched by Gödel's theorems. I shall briefly describe this later. And on this view it is precisely by virtue of the meaning of sign-configurations that mathematicians are so interested in some systems of sign-configurations and not in others. None of this of course means that formalization is not possible. Husserl was in fact quite favorably disposed toward efforts to formalize mathematical theories. It is only a matter of properly understanding what we are doing.[15] The effort to develop formalized mathematical languages can be viewed as an attempt to express what is essential to the form of the 'noematic nucleus' of our acts, depending on various background assumptions.

The way that meaning is built up in our mathematical acts is also linked to the applications of mathematics. Husserl has an analysis of the 'origins' of mathematical content according to which mathematical acts and contents are founded on more immediate perceptual acts and contents. To say they are 'founded' means that they depend on such underlying acts and contents, which could exist even if there were no mathematics.[16] Husserl says that

what we have are acts which . . . *set up new objects*, acts in which something *appears as actual and self-given*, which was not given, and could not have been given, as what it now appears to be, in these foundational acts alone. *On the other hand, the*

---

[14] On 'free variation' see "Philosophy as Rigorous Science"; *Experience and Judgment (EJ)*, §§ 87–93; and *Ideas I*, §§ 67–72.

[15] Husserl's views on formalization and formalism are scattered throughout the *Philosophy of Arithmetic, LI, Ideas, FTL,* the *Crisis,* and various unpublished writings and letters. For my remarks here, see especially *FTL,* §§ 73–93, 101–107, and *LI,* Investigation I, § 20.

[16] The themes of genetic analysis, founding and founded acts, and acts of reflection and abstraction are found throughout Husserl's work. See, for example, *LI,* Investigation VI, §§ 40–58; *FTL,* §§ 70–71, 82–91; *EJ;* and the *Crisis,* Appendix VI, "The Origin of Geometry." See also R. Tieszen 1989.

new objects are based on the older ones, they are related to what appears in the basic acts. (*LI*, Investigation VI, § 46)

Then it is a condition for the possibility of mathematical knowledge that there be acts of reflection on and abstraction from our basic sensory experience. Mathematical knowledge would not be possible without such acts. Whereas formalist and fictionalist views get stuck on the problem of applications, Husserl could argue that parts of mathematics have applications because mathematics has its origins in our everyday experience in the first place. It is just that some of our idealizations, abstractions, and formalizations in mathematics are quite far removed from their origins.

§ 4 Mathematical Objects

If mathematical language is to be taken at face value, then on Husserl's theory of intentionality our mathematical beliefs are evidently about objects such as numbers, sets, and functions, and not objects of any other type. This fact, taken together with what Husserl says about the 'ideal' nature of mathematical objects, suggests that we must confront the platonists' problem of explaining how it is possible to know about ideal or abstract objects. Husserl's solution to this vexing problem is quite interesting. Let us approach it by considering his view of mathematical objects.

Perhaps the single most important thing to say about the conception of objects of cognition in Husserl's transcendental phenomenology, whether the objects be mathematical or physical, is that they are to be understood in terms of the 'invariants' or 'identities' in our experience. Many aspects of our experience are variable and in a constant state of flux, and out of this flux, or against this background of variation, we find that certain invariants emerge.[17] Physical objects are identities that emerge for us through various sensory experiences or observations. Facts about such objects, obtained by empirical induction and expressed in empirical laws, are to be understood in the same way. Now Husserl argues that we also find invariants or regularities in our mathematical experience although in mathematics the facts are not established, strictly speaking, by empirical induction. Mathematicians are not satisfied, for example, with establishing the truth of Goldbach's conjecture or Fermat's last theorem on

---

[17] See, e.g., *FTL*, §§ 61–63; *Cartesian Meditations* (*CM*), §§ 17–18, 21; and *EJ*, §§ 13, 64–65.

the basis of inductive generalizations from specific numerical instances of these propositions. Rather, a mathematical proof is required.

In mathematics we suppose that different mathematicians are reasoning about the same number, or the same set, or the same function at different times and in different places, and if it were not the *same* object we were reasoning about in all these different circumstances, one would have considerable difficulty seeing how the science of mathematics would be possible at all. Mathematics would be utterly fragmented. No two statements of a theorem (or of theorems) could be about the same objects. But this is absurd. There are facts about mathematical objects, expressed in axioms and theorems, that constitute invariants across our experience with these objects. The view that an object is fundamentally just an invariant through different experiences does not beg questions about the nature of mathematics by identifying 'object' from the outset with 'empirical object' or 'sensible object'. Husserl is concerned that we not let our views be clouded from the beginning with various philosophical presuppositions.[18]

Invariants may simply emerge as we gain more experience with objects, much as this happens in ordinary sense experience, or we may try to make them emerge through conscious and systematic efforts of the sort that are embodied in the methods of the sciences. Husserl argues that although both mathematical and physical objects are 'invariants' in our experience, the meanings under which we think about such objects are quite different.[19] Mathematical objects are not meant as 'real'; rather, they are meant as 'abstract' or 'ideal', and this meaning is derived from several sources. They could not be objects (identities) of sense experience because objects of sense experience occur and change in space and time and interact causally with one another and with us. They could not be mental in nature because what is mental occurs and changes in time. If mathematical objects were objects of either of these types, there would be no stability in mathematics. The ground would constantly be shifting beneath us. Husserl's extensive arguments against psychologism, nominalism, and other forms of empiricistic reductionism in the *Logical Investigations* and many other works are meant to ward off efforts to assimilate mathematical objects to the objects of 'inner' or 'outer' perception. Mathematical objects also cannot be assimilated to social or cultural

---

[18] See, e.g., *Ideas I*, §§ 19–23; and "Philosophy as Rigorous Science."

[19] See, e.g., *LI*, "Prolegomena to Pure Logic," Investigation II, Investigation VI, §§ 40–58; *FTL*, §§ 55–68, 82–87; *Ideas I*, §§ 1–26; *EJ*, §§ 63–65.

objects, because social and cultural objects are bound to times and places. Thus, Husserl says that mathematical objects are not 'bound idealities' but, rather, 'free idealities'.[20]

Objects such as numbers are also identities that transcend consciousness in the sense that there are indefinitely many things we do not know about them at a given time, on the analogy with our knowledge of perceptual objects, but at the same time we can extend our knowledge of them by solving open problems, devising new methods, and so on. They transcend consciousness in the same way that physical objects do. And, similarly, we cannot will them to be anything we like; nor can we will anything to be true of them. They are mind independent. On Husserl's view, unlike some empiricist or pragmatist views, it is not a puzzle that experimental methodologies in the physical sciences are different from methodologies in pure mathematics. Husserl was attempting to account for the fact, for example, that theorems of mathematics are not expressed probabilistically, and that we do not find in pure mathematics any statements referring to the space-time properties of the objects under consideration. Husserl's view also explains differences in the dynamics of the growth of the natural sciences and pure mathematics.

The abstractness of mathematical objects, on Husserl's view, implies and is implied by the fact that mathematical objects are unchanging, omnitemporal, and acausal. As we just noted, there is also a sense in which mathematical objects are mind independent and transcendent. Husserl's position is nonetheless different from classical or naive platonism in a number of important respects. Husserl's is a phenomenological (and a transcendental) view of objects, not a classical metaphysical view. This means that objects are to be understood as invariants *in the phenomena* or in our actual experience in mathematics, and yet they are given as objects that would exist even if there were no subjects. They are meant or intended as objects that need not be grasped by any subject in order to exist. The response to the nominalist is that the existence of mathematics presupposes that *there are* identities through difference, except that these are now simply understood as identities through the multiplicities of our own cognitive acts and processes. They are not anything more ultimate than that, not anything 'metaphysical' lying behind the phenomena. The view is very much like what Kant has to say about empirical objects and empirical realism, except that now it is also applied to mathematical experience. On the *object* side of his analysis Husserl can still claim to be a

---

[20] See *EJ*, § 65.

kind of realist about mathematical objects, for mathematical objects are not our own ideas. On the analogy with Kant's view of empirical objects, mathematical objects are not mental entities; nor are they fictions.

This view deflates standard nominalist objections to abstract objects, for such objections are directed toward naive metaphysical platonism. But how can the nominalist object to a *phenomenological* account of the 'invariants' in our mathematical experience without failing to respect that experience? Nominalism is just a kind of skepticism about abstract objects or universals. Skepticism about metaphysical objects that are somehow supposed to lie behind given phenomena is indeed warranted, but to doubt that there are invariants in mathematical experience would simply be disingenuous. The latter kind of doubt is inconsistent with the daily practice of mathematicians. Let us go further in our response to nominalism by filling in the account of mathematical knowledge.

§ 5   Knowledge of Mathematical Objects

We have already said enough about how 'abstract' objects should be understood to go a considerable distance toward solving the problem of how knowledge of abstract objects is possible. This problem has figured centrally in the recent philosophy of mathematics, in part because of a particular formulation of it by Paul Benacerraf.[21] On Benacerraf's formulation, our best account of truth for mathematical statements, Tarski's, calls for the assignment of numbers to number terms, sets to set terms, functions to function terms, properties or relations to predicate terms, and the like. Since it is numbers, sets, functions, and so on, that we are assigning, and not other objects which might be understood as concrete, the concept of truth for mathematical languages seems to require objects to which subjects could not stand in causal relations. Our best hope in epistemology, Benacerraf argues, is some form of a causal theory of knowledge. A causal theory of knowledge requires that for M to know that S, there must be a causal relation between M and the referents of the names of S, or at least that there must have once been such a relation for someone. On these premises it is a deep puzzle as to how we could have any mathematical knowledge at all.

Almost no one now believes in the causal theory of knowledge in the exact form in which it was formulated by Benacerraf. That there is still

---

[21] Paul Benacerraf "Mathematical Truth" 1973.

a lingering problem, even for newer reliabilist accounts of knowledge, is shown by the literature in the philosophy of mathematics since Benacerraf's paper was published. Certain kinds of empiricists continue to think that some kind of causal link to objects is a condition at least for noninferential knowledge of objects. This has led some philosophers to go to rather extreme lengths, from a Husserlian perspective, to solve the puzzle. In recent work by Penelope Maddy and Jaegwon Kim, for example, it is argued that at least some abstract mathematical objects – sets of physical objects – exist in space and time and hence are themselves the kinds of objects to which we can be causally related.[22] On the views of Maddy and Kim we literally sense these *sets,* using the same kinds of neural mechanisms that we use to sense insects, even though we cannot sense other kinds of mathematical objects, for instance, the set of real numbers or the empty set. The problem for these views is that they equivocate on either the notion of sensory perception or the notion of what counts as a mathematical object. It would presumably be a mistake in mathematics itself, for example, to suppose that a set has a color. If sets never have colors, what could it mean to say that they are objects in the physical world that we sense through various neural mechanisms? Husserl would object that these views do not do justice to the nature of mathematical objects or to mathematical knowledge. It could not be the case that some kinds of abstract, mathematical objects are located in space and time, are subject to causal interactions with other physical objects, are changeable, and so on, whereas other abstract, mathematical objects do not have these characteristics. From Husserl's viewpoint, attempts to solve the puzzle along these lines amount to imposing on mathematics a predetermined philosophical position which simply does not fit the data. One ends up twisting things out of shape to make the theory fit.

We can make some progress on the problem, however, if we start with the relatively harmless idea of mathematical objects as invariants that persist across acts carried out by different mathematicians at different times and places. Now even in the case of ordinary sense perception it can be pointed out that the only way to identify a perceptual object to which we believe we are causally related at a stage in our experience is through ongoing perceptions which either correct or fail to correct that identification. So the focus of our account will be on the sequences of acts through which evidence for mathematical objects (invariants) is

[22] Penelope Maddy 1990, Chp. 2. Also, Jaegwon Kim 1981.

acquired, and on this basis we can give an account of how we come to reject some propositions in mathematics but to accept others. Let us first briefly consider the analysis of how we reject propositions.

Mathematical objects and facts have a stability over time in our mathematical experience. In mathematical experience, as in ordinary perceptual experience, it can happen that what is believed to be an object at one stage of our experience is later seen to be an illusion. Thus, suppose we have an axiom from which we can derive the existence of a particular object, such as the set that contains all and only those sets that do not contain themselves. Then at a later stage of our experience we see that in believing that such a set exists we are led to a contradiction. Thus, at the earlier stage of our experience we were under an illusion about the existence of this 'object', and about the 'axiom' or 'fact' from which its existence was derived. This is an 'object' which does not persist over time, which does not have any stability in our mathematical experience. Husserl says that the experience in which such an object is intended as existing 'explodes', as a result of the discovered contradiction.[23] (This is something that Frege may have experienced firsthand.) Many other examples could be used to illustrate the same point. For example, round squares are not regarded as mathematical objects and there are no positive existence statements about them in mathematics because the assumption that they exist leads to contradiction, even though the expression *round square* does of course have a meaning. (The expression is grammatically acceptable and meaningful but inconsistent. See my remarks on the first two levels of 'objective logic' in Chapter 1.) Likewise, there can be no (universal) Turing machine which determines whether an arbitrary Turing machine running on an arbirary input halts or not. We now know that there could be no such mathematical object, although there was uncertainty about this at earlier stages of our experience with the concept of a mechanical procedure or recursive function.

So much for the basic idea of how we come to reject certain kinds of propositions in mathematics. The account of how we come to accept propositions in mathematics is more complex, for there are, in effect, different types and degrees of acceptance depending on the kind of evidence that we might acquire in multiplicities of acts. A justified belief in the existence of an object toward which an act is directed will depend on what Husserl calls the *fulfillment*, partial or otherwise, of an (empty) intention directed toward the object, and one cannot have evidence for

[23] *Ideas I*, §§ 138, 151.

existence independently of this.[24] There is no room for any mysterious source of knowledge on this account. Husserl distinguished different types and degrees of evidence. Thus, one might or might not have evidence that is apodictic, adequate, clear, and distinct.[25] Briefly put, apodictic evidence is evidence of necessity, whereas adequate evidence is evidence in which every possible act required to make the knowledge of an object complete would be carried out. Since infinitely many acts would be required for perfect knowledge, Husserl regards perfect adequacy as an ideal. There are only degrees of adequacy/inadequacy. There are also degrees of clarity and distinctness of evidence.

Husserl uses the idea of the fulfillment of an empty intention to define intuition: we (partially) intuit an object when our intention toward the object is (partially) fulfilled. The basic distinction between an empty and a fulfilled intention is straightforward. An intention is fulfilled when the object intended is actually experienced and is fulfillable when the object could be experienced. If existence proofs in mathematics are supposed to be expressions of the *knowledge* that *objects* of our intentions exist, then we must evidently actually experience the object, or at least possess the means for coming to experience it, and not merely produce a contradiction from the assumption that the object does not exist. In the latter case we have only the bare contradiction. We are not presented with or do not see the object, even partially; nor are we given any means to find the object. It may be that our reasoning and knowledge in a particular domain of research stabilize over time as we build up more propositions on the basis of indirect proofs, and we may even hold that this is a way of 'filling in' our mathematical knowledge, but this is still not the same thing as being presented with an object, or with having an intention toward an object be fulfillable. Husserl's view of the presence or absence of particular objects such as numbers, sets, and functions thus takes on a 'constructivist' slant. This is already suggested earlier by our talk about objects' being invariants *in our experience.* Saying that *objects* are invariants in our experience must be counted as different from saying that there are stable sets of thoughts-about-objects in our experience that may be further clarified. (Note that we are distinguishing here the object toward which we are directed by a meaning or intention from the meaning or intention itself. The meaning or intention may itself become an object in an act of reflection, but it is not itself a number, set, and so on.)

---

[24] See *LI*, Investigation VI, §§ 1–29, 36–48; *Ideas I*, §§ 136–141.
[25] *FTL*, §§ 58–60, 69, 78–80, 82–91, 105–107; *CM*, § 6; *Ideas I*, §§ 67–71, 137–138.

Other elements of Husserl's transcendental view of cognitive acts and processes in mathematics, and of the genetic analysis of mathematical concepts, also point to an epistemology of mathematical objects that is constructivist or 'critical' in a Kantian sense. In fact, these ideas have struck a chord in a number of logicians and mathematicians who have studied Husserl's work, including Hermann Weyl, Oskar Becker, Arend Heyting, and Per Martin-Löf.[26] Indeed, Becker and Heyting straightforwardly identified fulfilled mathematical intentions with constructions. Intuitionism is also a form of constructivism that recognizes some abstract objects, such as objects that cannot be encoded in the natural numbers, and this is another respect in which it is similar to Husserl's view.[27] Although Husserl's views on mathematics are similar in some respects to those of traditional intuitionism, as a consequence of the fact that both views are concerned with mathematical cognition, it is nonetheless clear that Husserl would be critical of the strains of subjective idealism, solipsism, extreme antiformalism, and psychologism that can be found in Brouwerian intuitionism.

Husserl also has a more sophisticated philosophical view of meaning and mathematical objects than can be found in traditional intuitionism, and I do not think that one could argue for the actual rejection of parts of mathematics on phenomenological grounds.[28] The propositions of classical mathematics are not meaningless. One could account for nonconstructive parts of mathematics on the basis of Husserl's views about meaning clarification, idealization, reflection, abstraction, and formalization. We might, for example, attempt to clarify further the meaning of the general concept of set, for although we cannot construct large, impredicatively specified sets, we also do not know that this concept is contradictory. Failure to construct an object that falls under some meaningful concept does not preclude reflection on the meaning and a partial understanding of it. Gödel has also suggested a position like this in a number of his papers.

As does Kant's transcendental philosophy, Husserl's transcendental phenomenology contains a critique of knowledge. Unlike Kant, however, Husserl was in a position to see what this meant in the case of modern

---

[26] See Hermann Weyl *Das Kontinuum* 1918a, and "Comments on Hilbert's Second Lecture on the Foundations of Mathematics" 1928. Also, Oskar Becker 1927; Arend Heyting "The Intuitionist Foundations of Mathematics" 1983; and Per Martin-Löf 1987. See Chapter 13.

[27] See R. Tieszen 1984.

[28] See R. Tieszen 1994a.

mathematics and to carry the idea further in the case of mathematical knowledge. Husserl's view strikes a balance between criticisms of mathematical knowledge that are too restrictive, such as those of ultrafinitism or finitism, and completely uncritical views on mathematical knowledge, such as those of naive platonism or naive rationalism. On the phenomenological view, mathematics will be reliable to the extent that our intentions toward mathematical objects are fulfillable in different degrees. Indeed, an analysis of processes of fulfillment might even be offered as an account of the 'reliable processes' sought by reliabilism. Parts of mathematical experience in which we do not have fulfillable intentions to objects will be reliable to the extent that they have at least stabilized over time in such a way as to allow for the development of new propositions, new methods, and so on. It might also be possible to find relative consistency proofs for these parts of mathematics. None of this, however, gives us absolute security. There are only degrees of security.

As is the case in other areas of cognition, mathematical intentions are to be viewed as expectations about objects or states of affairs, where these expectations are a function of the knowledge we have acquired up to the present time. They have their associated 'horizons' in which the possibilities of filling in our knowledge are determined by the meaning of the act.[29] Husserl says that these possibilities depend on structures of cognition that are governed by rules. Our expectations then can either be realized, be partially realized, or fail to be realized. Now, as suggested previously, Husserl was well aware of the fact that there are limitations on intuition, and hence on the kind of direct evidence we can have in mathematics. We cannot, for example, complete an infinite number of intuitions. In a section of *Formal and Transcendental Logic (FTL)* on the "idealizing presuppositions of logic and the constitutive criticism of them" he says, for example, that the subjective correlate of "reiterational infinity" is that one can carry out an infinite sequence of acts. He says that "this is plainly an idealization" since in fact no one has such an ability (*FTL*, § 74). The limitations on human knowledge are clearly noted in his discussions of types and degrees of evidence, and in other works in which he recognizes the finiteness in time and space of human experience. Husserl therefore says that there is no such thing as absolute evidence which would correspond to absolute truth. From the viewpoint of epistemology, all truth is "truth within its horizons" (*FTL*, § 105).

[29] On the concept of the 'horizon' of an act see *Ideas I*, §§ 27, 44, 47, 63, 69, 82–83, 142; *EJ*, § 8; *CM*, § 22.

Now in a Tarskian theory of truth, invoked by Benacerraf, we are in effect dealing with a dead or frozen relationship between a formal language and mathematical objects. The relationship can be highly idealized if 'truth' is abstracted from the horizons of subjects who know about truth or if, as Husserl says, truth is detached from the 'living intentionality' of human subjects. The way to solve Benacerraf's problem, and indeed the problem of the relation of the subjective to the objective in mathematics posed even earlier by Husserl, can be summed up as follows: we can agree that mathematical objects are 'abstract' and that number terms, function terms, and so on, refer to such objects. But what we mean by this is that these terms refer to the invariants or identities found in the multiplicites of our mathematical acts and processes. We *know* about these invariants in the way indicated earlier, that is, through sequences of acts in which we have evidence for their existence, even if we know about them only partially or incompletely. Otherwise, they are merely intended, in the sense that an intention directed toward an object exists even if we do not experience the object itself, and these intentions may have some stability in our experience and admit of further clarification. Relative consistency proofs may be forthcoming in some cases. Note that at no point in this argument for how mathematical knowledge is possible do we suppose that we are causally related to mathematical objects.

§ 6   Against Fictionalism

Many basic objections to fictionalism about mathematical objects, and to assimilating mathematics to fiction, fall out of what has already been said. We have seen that Husserl does not wish to be understood as a 'naive' realist or platonist about mathematical objects, but he is also not a fictionalist. Husserl says that "It is naturally not our intention to put the *being of what is ideal* on a level with the *being-thought-of which characterizes the fictitious or the nonsensical*" (*LI*, Investigation II, § 8). Let us consider some of the objections that can be raised to fictionalism. Our intentions toward some mathematical objects are fulfillable, but it is not clear what it could mean for intentions to fictional objects to be fulfillable. Even where our intentions to mathematical objects are not presently fulfillable we should hesitate to assimilate these intentions to intentions expressed in fiction. Generally, the idea that mathematical objects are fictions runs contrary to the way that these objects are intended in the science of mathematics. For example, various kinds of existence statements abound in mathematics texts, and it is not the business of the philosopher of mathematics to explain these away. They are to be taken at face value, and it is the business of the philosopher to explain how they are possible. In texts concerning fictional objects we do not, however, find sets of statements asserting the existence of objects. If the concept of existence even appears in such a context, it is subject to various anomalies that are not present in mathematical language. A very substantial difference between mathematics and fiction is suggested by Husserl's discussion of reason and actuality:[30] a sustained belief in the existence of fictional objects, as distinct from mere imagining of such objects, is a sign of irrationality. It is a psychopathological phenomenon. On the other hand, belief in the existence of mathematical objects, in the sense in which one believes existence theorems one has read in a mathematics text, is not at all a sign of irrationality. On the contrary, mathematics is usually taken to be one of the finest achievements of reason.

Other differences are also manifest. The idea of obtaining contradictions from the assumption of the existence of certain objects, which shows how such 'objects' do not have any stability in our mathematical experience, has no analog in our understanding of fiction. There is also a disanalogy concerning questions of the development of knowledge and of the determinacy of mathematical intentions when compared to intentions in fiction. Thus, when we ask, "Was Hamlet over six feet tall or not?" we find that no determinate answer can be given. Of course we can extend the story of Hamlet however we like so that we get a determinate answer, but this is not at all how open problems are solved in the science of mathematics. It could also be argued that mathematics ought not to be assimilated to fiction because mathematics has extremely rich and fruitful applications. It is not clear how this would be possible if mathematics were nothing but fiction. We might generalize from the point Kant made about empirical objects: mathematical objects ought not to be construed as 'fictions' just because we have no God's-eye view of them, or because it would be wrong to understand them as noumenal 'things-in-themselves'.

§ 7 Descriptive Clarification of the Meaning of Mathematical Concepts

We just said that we do not solve problems in mathematics by extending the mathematical 'story' any way we like. On Husserl's view there is a kind

---

[30] *Ideas I*, §§ 136, 138.

of evidence that is intrinsic to mathematics. Husserl's mature conception of the phenomenological method includes the idea of descriptive clarification of the content (meaning) of our acts through the procedure of free variation in imagination and through the analysis of the 'origins' of content. I do not have the space to discuss Husserl's ideas on genetic analysis and free variation here, but I would like to note that what lies behind these ideas is the belief in the kind of informal rigor mentioned at the beginning of this chapter. The idea of informal rigor is that one obtains definitions, rules, and axioms by analyzing informal, intuitive concepts and writing down their properties. On Husserl's view informal, intuitive concepts in mathematics are assumed to be significant, and it is believed that we can extend our knowledge of such concepts, make them precise, and gain insight into the fundamental relations that hold among them through genetic analysis and free variation.

This helps us to account for the problem the pragmatist has of explaining why mathematicians are not gripped only by considerations of fruitfulness in their research. As the example of the continuum hypothesis shows, set theorists are concerned about whether proposed axioms are faithful to our understanding of the general concept of set. There is evidence that is unique to mathematics which supports our beliefs in certain axioms. Moreover, we can solve the Quinean problem of explaining the obviousness of elementary mathematics precisely on the basis of the concept of mathematical intuition described earlier. There is, within mathematics itself, a distinction between more observable and more theoretical parts of research, and between what is closer to and further from its origins in the 'lifeworld' of everyday perceptions and practices.[31]

Husserl's view also helps us to solve problems of conventionalism and strict formalism. We cannot make the invariants in our mathematical experience be anything we want them to be. Mathematical theories are not arbitrary creations, but rather we are 'forced' in certain ways in the development of mathematical knowledge. We do not, for example, solve open problems by convention.

It is important that Husserl is advocating a descriptive method for clarification of the meaning of mathematical concepts. A good deal of philosophical methodology, especially in the analytic tradition of philosophy, is concerned with offering arguments for the truth of various

---

[31] The concept of the 'lifeworld' in Husserl's later philosophy is an outgrowth of his earlier ideas on founding and founded acts, genetic analysis, and related themes. See especially the *Crisis*, §§ 9, 33–38, Appendix VII.

assertions. There is clearly a problem in trying to apply this methodology to elucidate primitive terms and to justify axioms of mathematical theories since with it one only succeeds in backing the problem up a step by depending on other unanalyzed concepts and propositions. As an alternative, Husserl's philosophy suggests that it is possible to understand the meaning of some mathematical assertions not only through argument but also through careful, detailed description of a concept, for example, the general concept of set. Husserl's work thus suggests that there is a way out of the old philosophical problem of how we should understand the meanings of the primitive terms and axioms of mathematical theories.

§ 8   Conclusion

The views described here are framed in Husserl's later philosophy by a variety of interconnected arguments against empiricism, naturalism, nominalism, and psychologism in mathematics. It is clear from comments in the "Prolegomena to Pure Logic" and other parts of the *Logical Investigations* that Husserl sees nominalism and psychologism as species of empiricism or naturalism about mathematics. They are forms of empiricistic reductionism, and Husserl is generally opposed to reductionism in the philosophy of mathematics. In his criticism in *Philosophy of Arithmetic* of Frege, who is certainly no empiricist about mathematics, we see Husserl pointing out that Frege's extensional definition of number also amounts to a kind of reductionism about the meaning of the concept of number.[32] Husserl is concerned that any of the views mentioned not do justice to our mathematical intentions, and to the way in which mathematical objects are intended in our acts. The views simply do not do justice to what we know in the science of mathematics. For the same general reasons, his view would be opposed to conventionalism, and to any kind of strict formalism about mathematics. His views on formalism do not entail that formalization in mathematics has no value. On the contrary, formalization of our mathematical concepts is quite important. However, from everything that Husserl says about intentionality, meaning, formal (as opposed to transcendental) logic, the idealizations and abstractions from experience required by formal systems, and the kind of informal rigor which makes mathematics possible, it is clear that the later Husserl is not a Hilbertian formalist.

---

[32] See R. Tieszen 1990 and Chapter 15.

Husserl thus gives us a way to solve the platonists' problem while avoiding the problems of nominalism and fictionalism. His call for a 'transcendental' foundation of mathematics is unaffected by Gödel's incompleteness theorems and the problems of 'meaningless' syntax, but it also avoids the problems that result from the pragmatic overemphasis on fruitfulness and from the supposition that mathematics is nothing but convention.

3

Free Variation and the Intuition of Geometric Essences

Some Reflections on Phenomenology and Modern Geometry

Edmund Husserl is perhaps the only philosopher of the past one hundred years or so who claims that we can intuit essences and, moreover, that it is possible to formulate a method for intuiting essences. Husserl calls this method 'free variation in imagination' or 'ideation'. It is explicated in some of his writings as the 'eidetic reduction'. His descriptions of ideation can be viewed as attempts to describe a method appropriate to the a priori sciences, a method that does not reduce to the methods of the empirical sciences. The best and clearest examples of this method, it seems to me, are to be found in mathematics. Pure mathematics, according to Husserl, is concerned with exact essences. Husserl's own examples of the method, however, are often concerned with the inexact (or 'morphological') essences of everyday sensory objects (e.g., color, sound) or of the phenomena that form the subject matter of phenomenology itself (consciousness, intentionality, and the like). It is unfortunate that Husserl does not give more examples involving mathematics. He seems to focus on the nonmathematical cases because he is very concerned to show that phenomenology itself can be a kind of science, a descriptive science of the essential structures of cognition. Perhaps he thinks it is not necessary to dwell on mathematics, which, unlike phenomenology, is already a firmly established science.

Husserl's views on intuiting essences have been subjected over the years to extensive discussion, criticism, and even ridicule. What I would like to do in this chapter is bring a fresh perspective to bear on these

I would like to thank a referee for *Philosophy and Phenomenological Research* for comments on this essay.

views by illustrating them in connection with some examples from modern pure geometry. One can see in these examples, I think, what Husserl was driving at in his descriptions of the method. Thus, I will first indicate what the components of the method of ideation are supposed to be and will illustrate them in the context of some ideas in modern geometry. Some of the specific ideas in geometry that I have in mind were under development in Husserl's time and even before. Indeed, in his writings on geometry Husserl refers to some of the central characters: Grassmann, Riemann, Lie, and Helmholtz (see Husserl 1973a, 1983). Looking back, I think we need to keep in mind a number of other central figures who were contemporaries of Husserl's or did their work later. The approach to geometry that began with Felix Klein in particular should be of special interest to those wishing to understand the examples of ideation I will discuss (Klein 1893, 1948; Torretti 1978; Greenberg 1993). (It is interesting to note that Husserl's ideas about *Wesenanalyse* influenced Carnap in his 1922 thesis on space (Carnap 1922) and Weyl's work on geometry and physics (in, e.g., Weyl 1918b).)

After presenting some examples and briefly discussing the many varieties of spatial ontology in modern geometry I will turn to some of Husserl's views on the origins of geometry. What does the genetic analysis of geometry show us about the various geometries and 'spaces' that are part of modern mathematics? The final section of the chapter is concerned with a few questions about applications of the method of ideation outside pure geometry.

The focal point of this chapter is free variation in imagination in connection with some examples from modern geometry. There are many other ideas in Husserl's writings on geometry that are not discussed here. It should also be noted that Husserl himself does not give the kinds of examples included here and that I will be appealing to some developments in geometry that occurred after his time.

§ 1   Ideation in Geometry: Some Examples

In an elementary geometry course one might be asked to imagine variations on given geometric figures. Imagine, for example, various translations or reflections of a triangle or of some other geometric figure, 'squishing' a circle in different ways, or transforming a circle into a triangle, and so on. If I start squishing a circle, for example, I might ask which properties of the circle change and which remain the same. So let us start in this vein with a simple example involving points and lines in Euclidean

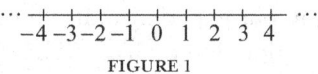

FIGURE 1

geometry. Consider the points on a number line in Figure 1. We can express transformations of points Px on this line in terms of their coordinates since there is an isomorphism between the points and their coordinates. Now perform the following (free) *variation* ('in imagination'): imagine each point moved 4 units to the right. Under this variation (technically known as a 'translation') all points are changed: -2 becomes 2, -1 becomes 3, 0 becomes 4, -1 becomes 5, 2 becomes 6, and the like. What is it that all of these particular changes have in common? We can write a little equation to express it: the coordinate x' of the point corresponding to the point Px is related to the coordinate x by the equation $x' = x + 4$. Under this variation *there are no invariant points.* If we like, we can immediately generalize from this basis: each equation of the form $x' = x + a$ represents a translation of the points Px to the points $Px' = Px + a$. If we imagine shifting or varying points in this way, does anything concerning the figure remain invariant? Yes. *Distance remains invariant,* in the sense that $x_1 - x_2 = x'_1 - x'_2$, for we see that this latter equation holds even though all of the points shift in accordance with the formula $x' = x + a$. Also, *direction remains invariant* under this kind of imagined variation, in the sense that $x_1 < x_2 \leftrightarrow x'_1 < x'_2$. Thus, under each translation distance and direction along the line remain invariant. They are in fact known to be invariants under the group (in the technical sense of group theory) of translations on the line. We can say that the formulas $x_1 - x_2 = x'_1 - x'_2$ and $x_1 < x_2 \leftrightarrow x'_1 < x'_2$ represent *necessities* through all the changes of a in $Px' = Px + a$.

With this example in mind let us consider how Husserl describes the method of ideation in a number of his writings (see especially Husserl 1973b, 1977). The method of ideation is composed of the following moments: (1) one starts with an example or 'model'; (2) one actively produces and runs through a multiplicity of variations of the example; (3) one finds that an overlapping coincidence occurs as a 'synthetic unity' through the formation of the variants; and (4) one actively identifies this synthetic unity as an invariant through the variations. It is at this final stage of the process that there is awareness of an essence. The essence is that which all of the variations have in common. It is that which remains invariant through all of the variations. The method is supposed to bring about an awareness of something that one might not have had otherwise or at least to make us *explicitly* aware of something that may have been only *implicit* in our experience. It is meant to be a method of a priori knowledge formation the results of which fix conditions that are necessary for things with a given essence.

Thus, in terms of Husserl's description of the procedure of ideation (1) we started with an example or 'model', and then (2) we actively produced and ran through a multiplicity of variations concerning the example. First we have the specific variation $x' = x + 4$ under which there are no invariant points. It is also immediately evident that there will be no invariant points under the variation $x' = x + a$ for any choice of a (other than 0). Husserl says that intuiting a universal or essence here, which is an awareness at a 'higher level', relates to the multiplicity of variations. If there is to be awareness of an essence, then there must be (3) an overlapping coincidence of the variants in which they all appear as variations of one another. $x_1 - x_2 = x'_1 - x'_2$ (distance) emerges as an invariant for us once we see, in the case where $x' = x + 4$, that there is an 'overlapping coincidence' of the variants $4 - 2$ and $8 - 6$, $2 - 1$ and $6 - 5$, and so on. There is something that these different pairs of expressions have in common even though we are aware of them at different times and we would evaluate the subtractions at different times. Against this background of variations, $x_1 - x_2 = x'_1 - x'_2$ emerges as a 'synthetic unity'. What we have here, in Husserl's language, is an 'identity synthesis'. The identity is 'synthetic' in the sense that it emerges or can be made to emerge against mental activities (forming variants) that *are taking place at different times*. Because these mental activities themselves have temporal duration and occur at different times there must be some cognitive 'synthesizing' that is taking place across them. Husserl adds that (4) there must be an active identification of this synthetic unity as an invariant through the variations. That is, the same universal or eidos must be 'singularized' against the background of the variations; it must be taken as an object (or state of affairs) in its own right that can be compared with other objects (states of affairs), be the subject of further predications, and so on. The idea is that in the midst of all of our free variations we will come up against certain constraints, as though we have a swirling sea of changes around some islands of permanence.

In order to forestall one possible misunderstanding of the method we should note that in referring to the role of imagination we are not claiming that forming an image in the mind is necessary. There need not be an image based on sense perception nor an image of any other type. Rather, in appealing to imagination we are appealing to the ability to think of logical possibilities where this need not be accompanied by any images at all. The same logical possibility might be accompanied by different mental images in different subjects. In short, variation in imagination may but need not be accompanied by mental images.

If we are to grasp an essence, then the multiplicity must be an object of consciousness *as a multiplicity* and it must be retained in consciousness as such. Otherwise we do not obtain an essence as that which is identical through the multiplicity. By the same token, the essence is precisely the one over the many. There must be an active identification of this 'oneness' or else we do not have intuition of an essence. We might be occupied merely with imagining a thing as different from a previous example, and then another variant, and so on. In doing this we would always have an awareness of something new and always only one thing, the last one imagined. In this case, however, no essence emerges. We would not see what, if anything, the variants have in common.

We must often actively look for what is invariant across the variations. It can sometimes be quite difficult to find this, as is apparent in cases where our knowledge of generalities is still under development. In cases where we do not have knowledge we might hypothesize or conjecture certain invariants, but this process is different from seeing that they actually hold or from actually finding them. Not just any invariants will be hypothesized in a given context. The possibilities (i.e., what is in the horizon) will be determined by the intentions that are at work in our thinking in a particular situation. Husserl also allows, as he should, that an invariant sometimes emerges for us in a purely passive way. It dawns on us, as it were, without an active search by us. In any case, it is at the final stage of the process that there is awareness of an essence. The essence is that which all of the variations have in common.

Now in order to press a little further with this illustration of Husserl's method we might imagine another variation in which we uniformly stretch our line about the origin. For example, transform each point Px at a directed distance x from the origin into the point P_{3x} at a directed distance $3x$ from the origin. The equation $x' = 3x$ thus represents such an imagined variation. Under this variation point 2 becomes point 6, point 3 becomes point 9, and so on. Now does $x_1 - x_2 = x'_1 - x'_2$ still hold? No. If I hypothesize that $x_1 - x_2 = x'_1 - x'_2$, then I see that my intention is now frustrated rather than fulfilled. Under this kind of variation it is no longer a necessary truth. What about $x_1 < x_2 \leftrightarrow x'_1 < x'_2$? This continues to hold. Direction remains invariant under this variation. We can generalize again: under the variations represented by the equation $x' = ax$, where a is positive, the origin remains invariant, as does direction, but distance does not remain invariant. Indeed, this can be made more precise in the language of modern geometry: the origin and directions are invariant under the group of transformations $x' = ax$ (where a is positive).

These simple examples show us that what remains invariant is a function of the kinds of transformations or variations we make. If the variations are radical enough, we might expect that fewer invariant properties will remain. Are there, for example, variations under which neither distance nor direction would remain an invariant property? In fact, there would be such variations, but then we can ask what would remain invariant under these more radical variations. There will be, as it were, deeper or more abstract invariants that we cannot eliminate even under the most extreme imagined variations. We should therefore distinguish different levels of invariants in relation to different groups of variations.

Husserl says in many works that the intuition of essences based on free variation is not at all mysterious. Indeed, if the reader understands the examples just presented, then he or she has just had such an experience. If you can grasp, that is, that distance is (not just hypothetically but actually) invariant in our first example, then you have grasped an essence. If you can see that direction is invariant in either example, then you have intuited another essence. The invariants we have spoken of just are *essences* in the sense that an 'essence' is a feature or property that remains the same through many variations. It is something that a multiplicity of particulars have in common and is in this sense a universal. If this is what we mean by 'essence' or 'universal', then we have in modern geometry and topology a domain of cognition in which there already exists a highly developed mapping out of essences (or universals) and their relations to one another (see § 2). A very sophisticated classification and unification within geometry is indeed brought about in this manner, including a sorting of geometric essences into species/genus hierarchies. The whole of modern geometry is full of the kinds of examples we have given.

We can thus see why Husserl speaks as though the experience of essences is or could be quite commonplace. In laying out these examples we have also not introduced any heavy metaphysical baggage into our discussion. Just imagine yourself engaging in the practice of geometry in this manner (as would be common in modern geometry) without getting especially philosophical about what you are doing. Then we might say with Husserl that you are apprehending an eidos or 'idea' (in something like a platonic sense) but in its purity and free of metaphysical interpretation. You are therefore apprehending it exactly as it is given immediately and intuitively in your experience. What we are emphasizing here is the *phenomenology* of your experience – the fact that an invariant stands out for you in the examples – without getting hung up about the metaphysics of essences. In this chapter I will make a point of suspending judgment about or 'bracketing' questions about the metaphysics of essences (i.e., questions of the form 'Do essences exist independently of us as eternally unchanging?' 'Are essences in some platonic heaven?' and so on).

Why should we call this an *intuition* of an essence? Husserl says it is an intuition because we are not merely hypothesizing or conjecturing some state of affairs. It is not a matter of merely speaking, thinking, or wishing without having the state of affairs itself before us. We are not merely expressing some opinion based on some vague source. There is a difference between referring to something indirectly or by means of empty symbols and referring to something that is given to us. It is in the latter case that we have knowledge in the best sense. What we have in our examples is geometric *knowledge* in which the affairs themselves are experienced. We have the invariant before us and we can then go on to compare it with other invariants, and so on. There is a mental overlapping in which some common thing emerges. Of course this is not a sensory grasping or intuition. It is a kind of rational intuition.

One can point to many aspects of the intuition of essences that are analogous to aspects of the intuition of sensory objects. For example, sensory objects are like essences in that they are given to us as invariants through multiplicities of acts, persons, times, places, properties, and so on. There are, of course, differences. Sensory objects are themselves temporal objects that are more or less stable. They are undergoing change. Relative to us, this change may be very slow or it may be very fast. The invariants given by sense perception are, it seems, characterized by their impermanence. If we say that what is characterized by temporality in this way is 'real', then we might hold that the notion of *permanence* through change represents an 'ideal' and that some kinds of invariants are 'ideal'. Some other analogies between the intuition of sensory objects and the intuition of essences seem to be as follows: (i) in both cases we do not merely think or speak about the object in an empty way: the object is present to us and plays a role in our *knowledge;* (ii) the intuition in both cases can be more or less clear and distinct; (iii) the intuition takes place through sequences of cognitive acts in time; (iv) the intuition is forced or constrained in certain respects; and (v) the intuition is in principle fallible. Much more could be said about all of this (see, e.g., Husserl 1982).

One additional point that might be mentioned here is that intuition of an essence should be distinguished from intuition of other kinds of objects that have sometimes been substituted for essences (especially in reductionist accounts of thinking and of mathematics). It is not, for example, intuition of a set. We make this point in order to be careful about distinguishing the language of essences from the language of sets. In the language of essences, for example, we can speak naturally about the difference between what is essential to certain kinds of spaces and what is incidental or accidental, but the language of sets is not used for this purpose. We use the two languages for different purposes. We do not want from the outset to confuse them or to adopt an eliminative reductionism about one of them hastily. In fact, the relationship between the two languages is itself a matter for further detailed investigation. Note, for example, that we can ask what the essential properties of sets are, but it is not clear how we could ask or answer such a question if essences were just sets. In phenomenology there is a view of the mind that recognizes both intensional and extensional aspects of consciousness in mathematics. There are reasons for taking essences to be intensional objects. We can think of them as the properties or relations expressed by (formal or material) predicates, not the extensions of these properties or relations. We can of course also consider extensions of properties or relations, that is, sets.

§ 2   Hierarchies of Geometric Essences

Let us now discuss in a little more detail the claim that in modern geometry and topology we already have a domain of cognition in which there exists a highly developed mapping out of essences (or universals) and their relations to one another. Since the inauguration of Klein's Erlanger program it has been possible to think of different geometries as defined by groups of transformations. (The term *group* here is used in the technical sense of group theory.) Each geometry investigates everything that is invariant under the transformations of the given group. Geometry will thus be the study of properties left invariant under groups of transformations. These properties will be expressed by the axioms, definitions, and theorems that are or could be set up for each particular geometry. What is invariant in one geometry is not necessarily invariant in another. Each different geometry had its own truths. These different geometries can, however, be systematically related to one another to a remarkable extent. What made this possible, historically speaking, was a substantial body of work in mathematics that had its origins in the theory of algebraic invariants of Cayley and Sylvester, and even earlier work centering on the concept of invariance, along with the subsequent developments in geometry due to Grassmann, Riemann, Lie, Helmholtz, and others. Husserl himself refers to these latter figures in his work on geometry. There is, in my view, a sense in which we can see a preoccupation with 'ideal' invariants in geometry right from the beginnings of this science, albeit in a much cruder form (see § 4 in this chapter).

What is nice about the approach to geometry developed by Klein, on my view, is that it meshes quite naturally with Husserl's descriptions of his method of ideation once we consider it in the case of geometry. (Husserl himself does not make this connection.) It provides a vivid illustration of the method and does so in connection with an existing body of mathematics. This does not mean that we should suppose that the method of free variation in imagination must be understood everywhere in terms of finding properties that remain invariant under groups of transformations. We need not necessarily use the group concept in all possible applications of ideation; nor need we import the specific technical aspects in the geometric case back into the use of the method in other domains (even in mathematics, to say nothing of 'persons', 'consciousness', and the like). Perhaps we should view the idea of finding invariants under groups of transformation in geometry as a *specific instance* of the method of free variation in imagination. It stands to the general method in a genus/species relation. It would be a variant of the method. We might in some contexts appeal instead to what remains invariant under a 'set' or a 'list' of variations. The main idea is to work out all of the a priori relations given whatever basic technical concepts we choose to employ at the outset.

To elaborate somewhat on how structures of variation and invariance are found throughout modern geometry, we can note that Euclidean geometry can be viewed as the study of those properties left invariant under so-called rigid motions: translations, rotations, and reflections. Projective geometry is concerned with the smaller class of essences that are a function of the rigid motions plus projections. Topology is concerned with the still smaller class of essences that we obtain if our variations are even more radical, including the most extreme stretchings and twistings. Thus, length and angle are Euclidean essences, but they are not invariant under projective variations. Linearity and triangularity are projective essences, but they are not invariant under the more radical topological variations. Connectedness and number of holes (to be more precise, 'genus'), for example, are topological essences (see, e.g., Meserve 1953). One might say that topological essences are quite abstract or deep, relative to other geometric essences. They quite literally result from greater levels of abstraction. We can relate this idea of greater *levels of abstraction* directly to the types of variations to which one subjects objects. In topology we do not, for example, lose the property of dimensionality of a geometric figure or object, or the property of having a boundary or not, but we do lose properties such as size and shape as these are understood in Euclidean geometry. Topological equivalence is much more abstract than equivalence in, say, Euclidean geometry.

We can thus think of topology as a generalization of Euclidean geometry, and indeed of projective geometry. The group of topological transformations has the group of projective transformations as a subgroup and therefore has the group of Euclidean transformations as a subgroup. The *generalization/specification relations* can in fact be made very precise here. We can say exactly what is involved in making the geometry more 'general' or more 'specific' by pointing to groups of transformations (variations). Some variations are more radical than others. We cannot perform them without changing a property that was invariant for a range of variations into a property that is no longer invariant. The idea is then to map out all of this, characterizing in each case the geometry obtained.

There are properties that are essential to Euclidean space, but different properties are essential to projective space, or affine space, hyperbolic space, elliptical space, and so on. As Husserl says, an essence proves to be that without which an object of a particular kind cannot be thought, that is, without which the object cannot be imagined as such. In this sense an essence is a kind of constraint that we cannot transgress. Thus, Euclidean space and the objects in Euclidean space cannot be thought without certain properties, projective space and the objects in projective space cannot be thought without another set of properties, and so on. A triangle or circle cannot be imagined to be just anything. Our imagination here is not free but is bounded or limited in certain respects. If the variations we make are too radical, we lose a property such as triangularity. The essential properties will, as we said, be expressed in the axioms, definitions, and theorems of each of the different geometries. We have, for example, the hierarchy of geometries in modern geometry shown in Figure 2 (viewing topology as a very abstract kind of geometry). An essence of a given geometry is also an essence of every special case of the geometry. A topological essence is also, for example, a Euclidean essence,

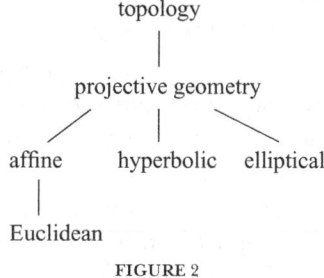

FIGURE 2

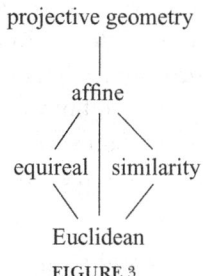

FIGURE 3

but there are many Euclidean essences that are not topological essences. There are many more constraints associated with Euclidean geometry than there are with the more 'abstract' geometries.

Note how this classification and organization extend to Euclidean and non-Euclidean geometries. One can start with Euclidean geometry and abstract until one obtains projective geometry or topological invariants, or one can specialize from the top down to Euclidean geometry. In the latter case the number of invariant properties increases by specializing the transformations under consideration until we have the invariant properties of Euclidean geometry. As we said, this is all made very precise in modern geometry by considering subgroups of a group of transformations.

To fill in the diagram in Figure 2 in even more detail, for example, we might start with Euclidean geometry and remove the requirement that area be invariant. Then we will necessarily lose some other properties as well, and we obtain the group of similarity transformations. If we remove the requirement that perpendicularity be invariant, we obtain equireal transformations. If we remove both of these (*abstract* from them, as it were) we obtain affine geometry. If we remove the restriction that parallelism be invariant, we obtain projective transformations. These remarks can be summarized as in Figure 3.

Topology was at the top of our earlier diagram. In topology we consider what remains invariant under arbitrary continuous variations. A sphere, a pyramid, and a cube, for example, can be transformed into one another by continuous variations, but a sphere cannot be transformed into a torus (a doughnut-shaped object) by such transformations. A sphere is an object of genus 0, whereas a torus is an object of genus 1. If we think of topology as the geometry of one-to-one and continuous point transformations, then set theory appears to be even more abstract. Cantor's results about one-to-one mappings of the unit interval to longer line segments, to surfaces, and to solids show us that all these are only different representatives (variants) of one set. In this sense set theory is even more abstract because now the notion of dimension no longer plays an essential role in the theory. Of course some properties remain, such as that of being a finite or an infinite point set, but we have left behind the whole rich world of geometric forms, retaining even fewer invariants. One could of course choose to focus on the set-theoretic invariants themselves.

§ 3   Spatial Ontologies and Unified, A Priori Science

Husserl is fond of saying that we can investigate different a priori ontologies and their logical relations to one another as part of a systematic, unified conception of the eidetic sciences (Husserl 1973c, 1982). Modern geometry can be seen as a model for at least part of what he has in mind. We think of each geometry as giving us a different spatial ontology. These spatial ontologies are then all related to one another in a systematic, interconnected whole, as indicated roughly in our diagrams. Each ontology is governed by a set of axioms that express properties that remain invariant (essences) under particular groups of transformations (variations). (We can also refine the notion of ontology somewhat by distinguishing regional from formal ontologies; see § 4.) These invariants hold for the space and the geometric character and behavior of objects in the space. This is just the kind of thing that Husserl envisions for the eidetic sciences in general. Indeed, he has a very broad conception of logic according to which all of the a priori sciences are to make up, ideally, a systematically connected body of truths. The view of logic and unified science here is very much like that of his great rationalist predecessor Leibniz. Such unification is, as it were, an ideal of reason. It is something we do not now have but that we should continue to work toward.

What does it mean to say that pure geometry, as an eidetic science, is a priori? Geometry, viewed as the study of what remains invariant through variations of different types, is quite independent of concerns about possible applications to nature. After the work of such mathematicians as Grassmann and Riemann in particular we know that there are many geometries – n-dimensional Euclidean and non-Euclidean geometries – for which there are no known applications. If we wish to consider the question of applications, then many additional considerations arise. In applications to the physical world there will be combinations of a priori and a posteriori elements. In relativistic physics, for example, we have the idea of space of variable curvature that is due to the distribution of mass in the universe. The geometry or manifold theory of general relativity theory reflects this fact. Some aspects of experience related to the distribution of mass, however, will evidently be a posteriori, and this will be reflected in empirical methodologies directed to this phenomenon.

With ideation itself, however, we have intuition of the a priori. It is a grasp of what *must* be the case regarding certain essences (invariants) in advance of any sense experience or applications to the physical world. Geometric essences of the sort mentioned in our examples envelop their extensions beforehand, not merely after sense experience. They are thereby capable of prescribing rules to any empirical particulars that might fall under them. The geometric invariants we find through free variation cannot be nullified by future sense experience. On the contrary, they prescribe rules for any possible application or experience involving such invariants.

Unlike the empirical generalizations based on sense experience, geometric invariants cannot be assumed to result (only) from empirical induction. Free variation in imagination is not an inductive procedure. It does not yield probabilities and contingencies. Moreover, we are not dealing with vague or inexact concepts here, as we are in parts of empirical science. In our first example, we could not find that $x_1 - x_2 = x_1' - x_2'$ fails for any particular choices of the x's that we might come across in the future, given the conditions indicated in the example. $x_1 - x_2 = x_1' - x_2'$ and $x_1 < x_2 \leftrightarrow x_1' < x_2'$ are *necessities* – not possibilities and not probabilities – through all the changes of a in $Px' = Px + a$. All future experience regarding instances of these formulas is fixed in advance. Determining properties that are invariant under groups of transformations, in other words, is not a matter of empirical induction. It is also not merely a matter of convention, for part of the very notion of a convention is that it is not in the first place forced to take precisely the form it takes. Moreover, a convention is something that we can change at will.

To think that geometric ideation is a form of empirical induction or a matter of convention is to fail to grasp the *meaning* of the geometric formulae in the practice of pure geometry. As a rough indication of whether the meaning is different we can use a substitution test: substitute the language of one method for the language of the other and see whether what was said in the original language about objects, their properties and relations, categories of objects, and so on, is preserved in the new language. Do the two languages direct our awareness and our projects in different ways? Do they have different purposes? If so, then we are dealing with different meanings.

Husserl says that if in the method of ideation we start with an example from sense experience, then we are to raise the example to the level of a pure possibility, treating it as merely one possibility among others. Free variation does not deal in sensory actualities at all. The spatial ontologies of modern geometry map out possible spaces. As pure geometers we need not be concerned with which one of them, if any, is the actual space of the physical world. This is why Husserl emphasizes the role of imagination in free variation. Sense experience (including its extensions by technology) is far more constrained than imagination. If we were confined only to sense experience, then seeing from whence all of these geometries could have originated would be difficult (see § 4). Ideation is supposed to be a method that is not limited to sense experience. Unlike some conceptions of methodology, it is not reductionistic in the direction of sense experience and it does not require that our variations be limited to what is indispensable to the natural sciences.

The general study of what is geometrically invariant through variations in imagination is radically underdetermined by sense experience and by the needs of the existing natural sciences, but there is an important methodological constraint on anything that could count as a spatial ontology. Each of these geometries must be logically *consistent*. Husserl had already built this into his views about manifold theory and the different levels of logic. He distinguishes three levels or strata of logic, consisting of grammar, 'consistency logic', and 'truth logic'. What Husserl calls consistency logic is concerned with the mere formal or material consistency of judgments independently of whether the judgments can be fulfilled in sense experience, or of whether we have evidence for their truth on the basis of intuiting the objects they are about (see Chapter 1). Husserl emphasizes considerations about the consistency of theories in connection with theories of various kinds of numbers (e.g., negative, real, complex) and with n-dimensional Euclidean and non-Euclidean manifolds where n > 3. The reason is that in these cases we cannot make the objects the theories are about present to ourselves in intuition in the same way that we can with natural numbers and objects of ordinary Euclidean geometry.

We should thus note that intuiting essences and their relations to one another, as we have been describing this, is very different from being able to have the geometric objects the ontologies are about present to us in sensory intuition. A cube, for example, is present to us in sensory intuition in a manner quite different from that of a hypercube (a four-dimensional object), and yet we can still develop the *conception* of four-dimensional space and of what would have to be true in such a geometry. This is simply to acknowledge that our sense of perceived space in dimensions beyond three falters quite dramatically. There is a shift to analytic or algebraic methods and to questions about conceptual impossibility, possibility, and necessity. This is what Husserl is emphasizing by highlighting the importance of the consistency of theories. It will be worthwhile to say more about how this is all part of a phenomenological account of the origins of geometry.

§ 4   The Origins of Geometry

How do we arrive at modern geometry with its many different 'spaces' and spatial ontologies? From the point of view of Husserl's philosophy, modern geometry is said to be *founded* on the most basic human experience of space in ordinary sense perception. We start with the space of the lived body in its immediate, everyday lifeworld experience. Here we encounter the space in which the human body moves about and orients itself toward objects, along with the spatial characteristics of those objects. How do we get from this basic experience of space up to the 'spaces' and 'geometries' that are part of modern mathematics? There is the basic *founding* experience of space, but then there are various *founded* acts of cognition (e.g., reflection, idealization, abstraction, imaginative variations, and formalization) that make modern geometry possible. Certain cognitive structures are required if we are to have modern geometry. So what we have in an account of the origins of geometry is an account of the a priori conditions for the possibility of modern geometry. I will sketch some of the elements of this account.

Husserl, Heidegger, Becker, Merleau-Ponty, and a number of other philosophers in the phenomenological tradition have described some basic features of the space of our everyday, prescientific practice and sense experience (Husserl 1970b, 1997; Heidegger 1962; Becker 1923;

Merleau-Ponty 1962). Space is a form of 'outer' perception. (This is different from time, in which case we can distinguish the time-consciousness involved in the perception of 'outer', transcendent temporal objects (such as chairs) from the temporality of 'inner' conscious experiences and acts themselves. In the case of 'outer', objective time we consider the temporal duration of events in space, the temporal duration of the existence of a spatial object, temporal relations between spatial objects, and so on. We set up systems for the measurement of time and temporal intervals.) In these descriptions the constitution of the space of everyday perception depends on bodily kinesthetic systems and the orientation of the body. The three-dimensional space of ordinary perception in particular depends on forms of bodily movement in relation to objects, for example, distancing and orbiting. Vision, touch, and other sensory modalities will have their own unique roles to play. The homogeneity of space is taken to have its origins in the relativization of the 'here' of the body. Any 'there' could in principle be a 'here'. One can go into all of these ideas in great detail (see, e.g., Drummond 1978–79, 1984; Ströker 1987b).

In Euclidean geometry, however, we do not have the shapes and the objects of everyday perceptual experience. Indeed, the shapes of objects in everyday sensory experience are imprecise or inexact in various ways. Quite generally, Husserl says that the essences of sensory objects are 'morphological' as opposed to exact (Husserl 1982). Sensory objects are 'notched', 'scalloped', 'umbelliform', and the like. We cannot obtain awareness of the objects of pure geometry by simply subjecting the bodies given to us in space and time to free variation in imagination. To obtain the 'pure' straight lines, planes, figures, and so on, of Euclidean geometry certain *idealizations* are required. The idealization of shapes involved in Euclidean plane geometry requires the limitation of our concern with objects generally to a theoretical concern with their shapes as measurable. One must focus attention on a side or aspect of an object in abstraction from the field and its horizons in which the side or aspect is presented and also from other sides and perspectives on the object. One limits attention to the two-dimensionality of the presentation of the object or its side. Here we already see how certain acts of *abstraction* (limitations of attention) and *reflection* (adoption of a more 'theoretical' attitude) are at work.

An important part of the idealization involves measurement. Our technical abilities to create a perfectly straight line, a perfectly flat surface, and so on, reach certain limits. The ideal of perfection, however, can be pushed beyond that, further and further. Out of the praxis of perfecting we can understand that in pressing toward the horizons of *conceivable* perfecting certain 'limit shapes' emerge toward which the series of perfectings tends, as toward invariant and never attainable poles. It is these *ideal* shapes that make up the subject matter of pure Euclidean geometry. Here we obtain an exactness that is denied to us in the intuitively given surrounding lifeworld. It is the measuring of shapes in the prescientific lifeworld that underlies these idealizations. The idealizations are a natural outcome of refining and perfecting measurement. Every measurement acquires the sense of an approximation to an unattainable but ideally identical pole, that is, to one of the definite mathematical idealities or to one of the numerical constructions belonging to them. It is in this way that we obtain the ideality of basic geometric notions of points, lines, planes, triangles, continuity, congruence, distance, direction, and so on. Euclidean geometry is concerned with exact essences, and the awareness of exact essences depends on idealization.

Euclidean geometry may thus be viewed as arising from the idealization of structures given to us in everyday sensory perception. This geometry existed for many years as a 'material' eidetic (a priori) science before it was formalized. The arithmetization of Euclidean geometry constitutes a very significant shift. Once numbers and the algebraic techniques of coordinate geometry are brought to bear, the possibility of *formalization* arises. We think of the *plane*, for example, as the set R^2 of ordered pairs of real numbers, $R^2 = \{\langle x, y \rangle : x, y \in R\}$, and a pair $\langle x, y \rangle$ represents a *point* in the plane. A *line* is the graph of an equation of the form $Ax + By = C$ where A and B are constants that are not both 0. Slope, distance, and so on, are all similarly characterized by formulae. With the subsequent formalization of Euclidean geometry we obtain for the first time what Husserl calls the 'Euclidean manifold' of two or three dimensions. Husserl's conception of a manifold depends on a form/matter distinction. We abstract away from the 'matter' or 'content' of geometric judgments to obtain the mere form of the judgments.

Once we can represent points by pairs or triples of real numbers, lines by algebraic formulas, and so on, it would be very natural to see whether we could start thinking of a four-dimensional 'space' in which points are given by quadruples of numbers $\langle w, x, y, z \rangle$, the formulas for distance and other properties are appropriately generalized, and so on. In other words, the foundation is now in place for generalizing and for constructing n-dimensional Euclidean or even non-Euclidean manifolds. One can start using free imaginative variations on both formal and material aspects of the existing geometry (or geometries) to obtain new geometries. In this manner Husserl wants to account for the origin and constitution of different 'spaces' and 'geometries', starting from the space of everyday perceptual intuition. He wants to allow complete freedom to devise new such constructions and to regard the resulting mathematics as perfectly legitimate, subject only to maintaining of consistency. We can thus see the role that ideation would play in the origins of all of the geometries that are part of modern mathematics. In § 2 we already noted how abstraction in pure geometry itself can be viewed as the result of looking for invariants through more and more extreme kinds of variations. We can spell out many of the technical details in terms of groups and subgroups of transformations.

In recent times it is common in mathematics to think of manifold theory as the study of geometric surfaces and generalizations of such surfaces. A manifold is a generalization of the notion of a surface to any number of dimensions. One-dimensional manifolds are curves, with the real line R as a special case. Two-dimensional manifolds are surfaces, with the flat plane R^2 as a special case, and so on. One can go on to distinguish many different kinds of manifolds and to link manifold theory in deep ways with real analysis and applications in physics. Husserl developed his own conception of 'manifold theory', which is even more general than the modern geometric or topological conception. He says that his conception was influenced by Grassmann's conception of an 'extension', Riemann's conception of a manifold, and some related work on manifolds by Lie and Cantor (Husserl 1973c, 1982; Riemann 1959), but he extends it beyond geometry and topology. In the most general sense a 'manifold' is said to be the purely formal ontological correlate of any consistent axiomatic formal system, whether it be a formal system concerning space or anything else.

Husserl distinguishes 'regional' from 'formal' ontologies, so that we have a refinement in our notion of ontology. Euclidean geometry, as part of exact eidetic science that already existed prior to formalization, evidently constitutes a regional spatial ontology. The Euclidean manifold, which then results from formalization of the Euclidean idealizations, constitutes a formal ontology. In the most general sense a manifold, as we said, is supposed to be the purely formal ontological correlate of a formal axiomatic system. With a manifold one leaves the content or interpretation behind and focuses only on the theory-form and its properties and relations to other theory-forms. The Euclidean manifold is then seen to be only one of indefinitely many manifolds.

Now as soon as we go beyond three-dimensional geometry it does indeed seem to be the case that the spatial intuitions that serve us so well in everyday sense experience begin to fail. But what if we could show that the resulting geometries were indeed consistent? Then there would be nothing illogical or incoherent about our conceptions in these cases. The existence of relative consistency proofs of different geometries is just the kind of thing Husserl would want to highlight in these cases.

Once we obtain new manifolds and formal systems through imaginative variations and formalization, we can reinterpret these formal structures any way we like. We are free to reinstantiate them with 'matter' or 'content' in a variety of ways. Although Husserl does not discuss the topic, one can say that this is just what happened in the case of the application of non-Euclidean manifolds in relativistic physics, where we see a surprising application of what had previously appeared to be merely conceptual or symbolic mathematics (Weyl 1918b; Becker 1923; Mancosu and Ryckman forthcoming). It is interesting to note, by the way, that at one time Einstein evidently considered calling relativity theory 'Invariententheorie'.

The Euclidean manifold and the Euclidean space for which it is the pure categorial form are, for Husserl, prior to non-Euclidean manifolds and spaces. So there is the already-characterized relationship between the three-dimensional Euclidean manifold and the 'space' of everyday sensory intuition. The three-dimensional Euclidean manifold can be applied in physics because it is a formalization of idealized geometry that was arrived at on the foundation of everyday sensory experience in the first place. Non-Euclidean manifolds might have applications to nature, as happens in the case of relativity theory, but it does not follow that our everyday sensory experience of the world is itself best characterized in terms of such manifolds. Both kinds of manifolds might have applications to nature and we need to separate the question of applicability from the question of characterizing of the space of everyday sense experience.

Given this sketch of the origins of modern geometry we can start from the bottom up or look from the top back down. In the one case we start from the founding and more practical, concrete, sensory, and particular and proceed to the founded and more theoretical, abstract, conceptual, and universal; in the other case we proceed in the opposite direction. At the deepest founding level one would consider the most basic, prescientific space of everyday human practice, whereas at the highest founded levels one would consider the most rarefied mathematical theories of space. There are levels in the constitution of 'space' in which the sedimentation of what comes before makes what comes after possible. There are already invariants for us in sense perception – the objects themselves and some of their properties and relations – but then we see that there are also invariants at 'higher' levels of cognition.

One might be tempted at this point to engage in what we might call 'transcendental geometry'. With the very general post-Riemannian conception of geometry in mind we might ask which of the indefinitely many n-dimensional Euclidean and non-Euclidean manifolds are a priori conditions for the possibility of our sense experience. Which are conditions for our everyday experience of space? Could they all be? Only Euclidean three-dimensional geometry? A related but different question is this: which geometries are a priori conditions for the possibility of modern physics?

We have seen in our characterization of the relationships of some of the geometries that some geometric invariants are very deep and that there are levels of invariants. Topological essences, for example, are also Euclidean essences, but not all Euclidean essences are topological essences. So one approach we might take, putting it very roughly, is just to plug the hierarchical arrangement of invariants we have indicated in our diagrams into our answer about the conditions for the possibility of sense experience. We would then have a layering of conditions, in which some are more abstract and general than others, and so on. The picture that results is thus quite a bit more elaborate than the view found in Kant. I do not, however, wish to go further into this subject in this chapter.

§ 5 Applications of Ideation Outside Geometry

There have been some skeptical challenges over the years to the very idea that there is a method of ideation. On the basis of the preceding geometric examples and the discussion, I would like to say that there is in fact such a method. One can see how it works in modern geometry. If this much is admitted, then there are additional questions about the possible use of the method outside pure geometry. In the first instance one might consider other parts of mathematics and logic. Here I think that one can also make a case for its application and its usefulness. One can see how mathematicians and logicians already are or could be making variations on the concepts of set, number, function, proposition, and the like, to find invariant properties of these kinds of objects. As I said, this need not always be cast in the particular technical group-theoretic framework in which it has been cast in much of modern geometry.

How about extending the method outside pure mathematics? Philosophers and scientists of a rationalist bent would certainly be tempted by the prospects. Indeed, one might come to think of it as a method within philosophy itself. Ideation not only is part of phenomenology but has, in effect, played a role in the 'analysis' of concepts in some schools of analytic philosophy. If Husserl is correct, then outside pure mathematics one will encounter inexact essences and also cases in which exact and inexact essences may be mixed. Suppose we think of a 'cognitive domain' or an 'ontology' quite generally as defined by a set of variations on some object(s) and as studying everything that is invariant under these variations. We take it as a goal of rational inquiry to map out all of this. We are clearly talking about a tremendous amount of work here. But would this even be possible if we have inexact essences in the mix? Progress might be possible. It seems, for example, that there would be variations under which a physical object would cease to exist, a priori conditions under which its existence would no longer be thinkable. Some of these conditions would be geometric, some would be temporal, and so on. One might, for example, think again of Weyl's use of essence analysis to find a priori elements in relativistic physics.

Would there not also be variations under which the existence of a person would, a priori, no longer be thinkable, or under which consciousness would no longer be thinkable? Even if *physical object, person,* and *consciousness* are terms that lack mathematical precision, should we not be able to make some progress in determining such conditions? Such a project is deeply connected with the most basic concerns of science and reason. Invariance and objectivity go hand in hand. Invariance is a cornerstone of rationality and science. Thinkers such as Leibniz, Husserl, and Weyl tell us not to give up on such questions, and it seems to me that we still do not have sufficient grounds for ignoring their advice.

KURT GÖDEL, PHENOMENOLOGY, AND THE PHILOSOPHY OF MATHEMATICS

4

Kurt Gödel and Phenomenology

From the available evidence we know that Kurt Gödel began to study Husserl's phenomenology in 1959 (Wang 1978, 1981, 1987, p. 28). This is an event of some significance for students of Gödel's work, for years later Gödel told Hao Wang that the three philosophers he found most congenial to his own way of thinking were Plato, Leibniz, and Husserl (Wang 1987, p. 74). Reports of Gödel's interest in Husserl have also surfaced in other sources. Gian-Carlo Rota has written that Gödel believed Husserl to be the greatest philosopher since Leibniz (Kac, Rota, and Schwartz 1986, p. 177). And Heinz Pagels has written that "during his later years he [Gödel] continued to pursue foundational questions and his vision of philosophy as an exact science. He became engaged in the philosophy of Edmund Husserl, an outlook that maintained that there is a first philosophy that could be grasped by introspective intuition into the transcendental structure of consciousness – the very ground of being" (Pagels 1988, p. 293). As part of his description, Pagels mentions how Gödel thought it meaningful to question the truth of axioms, and to ask about their philosophical foundations, and he then mentions Gödel's view on mathematical intuition. Georg Kreisel has also noted Gödel's

I would like to thank Hao Wang for comments, and for discussion and correspondence about Gödel's philosophical interests. I have also benefited from correspondence with Solomon Feferman, John Dawson, Jr., and Cheryl Dawson, and from comments by Lila Luce, Penelope Maddy, Pieranna Garavaso, Steven G. Crowell, J. N. Mohanty, Izchak Miller, and a referee for *Philosophy of Science*. Parts of this chapter were presented to the 1989 Eastern Division meeting of the American Philosophical Association, to the 1989 spring meeting of the Association for Symbolic Logic, and to the Philosophy Department Colloquium at the University of Iowa. I thank members of those audiences for comments.

interest in Husserl in his article on Gödel in the *Biographical Memoirs of Fellows of the Royal Society* (Kreisel 1980, pp. 218–219).

In the first part of this chapter I describe some of what is now known about Gödel's interest in Husserl's phenomenology. In particular, it appears that Gödel's comments in the famous 1963 supplement to "What Is Cantor's Continuum Problem?" ("WCCP"; Gödel 1964) may have been influenced by his reading of Husserl. Wang has remarked in connection with "WCCP?" that "presumably Husserl's elaborate analysis of our perception of a physical object can . . . be viewed as supporting G[ödel]'s conclusion" (Wang 1987, p. 303) about the objective existence of mathematical objects and about mathematical intuition. He comments in another place that "perhaps Husserl's considerations of *Wesenschau* can be borrowed to support G[ödel]'s belief in the objective existence of mathematical objects" (Wang 1987, p. 304). Also, Charles Parsons has conjectured that Husserl's conception of intuition is Gödel's model in "WCCP?" (Parsons 1983a, p. 24).

In the second part of the chapter I show that whatever Gödel's understanding of Husserl may have amounted to, his views on mathematical intuition and objectivity can be readily interpreted in a phenomenological theory of intuition and mathematical knowledge. Gödel's view on mathematical intuition has been discussed by the philosophers Benacerraf, Chihara, Maddy, Steiner, and many others, but little of this work has taken into account the nature of Gödel's own philosophical interests.

§ 1

According to Hao Wang, one of the logicians closest to the later Gödel, Gödel remained interested in Husserl's philosophy for a long time. Wang has said that during his meetings with Gödel in 1971 and 1973 Gödel often mentioned Husserl in conversation and urged him to study Husserl's post-1905 writings (Wang 1987, p. 120).[1] Gödel was apparently interested only in Husserl's later, transcendental phenomenology, although he had some reservations, to be discussed later, about Husserl's last published work, *The Crisis of the European Sciences and Transcendental Phenomenology (Crisis)*. Wang has also said that he and Gödel made a study of Husserl's "Philosophy as Rigorous Science." A number of ideas about philosophy as a rigorous science are included in Wang's *From Mathematics to Philosophy (FMP)* (1974, p. 6, pp. 352–356) and *Reflections of Kurt Gödel (RKG)* (1987,

---

[1] In Wang 1987, p. 120, Wang says post-1907 works.

pp. 219–221). In addition to Wang's reports, the Gödel *Nachlass* is known to contain many notes on Husserl. There are many notes on Kant and Leibniz as well, but virtually all of the notes on these philosophers are in Gabelsberger shorthand (the system of shorthand also used by Bernays, Zermelo, Schrödinger, and Husserl), and to date very little of this material has been transcribed.

Gödel's interest in Husserl developed relatively late in his career so it is virtually certain that his study of phenomenology had no impact on any of his important published logical or mathematical results. It is known that Gödel had studied Kant and Leibniz from early on, and that he was a "realist" or "platonist" long before he read Husserl. Whether his study of phenomenology had any impact on his published philosophical writing is not clear. The materials in the *Nachlass* will, we hope, shed light on this question and on many related questions concerning Gödel's interest in phenomenology. (See van Atten and Kennedy 2003 for more information on this.)

It appears, from Wang's writings, that Gödel was first drawn to phenomenology for methodological reasons. He was searching for a new method for thinking about the foundations of mathematics and for metaphysics in general, and he thought that phenomenology might offer such a method with its phenomenological reduction or epoché. Gödel evidently believed that the method of "bracketing," especially in connection with Husserl's conception of the "eidetic reduction," would enable a person to perceive "concepts" more clearly, or to arrive at essential characteristics of concepts (Wang 1987, p. 193). It would allow for this in part through its suspension of what Husserl called the "naive" or "natural standpoint," but also because of the sharp distinction between sciences of fact and sciences of essence that one finds in Husserl's conception of the eidetic reduction. The failure to make the latter distinction was, as Husserl saw it, the source of various empiricist or positivistic misunderstandings of sciences of essence such as mathematics and even of phenomenology itself. Let us briefly consider the idea of suspending the naive or natural standpoint, and also the idea of determining essential characteristics of concepts.

The "natural standpoint" is the naive, prereflective perspective on reality which is the starting point of philosophy. It will be useful, in what follows, to understand the suspension of the natural standpoint by way of a central concept in Husserl's philosophy, the concept of intentionality. Many cognitive scientists and philosophers of mind believe that intentionality is a basic, irreducible feature of cognition, certainly of the more theoretical forms of cognition. Intentionality is the characteristic of "aboutness" or "directedness" possessed by various kinds of mental acts. It has been formulated by saying that consciousness is always consciousness of an object or state of affairs. A standard way to analyze the concept of intentionality is to say that acts of cognition are directed toward, or refer to, objects (or states of affairs) by way of the "content" of each act, where the object of the act (or the state of affairs) may or may not exist. The canonical way to determine the content of an act is by "that" clauses in attributions of beliefs and other cognitive acts to persons. Historically, a variety of terms have been used in place of *content*, for example, *ideas, representations, concepts, intentions*. As indicated later, in some of his writings Gödel appears to use the term *concept* to stand for the notion of content, or for that part of the content which is expressed by predicates. We might picture the general structure of intentionality as:

$$\text{Act(Content)} \longrightarrow \text{[object]},$$

where we "bracket" the object because we do not assume that the object of an act always exists. Phenomenologists are famous for suggesting that we "bracket" the object, and that we then focus our attention on the act (noesis) and act-content (or noema), where we think of an act as directed toward a particular object by way of its content (or noema). Husserl thought it possible to do a good deal of phenomenology on the basis of simply reflecting on content without being concerned about whether objects of acts existed or not. However, he also developed a theory of knowledge on the basis of the notion of intentionality according to which the question whether an object exists or not depends on whether we have evidence for its existence, and such evidence is given in further acts carried out through time. The phenomenological reduction has the effect of making a belief in the existence of an object toward which an act is directed dependent on the fulfillment, partial or otherwise, of (empty) intentions directed toward the object, and one cannot impute existence independently of this. An intuition is understood as a fulfillment of an (empty) intention. It is the source of evidence.

Now from the "natural standpoint" one would assume that the object toward which an act is directed exists. So the idea of suspending the natural standpoint amounts to not making this assumption, allowing that there might not be an object, even though there is an intention directed toward an object, and then understanding beliefs about whether an object exists or not in terms of fulfillments of such intentions. This has many consequences which I believe are relevant to Gödel's views on mathematical intuition, but I mention here only the obvious consequence that one ought not assume (as one would from the natural standpoint) that the object toward which an act is directed at some stage of experience is necessarily one to which the subject is causally related at that stage. I will return to this point in the next section.

In addition to suspending the natural standpoint in this sense, Husserl's philosophy includes a sharp distinction between sciences of fact and sciences of essence which would no doubt have been appealing to Gödel. This distinction emerges most forcefully in Husserl's philosophy in such works as *Ideas I* and "Philosophy as a Rigorous Science," in which the view that there are no sciences of essence is seen as the curse of positivism, empiricism, and naturalism. It is reported by both Wang and Kreisel that Gödel rejected various forms of positivism, naturalism, and also conventionalism in the philosophy of mathematics. Gödel evidently thought that if we could shed empiricist or positivistic confusions about concepts, we could make great progress in mathematics, and even metaphysics. As Rota has put it, the phenomenological reduction (including the eidetic reduction), with its bracketing of the physical world, is supposed to have the effect of bringing out the experiential reality of "ideal" phenomena. Only when a phenomenon is taken seriously and studied at its own level, Rota says, can it reveal its own properties within its own eidetic domain (Kac, Rota, and Schwartz 1986, pp. 169–170).

Wang reports that "related to [Gödel's] conception of concepts is that part of his views that he attributes to Husserl" (Wang 1987, p. 161). In Wang's notes in *FMP* of his discussions with Gödel we are told that concepts are, in effect, abstract objects.[2] They are supposed to exist independently of our perceptions of them, and we simply have better or worse perceptions of them, much as we have a clearer perception of an animal when it is nearer to us than when it is farther away. According to Wang, Gödel thought of the idea of "seeing" a concept as very closely related to understanding the meaning of an expression. Much as Husserl does in "Philosophy as a Rigorous Science" (1965) and *Ideas I* (1982), he claimed that it was a prejudice of our time, a kind of naturalistic prejudice, to suppose that we do not "see" or "intuit" concepts. (Gödel even conjectured that some physical organ closely related to the neural center for language was necessary to make the handling of abstract impressions possible.)

---

[2] On the basis of the preceding remarks about intentionality, this parallels an interpretation of noemata due to Dagfinn Føllesdal.

These comments are remarkably similar to Husserl's views on reflection on noemata and on the intuition of essences in *Ideas I*, especially in the section on method where Husserl discusses the "method of clarification" and the "nearness" and "remoteness" of the data of intuition (Husserl *Ideas I*, §§ 67–70). On Husserl's view, reflection on concepts, or content, is itself a kind of intuition, governed by the same principles that govern other forms of intuition. Thus, a concept can itself become the object of an act, and one can then set about determining what kinds of properties are true with respect to the concept. Gödel applied his comments to the concept of a mechanical procedure, which he believed was finally brought into the correct perspective with the work of Turing. He told Wang that there were more similarities than differences between sense perceptions and perceptions of concepts, and that the analogue of perceiving physical objects from different perspectives is perceiving different logically equivalent concepts. This last remark refers to one of the hallmarks of intensionality, and intentionality, and it also reveals the connection with the idea of intuiting essences. For we have evidently arrived at the "essence" of the concept of mechanical procedure by way of the many different definitions of this concept that were subsequently proved to be logically equivalent. This is a particular example, about which more will be said later, of what intuiting the essence of something must be like.

Gödel believed, in particular, that the phenomenological method might allow us to perceive the meaning of the primitive concepts of our theories more clearly, so that our attempt to find axioms true of the concepts would be greatly enhanced. From remarks in Wang's *FMP* it appears that Gödel was attempting to apply the method for just such a purpose in the case of set theory, especially in connection with the concept of set and the axioms true of this concept that would decide the continuum hypothesis (CH) (Wang 1974, p. 189). Gödel must have had in mind the component of the method that Husserl referred to in *Ideas I* as the method of "clarification." Clarification is the method for making clearer the aspects of an essence which are only dimly perceived. It is implemented through what Husserl calls the process of free variation. The method of clarification and the process of free variation are described in Husserl's "Philosophy as Rigorous Science," *Ideas I, Phenomenological Psychology*, and *Experience and Judgment*. The basic idea of the method of free variation is to start with a concept and to vary it is as many ways as possible in imagination. The concept will not survive the addition, subtraction, or substitution of some properties, but it will survive in the case of others. The properties that cannot be subtracted, or for which substitutions cannot be made, are essential to the concept, and otherwise they are accidental. One might, for example, apply the method to the concept of a mechanical procedure, or to the concept of number or set, or to various geometrical concepts. Consider, for example, the concept of a triangle and the properties that are or are not essential to this concept, given a particular set of background assumptions. The method seems especially appropriate for mathematical thinking, although with suitable distinctions between material and formal essences, Husserl wished to apply it to concepts of any type.

What remains invariant through the multiplicity of free variations is supposed to be what is essential to a concept. The process of free variation is meant to foster the emergence of identities from or against a background of multiplicities, and "essences" are then understood in terms of such identities or invariants. Husserl believed that through the process of free variation one would uncover a "rule" governing the concept which would not have been seen prior to the process of variation. The rule, even if only partially understood or not fully determinate, would open up a "horizon" for further possible insights.

Hence, the idea would be to attempt to understand (enough of) the essence of the concept of set through the determination of new properties of sets to decide CH. In "WCCP?" Gödel had a particular concept of set in mind: the "iterative" concept. Applying the method of free variation to this concept could yield an axiom or axioms that would not be ad hoc or artificial for the concept. Certain axioms would be "forced" upon us in the sense that we could not imagine sets as objects in the cumulative hierarchy without the properties expressed in the axiom(s). One might look at the existing axioms of Zermelo-Fraenkel with Choice (ZFC) as axioms that would be shown by free variation to be true of the iterative conception of set.[3] This aspect of Husserl's philosophy may be the foundation for Gödel's claim that just as we are forced in certain ways in perceptual intuition, so certain axioms force themselves upon us as true of the concept. One would not of course want a "decision" of CH to depend on an

---

[3] One might try this as an exercise and a way to enter the frame of mind of using free variation. There will no doubt be nuances and differences in evidence concerning some of the axioms. In many of the expositions of the iterative concept, for example, it is argued that replacement does not have the kind of immediate evidence that one finds in the case of the other axioms. The axiom of choice is also set apart from the other axioms in some treatments. Extensionality, unlike the other axioms, may be viewed as a defining characteristic of sets (as opposed to properties). Much more could obviously be said about this.

accidental property of sets, a property one could imagine sets' not possessing. Hence, one might ask, for example, whether it is essential to the iterative concept of set that sets be well founded, constructible, that there be measurable cardinals, supercompact cardinals, that the axiom of determinacy hold, or that the axiom of quasi-projective determinacy hold. One can formulate axioms that decide CH, but at least at present none of these is (intersubjectively) perceived as essential; they do not force themselves upon us. How would we find new essential properties of sets? Husserl's method simply implies that further productive imagination is needed. Productive imagination would be guided by the uncovering of rules that would result from the process of free variation, given the appropriate depth of understanding and background belief. But, in any case, the idea that one can arrive at the essence of a phenomenon in this way may appear to be a rather optimistic attitude toward open problems that have been as vexing as the continuum problem. It appears to me to be closely related to what Wang has described as Gödel's "rationalistic optimism." Gödel's rationalistic optimism appears to go hand in hand with his realism or platonism, and with his view of the meaningfulness of undecided mathematical statements.

Gödel was apparently also impressed by the phenomenological claim that philosophy calls for a different method from science, and that it can provide a deeper foundation for science by reflecting on everyday concepts. Gödel was here presumably influenced by the role of what Husserl later came to call the *Lebenswelt* in our scientific thinking. Nonetheless, Gödel seems to have been very attracted to the idea that philosophy itself would become a new kind of "rigorous science," a science of essences. Gödel apparently saw in Husserl's idea of "philosophy as a rigorous science" a revitalization of the Leibnizian ideal of philosophy. The tenor of Gödel's interest in Husserl in fact reflects the irony in how much of Husserl's philosophy was lost on his Continental followers, especially once phenomenology took an "existentialist" turn at the hands of Husserl's successors. Husserl had of course been trained as a mathematician, and his philosophy originally developed out of a concern for problems in the philosophy of logic and mathematics. The sensibilities of a mathematician are apparent even in some of his very late writings. It is thus not very surprising that Gödel did not care for parts of Husserl's last published work, the *Crisis*. It has been suggested by such philosophers as Merleau-Ponty that the *Crisis* broke, at least tacitly, with the philosophy of essences (I disagree with this assessment). In the language of the existentialists, existence had begun to precede essence.

We can surmise from these comments that Gödel was primarily interested in the period of Husserl's work in which Husserl believed that philosophy could become a science. This is also the period in which Husserl was a Kantian transcendental idealist who was nonetheless an "objectivist" or a realist about mathematical objects. This last point is of some interest because, as Kreisel has noted, Gödel was successful at mixing realistic and idealistic conceptions in his own work in logic, mathematics, and physics (Kreisel 1980, pp. 209–213). Gödel's published and unpublished work shows interests in both realism and idealism. In mathematics the interests in idealism are very clear in the papers on constructive mathematics, whereas in physics one sees such interests in the papers on Kant and relativity theory. Gödel had a long-standing interest in Kant even though there were aspects of Kant's philosophy that he did not like. Gödel (Gödel 1949a) argued that the models of Einstein's field equations developed in Gödel (1949) support an idealistic view of time and change in the sense that these are to be viewed as a contribution of our own mind rather than as an objective aspect of the physical world. Kreisel remarks that in conversation Gödel did not view realism and idealism so much as conflicting philosophies; rather, "he was ready to treat them more like different branches of the subject, the former concentrating on the things considered, the latter on the processes of acquiring knowledge about these objects or about the processes" (Kreisel 1980, p. 209). This attitude toward realism and idealism is similar to Husserl's attitude, and I will say more about it later.

In the remaining part of this chapter I would like to pause over Gödel's comments on mathematical intuition in the 1963 supplement to "WCCP?" for they have perhaps been his most widely publicized philosophical views.

§ 2

Gödel's most extensive published comments about mathematical intuition are expressed in the supplement to "WCCP?." Gödel speaks more directly of intuiting sets than he does of intuiting or reflecting on the concept of set, although the main body of the paper does contain references to how reflections on the iterative concept of set "of a more profound nature than mathematics is used to giving" (Gödel 1964, p. 257) are required to solve open problems in set theory. In the supplement, Gödel is speaking about the intuition of transfinite sets, where one thinks of such sets as objects in the (a) cumulative hierarchy. Gödel says that "the objects of transfinite set theory... clearly do not belong to the physical world and even their indirect connection with physical experience is very loose" (Gödel 1964, p. 267). Later he says that

evidently the "given" underlying mathematics is closely related to the abstract elements contained in our empirical ideas. It by no means follows, however, that the data of this second kind, because they cannot be associated with actions of certain things upon our sense organs, are purely subjective, as Kant asserted. Rather they, too, may represent an aspect of objective reality, but, as opposed to the sensations, their presence in us may be due to another kind of relationship between ourselves and reality. (Gödel 1964, p. 268)

In the next sentence of the passage Gödel says that the question of the objective existence of the objects of mathematical intuition is nonetheless an exact replica of the question of the objective existence of the outer world.

In light of these kinds of comments, how is it that Husserl's analysis of intuition could be used to support or at least to help make sense of Gödel's views about mathematical intuition and the objective existence of mathematical objects? Let us start with an example concerning ordinary perceptual intuition. Perceptual acts, for Husserl, are paradigmatic of intentional acts. Thus, they are acts directed toward objects by way of their content, and the objects toward which they are directed need not exist. The following example will help us to make a useful distinction for a phenomenological understanding of mathematical intuition and mathematical knowledge. Let us agree to speak of stages in our experience of objects. We should evidently think of the stages of our experience as structured in linear time of type ω, although for my purposes at the moment this will not be crucial. Now suppose I am staying at a cabin in the mountains, look under my bed in the cabin, see a snake, scream, "Snake!" and flee the room. Then at that stage of my experience I "see" a snake. But suppose that what was really lying under the bed was a coiled rope. Then, in a different sense of the word *see*, what I "see" at that stage of my experience is a coiled rope. For the sake of continuity with Gödel's language let us use the terms *see* and *intuit* interchangeably. Then we can distinguish these two senses of "seeing" or "intuiting" by adopting some useful terminology that David Smith (Smith 1984) has introduced in his study of phenomenology and theories of reference. Let us say that I am *representationally related* to the coiled rope, but that, phenomenologically speaking, my experience has the *representational character* of being "about" a snake. What I "intuit" at that stage of my experience is a snake.

My representational relation to the coiled rope bears a kind of causal relationship to the rope which at that stage the character of my experience does not. The *character* of my experience at that stage may still have some perfectly coherent underlying causal explanation, but it is nonetheless an explanation of a kind different from the explanation of my *relation* to the coiled rope.

The fact that we can make this distinction opens up the possibility of developing a purely phenomenological conception of intuition. That is, in a phenomenological approach to intuition and knowledge we might just take the step of viewing *reference* to all objects in terms of representational *character*. If we took this perspective on the matter, then we would view the knowledge provided by intuition as something that could either be sedimented over time or corrected and refined in various ways in further acts of intuition. Hence, my belief that a snake is lying under the bed could very well be corrected through further experience. In fact, it probably would be, and exactly this kind of thing takes place in knowledge acquisition. Note that this also shows us that intuition in the phenomenological sense is corrigible, that it does not give us any special, certifiable kind of knowledge except perhaps when it meets special conditions. It might, for example, provide evidence that is a priori, or apodictic, or adequate, or intersubjective, or it might not provide such evidence. There are different types and degrees of evidence. If we are considering only the *character* of our experience, then whether there is some real object "out there" is also no longer an appropriate question, or, rather, we might say that this question gets reinterpreted. The question that is now important is whether or not we possess sequences of acts, let us say fulfillment or verification procedures, in which we have evidence for objects.

If we choose to view perception or reference from this perspective, we have made what Husserl called the "phenomenological reduction." Husserl made the interesting observation that if we do take this perspective, we might simply drop the idea of a causal relation to the object altogether. Instead, what comes to the fore is whether we do or do not have evidence for objects as this is provided in sequences of acts carried out through time. In fact, one might then point out that the only way we could even identify the perceptual object to which we (believe we) are causally or representationally related is through ongoing perceptions which either correct or fail to correct that identification. Thus, at a still later stage in my experience I might see that the object under my bed is not a coiled rope, but is a coiled garden hose. The evidence I have for the belief that it is a coiled garden hose is of course contingent on what future perceptions will yield. After some point the experience would presumably stabilize, just as our experience with sets will presumably stabilize in various ways as research progresses. From this perspective it would also be natural to speak of the objects of experience as "intentional objects." The intentional object in perception would be the object as it is presented to us at a stage in our experience given the sensory data and the way in which the sensory data are interpreted as a function of the intention, background beliefs, memory, attention, and so on, at that stage.

We can read Gödel's views on mathematical intuition as directly analogous to these views on perceptual intuition, with the qualification that sensory data will not play the same role in mathematical intuition that they play in perceptual intuition. Thus, Gödel does not say that we are causally or representationally related to transfinite sets but rather that our experience has the phenomenological character of being "about" such objects. In fact, we could think of such objects as "intentional objects" in the same way that we would think of ordinary perceptual objects as intentional objects once we had taken a phenomenological approach to perception. The question of the objective existence of objects of mathematical intuition would then be an exact replica of the question of the objective existence of the outer world: what would be relevant in both cases would be whether we have evidence for the objects of our cognitive acts as this would be provided in sequences of acts carried out through time, whether we have fulfillment or verification procedures. As in the case of the snake under the bed, our beliefs would be either verified or not in further acts, and knowledge would accrue accordingly. This is why Gödel could say in the 1963 supplement,

> I don't see any reason why we should have less confidence in this kind of perception, i.e., mathematical intuition, than in sense perception, which induces us to build up physical theories and to expect that future sense perceptions will agree with them and, moreover, to believe that a question not decidable now has meaning and may be decided in the future. (Gödel 1964, p. 268)

If Gödel was viewing matters this way, then he was, in effect, making the phenomenological reduction. This conception of intuition would explain, as Gödel remarks in another passage, the analogy between illusions of the senses and illusions in mathematics (Gödel 1964, p. 268). Gödel remarks, for example, that "the set-theoretical paradoxes are hardly any more troublesome for mathematics than deceptions of the senses are for physics" (Gödel 1964, p. 268). The discovery of the set-theoretic paradoxes showed that on the basis of the naive concept of set we were under an illusion about what sets were, which led to a correction or refinement of our knowledge, just as I am under an illusion about a snake's being under my bed at a certain stage in my experience, an illusion that could be corrected through further experience.

On the basis of the passages cited previously, Gödel evidently wants to say that we are not causally related to the objects of transfinite set theory, at least not in the way we are causally related to ordinary physical objects. Yet he says that from the fact that the abstract data underlying mathematics cannot be associated with the actions of things upon our sense organs it does not follow that they are something purely subjective. They too may represent an aspect of objective reality. We could read Gödel as suggesting that from a phenomenological perspective marks of objectivity occur in our experience of sets that are analogous in certain ways to marks of objectivity in our experience of perceptual objects. For example, our intuition of sets is constrained or "forced" in certain ways just as our intuition of physical objects is constrained in certain ways. We cannot change our intuition at will, or make an object be anything we want it to be. In the case of sets, numbers, mechanical procedures, geometrical objects, and so on, this is shown by the way our concepts of these objects have been sedimented over time, and by the process of free variation. Our experience with these objects has stabilized enough to permit, for example, of some degree of formalization, axiomatization, solution of open problems, and other research developments.

A consideration of Husserl's analysis of the perception of physical objects shows that Gödel may have had a related point in mind, one that is also connected with his view of the relation between realism and idealism. The point is that the phenomenology of perception shows that we do not "create" or "construct" perceptual objects, but that we do in a certain sense create or construct our knowledge of objects. The perceptual object is given as a whole of which we actually see only an aspect or part at a given stage in our experience. This basic fact of experience would not be possible if we had somehow to construct or create the object itself from the parts of it that we actually do see. We would not see objects in the way that we do. Instead, we would have something like fragmented, partial pieces of the object. We might see things, as Jaako Hintikka has suggested, as they are represented in cubist paintings, with various parts of the object pulled around front. Of course we do not see perceptual objects that way. Rather, at any particular stage they are given as complete, whole, determinate, even though we could not have had enough partial perceptions to see them that way. What is incomplete, partial, and indeterminate at a particular stage is our knowledge or experience of the object. The object itself is said to transcend our experience of it.

The phenomenology of perception thus shows that there is a robust kind of "objectivity" associated with perceptual objects, and one could say exactly the same thing about mathematical objects (or, if you like, essences). In reasoning about numbers and sets we do not, for example, suppose that these objects are somehow themselves indeterminate, incomplete, or partial, although our knowledge about numbers and sets certainly is in various respects indeterminate, incomplete, or partial. Numbers and sets, moreover, are not thought of as objects that have temporal characteristics, and they are not treated that way in mathematics, but the acts in which we come to know about numbers and sets certainly do have temporal characteristics. We can reason about the same number in different acts, different people can reason about the same number, and so on. Hence, the representational character of our experience presents us with "an aspect of objective reality," not something "purely subjective." This view also supports the platonist metaphor of knowledge as discovery, for we do make discoveries about objects. We are firmly convinced in many situations that "there is" another side to or aspect of an object even though we do not presently see it. Something is there that we can come to know about. We do not "create" what is there, so that in an important sense it is not our own production. Similarly, in the case of open problems of mathematics we may be convinced that "there is" a solution to a problem, or that there is an object of a certain type, even though we do not yet know what it is. The process of free variation would enable us to uncover a rule governing the concept(s) in question which would open up a horizon of further possible insights.

The belief that problems have solutions, or that objects exist, even though we do not yet know what they are, is connected with Gödel's "rationalistic optimism." Wang reports, for example, that Gödel believed Hilbert to be correct in rejecting the view that there are number-theoretic questions undecidable for the human mind. For if the view were true, it would mean that human reason is utterly irrational in asking questions that it cannot answer while emphatically asserting that only reason can answer them (Wang 1974, pp. 324–325).[4] Human reason would then be

---

[4] Gödel thought that one of the most important rigorously proved results about minds and machines was the disjunction that either the human mind surpasses all machines, in the sense that it can decide more number-theoretic questions than any machine, or there

very imperfect and even inconsistent in some sense, and that possibility contradicts the fact that some parts of mathematics have been systematically and completely developed using laws and procedures that were unexpected. Gödel cited the theory of first- and second-degree Diophantine equations (the latter with two unknowns) and noted that solution of all relevant problems in this area of mathematics supported rationalistic optimism. Gödel's rationalistic optimism goes hand in hand with his realism or objectivism, and with his view of the meaningfulness of undecided mathematical statements.

Our phenomenological observations also show how we could think of idealism and realism as compatible in the way that Gödel evidently thought they were. For unlike in naive forms of idealism or constructivism, we are not claiming that we construct the object. That would be a rather careless way of putting the matter. The correct way would be to say that constructivism is concerned with the way the knowledge of the object, not the object itself, is created or constructed. Idealism and realism evidently are incompatible if idealism claims we construct the object and realism claims we do not, or if idealism claims we construct our knowledge of the object and realism claims we do not. I see this view of the relationship between idealism and realism as reflecting just the kind of twist on Kantian philosophy that one finds in Husserl's transcendental phenomenology: transcendental idealism coupled with mathematical (as distinct from empirical) realism. On the analogy with the treatment of empirical objects in Kant, numbers and sets are not mental entities; nor are they fictions; nor are they some kind of radically mind-independent things-in-themselves.

Gödel says that the presence in us of the abstract elements underlying mathematics may be due to "another kind of relationship between ourselves and reality" (Gödel 1964, p. 268). We might read this as meaning that the character of our experience of sets may have a coherent causal explanation but that it is of a different, probably more complex kind from the causal explanation involved in my representational relation to the coiled rope. The explanation would have to be different because Gödel does not take sets, unlike coiled ropes, to be objects in the external world. Thus we could not be causally related to sets in the way that we

exist number-theoretic questions undecidable for the human mind (Wang 1974, p. 234). Gödel evidently believed that the second disjunct ought to be rejected. At the end of the chapter I will briefly mention how Gödel connected the idea that the human mind might surpass all machines to the notion of intuition.

are causally related to coiled ropes. What would presumably be needed, contrary to Penelope Maddy's (Maddy 1980) account, is not some kind of mechanism for detecting sets in the external, physical world, but rather an "internal," intentional mechanism for "constructing" our directedness toward sets. Gödel's conjecture that a physical organ closely related to the neural center for language is necessary to make the handling of abstract impressions possible might be explicated in terms of our best connectionist or neuroscientific hypotheses about "internal" cell-assemblies in the brain, where internal cell-assemblies are not directly connected to the environment but are activated and inhibited by complex causal interactions with other internal and possibly also external cell-assemblies.

Of course this reading of Gödel's views on mathematical intuition only holds up if we do have evidence for objects such as transfinite sets in acts carried out through time, or if we do have fulfillment procedures for these objects. Do we? In order to make sense of Gödel's views on mathematical intuition we might say, with some important reservations, that we do. One could read Gödel's emphasis on a "genetic" or "iterative" conception of set as bearing out the claim that, relatively speaking, we do have some kind of verification procedure for these objects, for sets are viewed as objects for which we (could) have evidence at various stages of our (possible) experience. An explicit development of a "stage theory" of the type that so often underlies descriptions of the iterative conception of set can be found, for example, in Boolos 1971.

Thus, Gödel draws a sharp distinction between a "logical" conception of sets according to which a set is obtained by "dividing the totality of all existing things into two categories" given a particular predicate, and the "mathematical" conception of sets according to which a set is obtained in stages by iterated applications of the operation "set of" to some given objects. In the case of the "logical" conception of set we have nothing that even approaches a verification procedure in this sense. The mathematical conception, on the other hand, has even been described as "quasi-combinatorial" by Bernays. Gödel points out that our experience with the latter conception has not led to antinomies, whereas the same is not true in the case of the logical conception. The idea of having a "procedure" for obtaining objects comes out somewhat more persuasively in the case of the "constructible" sets in Gödel's consistency proof of axiomatic set theory with choice and the continuum hypothesis, for the constructible sets are those which can be obtained by iterated application of the operations given by the axioms presented by Gödel (1940). Gödel thus speaks of the constructible sets as formed at stages in a "generating process," and of the generating process as continuing into the transfinite. Gödel made the interesting remark to Kreisel that he used L in the axiom $V = L$ (i.e., every set is constructible) to stand for *lawlike*. The idea is even more plausible, for example, in the case of hereditarily finite sets.

Nonetheless, Gödel is speaking of transfinite sets, and this puts a great strain on the notion of having evidence for objects in acts carried out through time, or on the notion of a fulfillment procedure, especially if the acts involved are supposed to be structured in linear time of type ω. Certainly it extends the concepts of evidence, of verification, and of a "procedure" far beyond what is presently recognized as constructive in mathematics. Very strong idealizations or abstractions from human experience would be involved, as Charles Parsons (Parsons 1977) has emphasized. Perhaps, much as in Husserl's writings, Gödel is only suggesting that we have some type or degree of evidence for these objects that we would not possess at all with the logical conception of set. For although the notion of a process or procedure is greatly extended in the case of the sets Gödel has in mind, there is still a significant difference from the logical conception of set, in which case there is no procedure of any kind for obtaining the objects at various stages from given objects. Surely there is a difference in evidence between having a concept, for example, the naive concept of set, which is provably contradictory and having a concept of set which at this stage of our experience is not provably contradictory but which, on the contrary, has some stability in our experience. And if the present reading is correct, Gödel would certainly not be suggesting that mathematical intuition is incorrigible, given the nature of the evidence at these levels. Our experience with the iterative concept of set might even prove illusory in some respects. Gödel's papers on constructive mathematics show that he does distinguish different degrees of evidence for mathematical objects. If we consider the contrast between constructive number theory and Zermelo-Fraenkel (ZF) set theory, for example, it is clear that in the former case we have well-defined procedures for obtaining the objects of the theory so that there is a kind of evidence associated with acts directed to natural numbers at a given stage that is not present in the case of intuition of sets. Thus we might say that at a given stage of our experience we do have evidence for objects such as transfinite sets, as these objects are understood on an iterative conception, but that, on the basis of several criteria, it is significantly different from the kind of evidence we have at a given stage of our experience for finite objects such as natural numbers.

§ 3

I believe that this reading helps to make sense of Gödel's views about mathematical intuition, even though epistemological difficulties about how best to understand the notion of evidence in connection with the idea of transfinite objects or processes remain. The notion of mathematical intuition is of course also connected with other themes in Gödel's work of which we have not taken note. For example, Gödel apparently believed that something about the notion of intuition might suggest that the mind is not mechanical, in the sense that it ought not to be viewed as a finite, discrete, combinatorial, syntax manipulator. This theme emerges clearly in Wang's notes of his discussions with Gödel in *FMP* and *RKG*, and in Gödel's very interesting but unpublished "Is Mathematics Syntax of Language?" (now published in two versions as Gödel *1953/59). It is perhaps again related to the basic fact that in our experience we intuit objects as complete, whole, and determinate, even though what we actually see at a given stage in our perception is only one aspect or part of an object. Gödel may have had in mind that it is possible that if intuition were nothing more than combinatorial operations on discrete, finite configurations of such "aspects" or "parts," we would not have experience as we know it. But intuition is not like that, and hence it is possible that mind is not mechanical. In a note published in *FMP* Gödel claims, among other things, that Turing's claim that mental procedures cannot carry any further than mechanical procedures is inconclusive because it depends on the supposition that a finite mind is capable of only a finite number of distinguishable states (Wang 1974, pp. 325–326). He says that "what Turing disregards completely is the fact that mind, in its use, is not static, but constantly developing" (Wang 1974, p. 325). There is much more to be said about this aspect of Gödel's philosophical view, but we cannot go into it now.

It seems to me that a phenomenological conception of reference and intuition is worth exploring in its own right for the insights it might provide into difficult problems about mathematical knowledge. Indeed, I have discussed many of the issues in this chapter in detail, but limited to finite sets, in Tieszen (Tieszen 1989). The notion of representational character, and the role that it has in a theory of intentionality, is especially interesting in the case of mathematics, for with it we are given some leeway, of a type not furnished by other views of mathematical knowledge, in dealing with the question of how knowledge of mathematical objects is possible.

Postscript (1992)

After this essay was written I learned from Solomon Feferman that a manuscript in which Gödel discusses Husserl and phenomenology had been recently transcribed and that it would appear in Volume III of Gödel's *Collected Works*. (It is now published as Gödel *1961/?.) The manuscript, written around 1961, shows that Gödel had by that time assimilated many Husserlian ideas and that any later work could have been influenced by his study of phenomenology. I also discovered that in the present essay I was led to anticipate a number of the ideas that Gödel explicitly discusses in Gödel *1961/?.

5

Gödel's Philosophical Remarks on Logic and Mathematics

The publication of Volumes I, II, and III of *Kurt Gödel: Collected Works* (*KG:CW*) (Feferman 1986, 1990, 1995) marks a major event in the history of logic and the foundations of mathematics. The material included in the volumes not only presents us with a picture of the great scope, depth, and significance of Gödel's accomplishments, but will also open up new avenues of thought and research for future generations of logicians, mathematicians, philosophers, computer scientists, and others who will find Gödel's ideas on various subjects to be of substantial interest. There is an abundance of material to be considered in studying these books, including work on relativistic cosmology, Kant and the philosophy of time, an ontological proof for the existence of God, and Gödel's abortive but interesting efforts late in his career to settle the continuum hypothesis. In this chapter I will focus only on indicating Gödel's main philosophical theses about mathematics and logic; even with this limitation, it will only be possible to scratch the surface.

The bulk of Gödel's technical work in logic and foundations was completed between 1929 and 1943. The most important results are the completeness theorem for first-order logic (1929), the incompleteness theorems (1931), and the theorems on the relative consistency of the axiom of choice (AC) and the continuum hypothesis (CH) (1938–40). From 1943 on, Gödel devoted himself almost entirely to philosophy, but he did not publish most of his work. Volume III of *KG:CW* contains

I thank Solomon Feferman and Charles Parsons for comments on an earlier draft of this essay.

some of the most interesting philosophical items, and I will comment on a number of these later.

Prior to the publication of these three volumes, Gödel's philosophical views were known only from the few essays he published and from sources such as Wang and Kreisel. The picture is now filled out considerably and we are able to see how Gödel assessed the philosophical implications of his incompleteness theorems and related results on decidability and consistency. The work in the thirties after the discovery of the incompleteness theorems (see especially *1933o, *193?, *1938a) shows that Gödel was already thinking about the philosophical consequences of incompleteness and undecidability. It is noteworthy that in the introductory section of the 1929 dissertation (which was omitted by Gödel in the published version of the completeness proof in 1930) Gödel mentions the possibility of the incompleteness of mathematical axiom systems. He says that

we cannot at all exclude out of hand . . . a proof of the unsolvability of a problem if we observe that what is at issue here is only unsolvability by certain *precisely stated formal* means of inference. (Gödel 1929, p. 63)

Much of this early work shows a great deal of concern for constructivity and for degrees of constructivity in mathematics. Indeed, Gödel had already published some important results on intuitionistic logic and arithmetic in 1933 (see 1933e, 1933f). He is, however, also critical of intuitionistic notions (e.g., the intuitionistic notion of proof) as not being "constructive and evident to a higher degree" (see *1933o, *1938a, *1941, 1958, and 1972). In this early period Gödel is especially interested in finding ways to establish the consistency of arithmetic and other parts of mathematics on grounds that are constructive but that extend beyond Hilbert's finitism. (It is clear, especially in *1933o, that he already appreciated the difference between finitism and intuitionism.) The "Lecture at Zilsel's" (*1938a) gives a fascinating glimpse into this effort and prefigures a number of important ideas and results. For example, the idea of using higher-type functionals to obtain a consistency proof for classical arithmetic is already discussed in the lecture. By 1941 Gödel (see *1941) had developed the so-called *Dialectica*-interpretation of arithmetic. He did not publish this material until 1958 (see Gödel 1958), and he continued to add philosophical comments to an English translation of it as late as 1972 (see Gödel 1972). Remarkably, the no-counterexample interpretation of theorems of arithmetic also appears in *1938a.

Gödel draws a number of basic consequences from the incompleteness theorems in connection with these lines of research. By the second incompleteness theorem, it is necessary to go beyond finitary mathematics in Hilbert's sense in order to obtain consistency proofs for number theory. Since finitary mathematics is supposed to be the mathematics of concrete intuition, it appears that abstract concepts are needed for the consistency proof for number theory (see 1972). The concept of 'computable functional of finite type' is said to be such an abstract concept, but it is nonetheless constructive and meets conditions on constructivity that are more strict than those recognized by intuitionism. In a comment about abstract concepts that combines a number of the ideas and themes of his later philosophical reflections on mathematics, Gödel says that

> by abstract concepts, in this context, are meant concepts which are essentially of the second or higher level, i.e., which do not have as their content properties or relations of *concrete objects* (such as combinations of symbols), but rather of *thought structures* or *thought contents* (e.g., proofs, meaningful propositions, and so on), where in the proofs of propositions about these mental objects insights are needed which are not derived from a reflection upon the combinatorial (space-time) properties of the symbols representing them, but rather from a reflection upon the *meanings* involved. (Gödel 1972, pp. 272–273)

Gödel also argues in other works that reflection on the thought structures or contents associated with mathematical symbols is needed in order to find consistency proofs and to decide meaningful, well-defined mathematical problems. He even explores 'abstract', nonformalistic concepts of provability and definability in some of his writings.

In another paper from the thirties (*193?) Gödel says that Hilbert's belief in the decidability of every clearly posed mathematical question is not shaken by the proof of the incompleteness theorems. The incompleteness theorems show only that something was lost in translating the concept of proof as "that which provides evidence" into a purely formalistic concept. Gödel concludes that it is not possible to formalize mathematical evidence even in the domain of number theory. The problem of finding a mechanical procedure for deciding every proposition of a class for certain classes of mathematical propositions, however, *is* absolutely unsolvable. Another way to put this, according to Gödel, is to say that it is not possible to mechanize mathematical reasoning completely. The claim that the bounds of mechanism and formalism are not to be identified with the bounds of human reason appears in many of Gödel's philosophical papers. Similarly, Gödel suggests in many passages that the incompleteness theorems show that (the bounds on) what can be known by appeals to concrete, finitary, immediate intuition should not be identified with (the bounds on) what can be known by human reason or 'rational intuition'.

It is this kind of view that underlies the late note on a "philosophical error in Turing's work" (1972a) that Gödel appended to "On an Extension of Finitary Mathematics" (1972). Gödel says that Turing at one point presents an argument to show that mental procedures cannot go beyond mechanical procedures. The problem is that there may be finite, nonmechanical procedures that make use of the meaning of terms. Turing does not recognize that

mind, in its use, is not static, but constantly developing, i.e., that we understand abstract terms more and more precisely as we go on using them, and that more and more abstract terms enter the sphere of our understanding. (Gödel 1972a, p. 306)

Therefore, although at each stage the number and precision of the abstract terms at our disposal may be *finite*, both (and therefore, also Turing's number of *distinguishable states of mind*) may *converge toward infinity* in the course of the application of the procedure. (Gödel 1972a, p. 306)

Although there will always in principle be some well-defined or clearly posed but undecidable problem for any machine, the same claim does not necessarily follow for human reason. There can be no consistent machine that is constantly developing its 'understanding' with respect to more and more well-defined yes-or-no questions P about a given mathematical concept. Human reason, however, can evidently achieve such a constant development by virtue of its ability to reflect on the abstract meaning of the terms involved in the problems. Perhaps the most sympathetic interpretation of Gödel's remarks is not that the constant development itself bars mechanization but rather that the development might have a nonrecursive character because it involves reflection on *abstract meanings* or concepts. A person would not have to stop or become static at some point in deciding mathematical propositions that are clearly formulated but presently undecidable. (For example, human reason would not necessarily have to become static in the presence of CH. Gödel is at pains in 1947 to show that the formulation of CH is determinate enough for us to expect that a solution should be forthcoming in an extension of set theory.)

The material in "Remarks Before the Princeton Bicentennial Conference on Problems in Mathematics" (1946) is related to these ideas. Gödel explores ideas of demonstrability and definability that are 'absolute' in the sense that they would be independent of any formal languages or theories. He says that, thus far, demonstrability and definability have been defined only relative to a given language, and "for each individual language it is clear that the one thus obtained is not the one looked for." With the concept of Turing computability, however, "one has for the first time succeeded in giving an absolute definition of an interesting epistemological notion, i.e., one not depending on the formalism chosen." Although Turing computability is merely a special kind of demonstrability or definability, it is "by a kind of miracle" not necessary to distinguish orders. The diagonal procedure does not lead outside the defined notion. It is this fact that should encourage one to expect something similar in the case of demonstrability and definability in spite of such negative results as the incompleteness theorems and the Richard paradox. Concerning demonstrability, Gödel is evidently thinking of the nonformal, abstract concept of proof he mentions in other work, where 'provable' is understood in the sense of 'knowable to be true'. In the incompleteness theorems for arithmetic, for example, it is not the case that one diagonalizes outside what is knowable to be true in the natural numbers with the formation of each new Gödel sentence. One does, however, diagonalize outside each given formal system.

Gödel published his set-theoretic results between 1938 and 1940. "Russell's Mathematical Logic" (1944), which contains the first published expression of Gödel's mathematical and logical realism, appeared a few years later. Gödel remarks on the realistic attitude that Russell adopted in some of his work, and he develops some analogies between mathematics and natural science that are suggested by Russell. In particular, Gödel notes how Russell compares axioms of mathematics and logic with laws of nature, and mathematical evidence with sense perception. Axioms need not be evident in themselves. Their justification could lie in the fact that they make it possible for the 'sense perceptions' to be deduced. Gödel thinks this view has been largely justified by subsequent developments, and that in the future we will find it even more convincing. He writes that

it has turned out that (under the assumption that modern mathematics is consistent) the solution of certain arithmetical problems requires the use of assumptions essentially transcending arithmetic, i.e., the domain of the kind of elementary indisputable evidence that may be most fittingly compared with sense perception. (Gödel 1944, p. 121)

The discussion of realism is deepened when Gödel comes to Russell's approach to the paradoxes. Gödel notes that one of Russell's formulations of the vicious circle principle (VCP) (i.e., no totality can contain members definable only in terms of this totality) makes impredicative definitions impossible and forces a kind of constructivity. If classical mathematics violates the VCP, however, then this is reason to take the VCP rather than classical mathematics to be false. One of Gödel's most widely quoted philosophical comments is made in this connection:

> Classes and concepts may, however, also be conceived as real objects, namely classes as "pluralities of things" or as structures consisting of a plurality of things and concepts as the properties and relations of things existing independently of our definitions and constructions. (Gödel 1944, p. 128)

Gödel adds that "the assumption of such objects is quite as legitimate as the assumption of physical bodies and there is quite as much reason to believe in their existence." They are just as necessary in obtaining a satisfactory system of mathematics as physical bodies are in obtaining a satisfactory theory of sense perceptions.

Gödel evidently feels the need at this point to adopt a view of abstract concepts and classes that is more robust and platonistic than the view he held in his earlier work. Indeed, in his discussion of the ramified theory of types he mentions that a transfinite extension of the ramified theory has technical value since this is what he used to define the constructible sets in his work on set theory. It is known that Gödel had been trying to obtain his consistency proof by using constructive ordinals but found that the problem yielded if he took the classical ordinals as given. It is at this point that Gödel might very well have arrived at the following moral: a platonistic view yields important (metamathematical) results that would otherwise be overlooked!

The paper on Russell also contains Gödel's earliest published comments on the analyticity of mathematics. He says that 'analyticity' may be understood in two senses: (i) "it may have the purely formal sense that the terms occurring can be defined (either explicitly or by rules for eliminating them from sentences containing them) in such a way that the axioms and theorems become special cases of the law of identity and disprovable propositions become negations of this law"; or (ii) "a proposition is called analytic if it holds 'owing to the meaning of the concepts occurring in it', where this meaning may perhaps be undefinable (i.e., irreducible to anything more fundamental)" (p. 139). In this and other writing (especially *1951), Gödel suggests that mathematics cannot be understood as analytic in the sense of (i) but that it can in the sense of (ii).

"What Is Cantor's Continuum Problem?" (1947, 1964) is widely known and the 1964 version is especially noteworthy for its expression of some central themes in Gödel's later philosophy of mathematics. It includes various theses of Gödel's set-theoretic realism, along with the claim that there is a kind of mathematical intuition of the objects of set theory and of the concepts expressed by basic terms of set theory, and the claim that there is an analogy between mathematical intuition and sense perception. Gödel is at pains in the early part of the essay to show that CH is meaningful and determinate enough that one can expect an unambiguous answer to be forthcoming in a natural extension of set theory. He does of course say that further clarification of the meaning of the basic terms of set theory is needed in order to obtain the extension and to solve the problem, but he rules out the meaning analyses proposed by Brouwer, Weyl, and Poincaré on the grounds that they change the original meaning of the problem. Instead, new axioms that only analytically unfold the content of the given (abstract) iterative concept of set are needed. Gödel suggests the use of large cardinal axioms, but he also says that other unknown axioms may play a role.

In "Some Basic Theorems on the Foundations of Mathematics and Their Implications" (*1951), Gödel discusses the philosophical implications of the incompleteness theorems and related results. He says that the results he wants to discuss are all centered around one basic fact, which might be called the incompletability or inexhaustibility of mathematics. The point is first illustrated in connection with the effort to axiomatize set theory. Instead of finding a finite set of axioms, as in geometry, one is faced with an infinite series of axioms that can be extended further and further but without the possibility of generating all of these axioms by a finite rule. The incompletability of mathematics is shown in a general way, however, by the incompleteness theorems. Gödel says it is the second incompleteness theorem that makes the incompletability of mathematics particularly evident. The theorem shows that it is impossible for someone to set up a well-defined system of rules and axioms and consistently make the following assertion about it: I perceive (with mathematical certitude) that all of these axioms and rules are correct and, moreover, I believe that they contain all of mathematics. Anyone who makes such an assertion contradicts himself or herself. If the axioms are perceived to be correct, then they are perceived with the same certitude to be consistent. One then has a mathematical insight not derivable from the axioms.

Gödel asks whether this means that no well-defined system of correct axioms can contain all of mathematics. His answer is that it does have this implication for mathematics in the *objective* sense (i.e., viewed as the system of all true mathematical propositions), but it does not for mathematics in the *subjective* sense (i.e., viewed as the system of all demonstrable mathematical propositions). If there exists a finite rule producing all of the evident axioms of 'subjective mathematics', then, Gödel says, the mind would be equivalent to a finite machine. It would, however, be a finite machine unable to understand completely its own functioning. Furthermore, if the mind were equivalent to a finite machine, then not only would objective mathematics be incompletable in the sense of not being contained in any well-defined axiom system, but there would exist *absolutely* unsolvable Diophantine problems. *Absolutely* here means unsolvable by *any* mathematical proof the human mind could conceive, and not just undecidable within some particular axiom system. According to Gödel, the following disjunctive 'theorem' about the incompletability of mathematics is inevitable:

> Either mathematics is incompleteable in this sense, that its evident axioms can never be comprised in a finite rule, that is to say, the human mind (even within the realm of pure mathematics) infinitely surpasses the powers of any finite machine, or else there exist absolutely unsolvable diophantine problems. (Gödel *1951, p. 310)

The philosophical implications of this conclusion are also disjunctive, but Gödel says that under either alternative they are decidedly opposed to materialistic philosophy. Under the first alternative, it seems to be implied that the functioning of the human mind cannot be reduced to the functioning of the human brain. The brain, however, is to all appearances a finite machine with a finite number of parts. One is apparently driven to a 'vitalistic' viewpoint. The second alternative appears to disprove the view that mathematics is only our own creation. It seems to imply that mathematical objects and facts exist objectively and independently of our mental acts and decisions. In other words, this alternative seems to imply a platonistic view. (We have already seen that Gödel suggests elsewhere (e.g., *193?, 1972a) that it is not necessary to embrace the view that there are absolutely unsolvable Diophantine problems, and that he explores an 'abstract' concept of proof. See also 1972a, remark 2.)

Gödel argues that modern developments in the foundations of mathematics, in any case, support platonism. He presents three short arguments for this claim (*1951, p. 314). The arguments are supposed to show that mathematics could not be our own free creation. This provides an occasion for Gödel to launch an attack on syntactical or linguistic conventionalism about mathematics, for these forms of conventionalism represent attempts to give a more precise meaning to 'free creation' or 'free invention'. Gödel argues at length that mathematical propositions are not void of content. On the contrary, they have their own content and refer to distinctively mathematical objects and facts. Gödel argues that the incompleteness theorems and related results are incompatible with conventionalism, and he returns to this theme in his essay "Is Mathematics Syntax of Language?" (*1953/59). Gödel mentions that certain axioms of set theory are valid because of the meaning of the term 'set' and might even be called analytic, but they are certainly not tautological or void of content. In this case, 'analytic' does not mean 'true owing to our definitions', but rather 'true owing to the nature of the concepts occurring therein' (p. 321). This concept of analyticity is so far from meaning 'void of content' that an analytic proposition can possibly be undecidable. Gödel concludes this essay by saying that he also thinks nominalism, psychologism, and Aristotelian realism are untenable. Only the platonistic view is acceptable.

Two versions of "Is Mathematics Syntax of Language?" (*1953/9, versions III and V) are included in Volume III. In these papers Gödel argues at length against the empiricist or positivist claim that mathematics is syntax of language. His targets are some early views of Carnap and the positivist program that attempt to reconcile strict empiricism with the a priori certainty of mathematics. Empiricism is the view that all knowledge is based on external or internal sense perception. Empiricists hold that we do not possess an intuition into some realm of abstract mathematical objects. Mathematics is thought to have an a priori certainty, but because such a mathematical realm cannot be known empirically, it must not exist at all. The objective of these logical empiricists (or positivists) is to build up mathematics as a system of sentences that are valid independently of sense experience without using mathematical intuition or referring to any mathematical objects or facts. This is to be accomplished by showing that mathematics can be completely reduced to syntax of language.

Gödel uses an interesting application of the second incompleteness theorem to show that mathematics could not, in the relevant sense, be syntax of language. Thus, mathematical statements must have their own kind of content or meaning that is abstract, relative to the rules of syntax. Mathematical intuition is not eliminable. Empiricists suppose that mathematical propositions are tautologies and are empty because, unlike empirical propositions, they fit all of the facts. This only shows, however, that empiricists commit a petitio principii by taking 'fact' to mean 'empirical fact'. It is not the case that mathematical propositions fit all of the *mathematical* facts. Rather, they are true or false relative to given concepts, and there are different mathematical concepts. Indeed, there are different sets of axioms for different concepts. Empiricists typically run roughshod over these matters of mathematical practice. Thus, instead of clarifying the meaning of abstract and nonfinitary mathematical concepts by explaining them in terms of syntactical rules, we must use abstract and nonfinitary concepts to formulate the syntactical rules. Instead of justifying mathematical axioms by reducing them to syntactical rules, we must use some of these axioms in order to justify the syntactical rules as consistent.

Finally, many themes of Gödel's later philosophy of mathematics are brought together in an interesting lecture manuscript that presents a rather broad perspective on the issues, "The Modern Development of the Foundations of Mathematics in the Light of Philosophy" (*1961/?). Gödel describes the development of foundational research in mathematics since the beginning of the twentieth century in terms of a general schema of possible worldviews that he divides into two groups. One group (the 'leftward') consists of skepticism, materialism, and positivism, and the other ('rightward') group consists of spiritualism, idealism, and theology. Gödel places 'apriorism' on the right side of this schema, and empiricism on the left. He also places optimism on the right, and pessimism on the left, describing skepticism as a form of pessimism about knowledge.

Gödel notes that philosophy, including the philosophy of mathematics, has moved in a leftward direction since the Renaissance. Hilbert's formalism was an attempt to do justice both to the leftward spirit of our time and to the nature of mathematics, which has always inclined toward the rightward view. Hilbert still held to the rightward ideas that a mathematical proof should provide a secure grounding for a proposition, and that every precisely formulated yes-or-no question in mathematics must have a clear-cut answer. The incompleteness theorems show, however, that Hilbert's attempted reconciliation fails. As Gödel sees it, this failure implies that an adjustment is needed in our philosophical view. It does not imply that giving up on the rightward aspects mentioned is necessary. A new combination of leftward and rightward views is needed, and Gödel suggests that this can be found in Husserl's phenomenology. Phenomenology purports to offer a way to clarify and cultivate knowledge of the abstract concepts that underlie mechanical systems of mathematics without succumbing to the excesses or mistakes of earlier rightward views. Gödel defends the phenomenological approach to meaning clarification against empiricist views. He notes that there are examples that show how we come to a better understanding of concepts in a way that is not based on sense perception. We do this even without the application of a conscious and systematic procedure. Once again, he cites the fact that new axioms become evident to us again and again, even though they do not follow by formal logic alone from previously established axioms. It is not excluded by the incompleteness theorems, Gödel says, that every clearly posed mathematical yes-or-no question may be solvable. More and more new axioms become evident to us on the basis of the meaning of the primitive notions, and this process is not something that a machine can imitate. The paper concludes with some sketchy but interesting comments on the relation of phenomenology to Kantian philosophy. Gödel says that it is Husserl's (transcendental) phenomenology that for the first time does justice to the core of Kantian thought.

It is evident from this overview that Gödel did not develop a systematic philosophy and that many of his claims are in need of clarification. There have been and will no doubt continue to be difficulties and controversies in the interpretation of some of his views. At least in outline, however, Gödel's later views on mathematics arguably do form a coherent whole. He holds that mathematical terms are expressions of abstract concepts and mathematical sentences are expressions of relations between these concepts. Sentences of pure mathematics have their own content or meaning, and they refer to mathematical objects and facts, just as sentences about ordinary sense experience have their own content or meaning and refer to sensory objects and facts. Indeed, Gödel thinks there is a kind of analogy between mathematics and physics, although the objects and facts in each case are of different types. Gödel evidently wants to accept mathematical practice as it is given in its 'whole original extent'. He is a platonist or realist about mathematical concepts, meanings, objects, and facts. He says that these entities are not human-made, they are abstract and not concrete, they are not located in space-time, and so on. We know about these entities through reflection, or through a type of rational intuition that is analogous to sensory perception. In the one case it is particular objects and their properties and relations that are known; in the other case we perceive the most general concepts and their relations. Gödel holds that both rational intuition and sensory perception are inexhaustible, both are constrained or forced in certain respects, and both are susceptible to illusion. In his later work, Gödel suggests that the incompleteness theorems and related results about decidability and consistency support this kind of platonistic view. They are also, in a sense, consequences of it. These metamathematical results are difficult to reconcile with various reductionistic viewpoints.

The use of rational intuition in mathematics provides a kind of evidence that is unique to mathematics. Gödel even allows that there may be different types or degrees of evidence within mathematics. For example, he seems in some passages (e.g., Gödel 1947) to suggest that in constructive mathematics the objects are completely given, or they can be intuited individually, whereas this is not the case in higher, impredicative set theory. It is, however, still possible to reflect on and develop a grasp of the meanings of sentences of impredicative set theory in such a way as to solve meaningful mathematical problems. We can hope to clarify (our understanding of) the meaning of these sentences if the sentences are approached in the right way. It is important not to approach them in a reductionistic way. There are many unacceptable versions of reductionism about mathematical meaning and objects that are associated with leftward philosophical views. All of these views lack the proper balance between empiricism and rationalism. Gödel mentions specifically Hilbertian formalism, mechanism, finitism, intuitionism (relative to higher set theory), psychologism, nominalism, positivism, conventionalism, empiricism, and Aristotelian realism. It is clear that Gödel's comments would extend to more recent forms of empiricism and naturalism.

Gödel says that an alternative to these views is to be found in Husserl's phenomenology. Of course Gödel never published the 1961 paper in which he discusses Husserl, and the remarks in the paper need not represent his final position. There is, however, other explicit and implicit evidence for his sympathy with Husserl's views, and Husserl does recognize the possibility of nonreductionistic meaning clarification in mathematics. Meaning clarification in mathematics can be regarded as analytic in a broad sense, and one can argue that mathematics could not be analytic according to the narrower formalistic or positivistic criteria that Gödel discusses in some of his papers. The relationship between the interpreted theorems of a formal system and the interpreted Gödel sentence for the system would be analytic in the broad sense. One sees examples of meaning clarification (even if it is not completely conscious and systematic) in the phenomena associated with the incompleteness theorems and, analogously, in set theory. In each case there is an unfolding of the content of an abstract concept in rational intuition. In each case we ascend to a higher form of awareness about the relevant concept(s). For example, we ascend to higher types, or new axioms of infinity, to decide problems that were previously undecidable in set theory. This is not done on the basis of sensory intuition nor even with an eye to rounding out our theories of sensory objects. Mathematics may be inexhaustible, but negative results such as the incompleteness theorems do not imply that there are absolutely unsolvable mathematical problems. These results do not exhibit a definite problem that cannot be solved. Rather, they establish the nonexistence of a certain kind of general procedure for deciding a class of problems. To a rationalistic optimist such as Gödel, this does not imply absolute undecidability. It is empiricists, skeptics, and materialists who are more likely to interpret it as an absolute undecidability result. Gödel thinks it would imply absolute undecidability only if the mind were a finite machine, but he gives arguments in various writings to show that the mind is not a finite machine. We should keep in mind Gödel's explorations of nonformal, absolute notions of demonstrability and definability.

There is no doubt room to attack some of these ideas, but there is just as much room to develop and defend many of them. In my view, Gödel has a better philosophical perspective on mathematics than can be found in much of the contemporary work on the subject. Of course there are certain perplexing aspects of his views on mathematical intuition and realism, but the perplexity seems to me to embody more insight than many of the supposed (reductionistic) 'solutions' to these problems. I agree with Thomas Nagel's observation that there is a persistent temptation to turn philosophy into something less difficult and more shallow than it is. Part of the perplexity engendered by these views of Gödel probably results from the fact that certain areas of philosophy have been neglected in our time. For example, there has been almost no work at all on elucidating a nonlogicist, nonreductionistic theory of concepts or properties in connection with mathematics. As another example, one could point to the fact that generations of analytic philosophers have ignored the work of thinkers such as Husserl.

After studying these papers, it is easy to believe that Gödel's ideas will open up some new directions in the philosophy and foundations of mathematics.

6

Gödel's Path from the Incompleteness Theorems
(1931) to Phenomenology (1961)

In a lecture manuscript written around 1961 (Gödel *1961/?), Gödel describes a philosophical path from the incompleteness theorems to Husserl's phenomenology. It is known that Gödel began to study Husserl's work in 1959 and that he continued to do so for many years. During the 1960s, for example, he recommended the Sixth Investigation of Husserl's *Logical Investigations* to several logicians for its treatment of categorial intuition (Wang 1996, p. 164). Although Gödel may not have been satisfied with what he was able to obtain from philosophy and Husserl's phenomenology, he nonetheless continued to recommend Husserl's work to logicians as late as the 1970s. In this chapter I present and discuss the kinds of arguments that led Gödel to the work of Husserl. This should help to shed additional light on Gödel's philosophical and scientific ideas and to show to what extent these ideas can be viewed as part of a unified philosophical outlook. Some of the arguments that led Gödel to Husserl's work are only hinted at in Gödel's 1961 paper; they are developed in much more detail in Gödel's earlier philosophical papers (see especially 1934, *193?, 1944, 1947, *1951, *1953/59). In particular, I focus on arguments concerning Hilbert's program and an early version of Carnap's program.

I would like to thank the spring 1997 Logic Lunch group at Stanford, especially Solomon Feferman, Grisha Mints, Johan van Benthem, Aldo Antonelli, and Ed Zalta, for helpful comments. I would also like to thank Andreas Blass, Albert Visser, and an anonymous referee for the *Bulletin of Symbolic Logic* for comments on an earlier draft of this essay.

§ 1   Some Ideas from Phenomenology

Since Husserl's work is not generally known to mathematical logicians, a brief mention of a few details about his background may be helpful. Husserl received his doctorate in mathematics in 1883 with a thesis on the calculus of variations. He then served for a brief period as an assistant to Weierstrass in Berlin. His interests soon turned to the philosophy of logic and mathematics. Eventually his philosophical interests widened considerably, although he continued to write on philosophical issues about logic. He was personally acquainted with Cantor, Zermelo, Hilbert, and a number of other major figures in logic. He also corresponded with Frege, Schröder, and others. Hilbert was evidently personally involved in helping Husserl to obtain a position in Göttingen in 1901. Husserl's philosophical work was deeply influenced by his mathematical background, as will become apparent later. Husserl exerted some influence on Hermann Weyl, on Arend Heyting by way of Oskar Becker, and on some of the Polish logicians who were active in the early part of the twentieth century. Heyting and Becker, for example, established an interesting connection with Husserl's work by identifying mathematical constructions (in the sense of constructive mathematics) with fulfilled mathematical intentions, where the notion of 'intention' is understood in terms of the theory of intentionality.

I begin with a description of some of Husserl's ideas that are directly relevant to Gödel's comments in 1961/? (see also Chapters 2, 4, and 5 of this book; Føllesdal 1995; Wang 1996). The most important ideas are related to the fact that human cognition, including mathematical cognition, exhibits intentionality. This means that cognition is always *about* something. Consciousness is always consciousness *of*. One can see this clearly in acts of believing, knowing, willing, desiring, remembering, hoping, imagining, and so on. For example, an act of belief is always directed toward a particular object or state of affairs, even if this directedness is indeterminate in some respects. Try to imagine for a moment what it would be like for a belief not to be about anything in particular.

That an act is directed in a particular way means that it is not directed in other ways. Cognitive acts are perspectival and we cannot take all possible perspectives on an object or domain. We do not experience everything all at once. Perhaps only a god could do that; we cannot. Our beliefs and other cognitive acts at any given time are, in other words, always about certain *categories* of objects. The mind always categorizes in this way. It cannot help doing so, for our experience would have to be very different otherwise. Thus, many types or categories are always at work in our experience, even in everyday sense experience. I shall also refer to these categories as 'concepts'.

In order to introduce some specific texts of Husserl that clearly influenced Gödel, I now introduce a new term for what I have been calling 'categories' or 'concepts'. I will call these categories 'essences'. It is important not to let your mind wander to other associations you may have with the term 'essence'. I only mean for it to be another way of speaking of these categories in our experience. There are several reasons why it makes sense to use this term. Thus, at least the following associations are permissible at this point. First, we are concerned with *what* the experience is about at a given stage, and 'whatness' has been associated with the notion of essence at least since the time of Aristotle. Second, we can think of an essence in this sense as a universal. For example, I can see different instances of the category 'chair'. I can believe that there are different instances of the category 'natural number' or of the category 'function' or 'set'. I cannot believe that just anything is an instance of the category 'natural number'. The essences are not arbitrary. There are constraints on them and they are not subject to being changed at will. They are not freely variable.

It is safe to say we *know* that certain things are not instances of the category 'natural number' and that others are instances of this category. Thus, we must have some grasp of this category. Of course our grasp of this essence has become more precise over time as a result of various refinements, including informal and even formal axiomatization. Similar remarks can be made about the categories 'function' and 'set'. Different sets of axioms are established to capture and clarify our grasp of different mathematical essences. Nothing about this claim to knowledge precludes the possibility of indeterminateness in the grasp of the essence or of a subsequent splitting or refining of the concepts involved. Indeed, history shows that this has happened with the categories 'function' and 'set'. We can be more specific and distinguish different types of functions and sets. For example, we can distinguish and investigate the essence 'mechanically computable function'. In this case it is possible to reach such a degree of clarification in understanding that we can say with great precision which things are instances of this essence and which things are not. No matter what ambiguities lurked in the early conceptions of functions or sets, however, it was still possible to know that certain things were not instances of these types. It should be noted that Husserl also calls for the genetic investigation of our experience of mathematical essences in order to reacquaint ourselves with the different meanings and refinements that have been sedimented across different times, places and persons.

Husserl thinks these essences always play a role in our experience even if we do not usually attend to them or we do not reflect on them in a conscious and systematic way. Essences are presupposed in any cognition of an object. We can reflect on essences, however, especially if we are interested in clarifying our grasp of them. We might, for example, try to arrive at a better understanding of their bounds, or of their relations to other essences. (Such relations could include species/genus relations, consistency, implication.) Husserl distinguishes the study of formal essences from the study of 'material' essences. In logic and mathematics, broadly construed, the study of essences could be and is regimented in various ways. One must turn to mathematical practice to see what has actually been established with respect to mathematical essences.

A few additional words of warning about this idea of essences are in order. First, it is clear that we need not have a complete or perfect grasp of essences. Typically we would not have such a grasp. That is why we speak of clarification in the first place. The fact that we do not have a complete grasp of them, however, does not render them useless in our experience. On the contrary, it only accurately reflects the fact that we are finite, limited beings and that we are not omniscient. At the same time, one cannot deny that we do think in terms of categories. Moreover, a mathematical essence has an intrinsic unity that could not be replaced with some notion of family resemblance or similarity. It is not the case that the number 9 I use on one occasion *resembles* the number 9 I use on another occasion. It is *identical* to it. A definition or lemma I use on one occasion cannot merely resemble the definition or lemma I use on another occasion. It cannot differ essentially from it. If mathematical definitions, objects or theorems resembled one another only from occasion to occasion or from person to person, then mathematical practice would be very different from the way it actually is.

I now introduce another term. Instead of saying we have a partial 'grasp' of these essences I will say we *intuit* them. Thus, 'grasping the categories' involved in our experience means 'intuiting essences'. It is 'categorial intuition'. Once again, please do not let your mind wander to all sorts of other meanings that 'intuiting essences' could have. The expression, as it is being used here, is quite innocuous. The term 'intuition' is used because an essence is immediately given as a datum in reflection on our experience. It is given before we can begin to analyze it or to consider its relations to other essences. Another important point that should be clear from our description is that essences can be grasped prior to knowing whether or not there are individual physical instances of those essences. Some essences may be instanced in the physical world and some may not. Given what they are about, mathematical essences (e.g., the essence 'natural number') could not have individual physical instances. Generally, it does not follow that if there is intuition of an essence, then there is intuition of the existence of an individual physical instance of the essence.

With all of this said, I would now like to quote and briefly comment on a few passages from an essay by Husserl that was one of Gödel's favorites: "Philosophy as Rigorous Science" (Husserl 1911). It should now be possible to read these passages with some understanding. Gödel was evidently quite taken with Husserl's ideal of philosophy as a rigorous science, perhaps because he saw in it a continuation of the spirit of Leibniz's work. The themes in these passages are repeated and developed extensively in other works by Husserl that were among Gödel's favorites (e.g., *Logical Investigations, Ideas I, Cartesian Meditations*). First, Husserl (Husserl 1965, pp. 90–91) says that

to study any kind of objectivity whatever according to its general essence . . . means to concern oneself with objectivity's modes of givenness and to exhaust its essential content in the process of "clarification" proper to it. . . . With this we meet a science of whose extraordinary extent our contemporaries have as yet no concept: . . . a phenomenology of consciousness as opposed to a natural science about consciousness.

Husserl distinguishes sciences of fact from sciences of essence. Phenomenology itself is not supposed to be a natural science since it is concerned with essences. Essences cannot be adequately understood in terms of natural science. Indeed,

the spell of the naturalistic point of view . . . has blocked the road to a great science unparalleled in its fecundity. . . . The spell of inborn naturalism also consists in the fact that it makes it so difficult for all of us to see "essences", or "ideas" – or rather, since in fact we do, so to speak, constantly see them, for us to let them have the peculiar value which is theirs instead of absurdly naturalizing them. Intuiting essences conceals no more difficulties or "mystical" secrets than does perception. (Husserl 1965, p. 110)

Intuiting of essences conceals no more difficulties than perception because, as I have been saying, it refers only to grasping categories in our experience.

But one must in no instance abandon one's radical lack of prejudice, prematurely identifying, so to speak, "things" with empirical "facts." To do this is to stand like a blind man before ideas, which are, after all, to such a great extent absolutely given in immediate intuition. (Husserl 1965, p. 146)

And, for just this reason,

it is important today to engage in a radical criticism of naturalistic philosophy. (Husserl 1965, p. 78)

According to Husserl, there are many ways in which essences have been 'absurdly naturalized'. It is important to note that much of what has gone by the name *phenomenology* since Husserl has itself naturalized essences and shunned the ideal of philosophy as rigorous science. This development began immediately with Husserl's most famous and influential student, Heidegger. It goes without saying that Husserl was not pleased with this development.

Interestingly, Husserl laments many of the same reductionistic attitudes about essences that are mentioned in Gödel's discussions of mathematical content and abstract concepts: empiricism, naturalism, psychologism, nominalism, and forms of conventionalism and formalism that are coupled with these views. Aristotelian realism may also be added to this list. It is precisely this worry about reductionist attitudes that is a central theme in Gödel's 1961 paper. I now turn to Gödel's remarks.

§ 2   'Leftward' and 'Rightward' Viewpoints in Philosophy

Gödel (Gödel *1961/?) seeks to describe the development of foundational research in mathematics since the beginning of the twentieth century in terms of a general schema of possible worldviews. He divides these worldviews according to the degree and manner of their affinity to, or renunciation of, metaphysics (or religion). He thinks we then obtain a division into two groups, consisting of skepticism, materialism, and positivism on one ('leftward') side, and spiritualism, idealism, and theology on the other ('rightward') side. There are also mixed cases. Gödel places 'apriorism' on the right side of this schema, and empiricism on the left. He also places optimism on the right, and pessimism on the left, describing skepticism as a form of pessimism. Idealists tend to see meaning, reason, and purpose in everything in the universe, whereas strict materialists do not see meaning, reason, or purpose in anything.

The development of philosophy since the Renaissance has, on the whole, gone from right to left, and Gödel says that it would be a miracle if this development had not also begun to prevail in the conception of mathematics. One sees this development, for example, in the work of Mill in the nineteenth century. At the turn of the twentieth century the antinomies of set theory were discovered. The significance of the antinomies was exaggerated by skeptics and empiricists as a pretext for a leftward upheaval. Mathematicians began to deny that mathematics depicts a system of truths. Instead, they acknowledged this for some part of mathematics and retained the remainder in, at best, a hypothetical sense.

§ 3   The Incompleteness Theorems and Hilbert's Program

Hilbert's formalism is an example of this development. It seeks to do justice both to the leftward spirit of the time and to the nature of mathematics. In accordance with the spirit of the time, it acknowledges that the truth of axioms cannot be justified or recognized in any way and that the deduction of consequences from them therefore has meaning only in a hypothetical sense. This deduction itself is construed as a mere game with symbols according to certain rules. On the other hand, in accordance with the rightward character of mathematics, it holds that a proof of a proposition must provide a secure grounding for the proposition, and that every precisely formulated yes-or-no question in mathematics must have a clear-cut answer. A certain part of mathematics must be acknowledged to be true in the older rightward sense if we are to have mathematical certainty, but this part is much less opposed to the leftward spirit of the time. It is the part that refers to concrete and finite objects in space in the form of combinations of symbols.

Hilbert hoped to show, in response to the foundational crisis brought on by the discovery of the paradoxes, that the Peano axioms for arithmetic (PA), and indeed the formalization of all of higher mathematics, could be proved consistent by using only concrete, finitist means. The hope was that 'Cantor's paradise' itself could be secured in this way. A concrete, finitist consistency proof would depend only on a finite number of discrete objects – sign-configurations – that would be immediately intuitable in space-time, and on the combinatorial properties of these objects. (The notion of intuition here is clearly much narrower than the Husserlian notion described earlier.) It was part of Hilbert's formalism that there should be no need to consider the meaning of the sign-configurations involved in the formalizations. Only syntactical properties and relations should figure into the consistency proof. Hilbert (Hilbert 1926, p. 376 of English translation) puts it the following way:

As a condition for the use of logical inferences... something must already be given to our faculty of representation, certain extralogical concrete objects that are intuitively present as immediate experience prior to all thought. If logical inference is to be reliable it must be possible to survey these objects completely in all their parts, and the fact that they occur, that they differ from one another, and that they follow each other, or are concatenated, is immediately intuitively given, together with the objects, as something that neither can be reduced to anything else nor requires reduction.... And in mathematics, in particular, what we consider is the concrete signs themselves, whose shape, according to the conception we have adopted, is immediately clear and recognizable.

The concern for reliability in Hilbert's conception of proof theory precluded any appeal to the thoughts or meanings associated with the concrete signs. The concrete signs should be given to our faculty of representation prior to all thought.

A finitist consistency proof could presumably be carried out in a system such as primitive recursive arithmetic (PRA), which may be viewed as a subsystem of PA. This constructive, finitist part of mathematics would, in Hilbert's view, contain only 'real' or 'contentual' propositions. The aims would be to formalize mathematics and to show that the formalism for higher parts of mathematics could be proved consistent by using only the real, contentual part of mathematics. All the parts of mathematics in which purportedly abstract elements (such as abstract 'meanings' of semantics) were found would thus be shown to be secure. Hilbert's program is thus, in effect, a conservation program: the formalizations that included ideal elements in mathematics would be shown to be conservative extensions of the real part of mathematics, thereby effectively eliminating the dependence on the ideal elements. The reference to ideal or 'abstract' elements would merely serve to shorten proofs, or to simplify a system of reasoning, or to make it more perspicuous.

Now consider the meaning of Gödel's incompleteness theorems against this background. Gödel says they show, against Hilbert, that it is impossible to rescue the old rightward aspects of mathematics in such a manner as to be in accord with the spirit of our time. This is because it is impossible even to find an axiomatic system of number theory from which, for every number-theoretic proposition A, either A or ¬A would always be derivable. Furthermore, the second theorem tells us that for reasonably comprehensive systems of mathematics it is impossible to carry out a consistency proof merely by reflecting on the concrete combinations of symbols. It is necessary to introduce abstract elements.

Suppose the objects or concepts that can be represented in PRA (or even in PA) are completely representable in space-time as finite, discrete, sign-configurations that are in principle amenable to concrete intuition in Hilbert's sense. Let us further suppose that any objects or concepts that can be represented in PRA (or even PA) can, as Hilbert evidently thought, be considered 'concrete' by virtue of their representability in space-time as finite, discrete sign-configurations. There are some arguments in the literature about whether the exact bounds of the 'concrete' in finitist mathematics should be drawn more narrowly or more widely than PRA, but this need not detain us now. It seems unlikely that finitists would have extended the bounds beyond PA, given their concern for reliability in the face of the paradoxes. (If the bounds were extended beyond PA, then there would still have to be some limit. We could then bring the form of our argument to bear at that point.)

Given these suppositions, it follows, by the second incompleteness theorem, that the objects or concepts needed for the proof of CON(PA) must *not* be completely representable in space-time as meaningless, finite, discrete sign-configurations which are amenable to concrete intuition. In other words, starting from Hilbert's original philosophical viewpoint, it appears that a proof of CON(PA) requires appeals to the *meaning* of the sign-configurations and to *abstract* objects or concepts that are in some sense *nonfinite*. It will also involve a form of intuition that is not restricted to Hilbert's concrete intuition.

These kinds of remarks on the incompleteness theorems are expressed in many of Gödel's writings. For example, Gödel (Gödel 1972, pp. 271–272) says that

P. Bernays has pointed out on several occasions that, in view of the fact that the consistency of a formal system cannot be proved by any deduction procedures available in the system itself, it is necessary to go beyond the framework of finitary mathematics in Hilbert's sense in order to prove the consistency of classical mathematics or even of classical number theory. Since finitary mathematics is defined as the mathematics of *concrete intuition*, this seems to imply that *abstract concepts* are needed for the proof of consistency of number theory.... [What Hilbert means by *Anschauung* is substantially Kant's space-time intuition, confined, however, to configurations of a finite number of discrete objects.] By abstract concepts, in this context, are meant concepts which are essentially of the second or higher level, i.e., which do not have as their content properties or relations of *concrete objects* (such as combinations of symbols), but rather of *thought structures* or *thought contents* (e.g., proofs, meaningful propositions, and so on), where in the proofs of propositions about these mental objects insights are needed which are not derived from a reflection upon the combinatorial (space-time) properties of the symbols representing them, but rather from a reflection upon the *meanings* involved.

The idea that it is necessary to reflect upon meaning plays a central role in Gödel's 1961/? paper, as does the idea that reflection on meaning (or intuition of essence) is of a 'higher level' than reflection on the combinatorial properties of concrete symbols.

There are proofs of CON(PA) and these proofs must therefore require objects or concepts of the sort that would be recognized by mathematical or phenomenological realists (see also Tieszen 1994b). That is, the proofs must require abstract concepts and/or meanings that are not available to concrete, sensible intuition. In addition, these objects must in some sense be nonfinite. The requisite sense in which the objects or concepts needed for the proof of CON(PA) must be abstract and nonfinite is seen in Gentzen's consistency proof, since the proof requires induction on the transfinite ordinals $<\varepsilon_0$. It is also seen in Gödel's consistency proof, since the theory of primitive recursive functionals requires the abstract concept of a "computable function of type t." If we combine the conclusion drawn about meaning with the conclusion about abstract elements, it appears that the meaning associated with arithmetic expressions must be 'abstract', and that abstract elements cannot be eliminated from mathematics in the way that Hilbert had hoped. Another way of putting this would be to say that Hilbert's appeal to finite, concrete *particulars* will not suffice for consistency proofs for interesting parts of mathematics. In addition, the abstract elements involved could not be given by Hilbert's concrete intuition since concrete intuition is restricted to finite sign-configurations. There must be, by the second incompleteness theorem and the consistency proofs for PA, a less restricted kind of mathematical intuition or insight that accounts for our mathematical knowledge. In Husserlian language, there must be an intuition of mathematical essences. (This is no doubt what lies behind Gödel's discussions with Wang (Wang 1974, pp. 84–86; 1987, pp. 188–192, 301–304) about how we perceive or intuit abstract concepts, i.e., intuit categories or essences.) It need not be claimed, however, that this kind of intuition is infallible (see later discussion), even if it is taken to be a basic source of evidence.

If the Gentzen or Gödel proof of CON(PA) is evident to us on the basis of the meaning of the terms involved, then there is reason to believe that there are such proofs elsewhere in mathematics. There must be proofs that are not fully formalizable at a given stage in our mathematical experience but that are evident to us at that stage on the basis of the *meanings* involved in the terms of the proofs. There must be, in other words, a kind of 'informal rigor' in mathematics. This suggests the possibility that human minds might surpass machines in solving problems or in obtaining proofs of statements based on an understanding of the abstract meaning of the statements involved. Gödel has noted this implication in many places in his writings and I will come back to it later. In some writings he discusses an 'absolute' concept of proof that is abstract and nonformal. This is not a concept of proof that is always relative to some particular formal system or machine. It is a concept according to which *proof* is to be understood as 'that which provides evidence' (Gödel *193?, 1946). What is 'provable' in this sense is what is 'knowable to be true' (Gödel *1951, p. 318; *1953/59, p. 341).

Since the Hilbertian combination of materialism and aspects of classical mathematics proves to be impossible, Gödel (Gödel 1961/?) says that only two possibilities remain: either one must give up the rightward aspects of mathematics or one must attempt to uphold them in contradiction to the prevailing spirit of our time. Gödel says that nothing about the mathematical results we achieve requires us to give up the idea of the certainty of mathematical knowledge or the belief that for clear questions posed by reason, reason can also find clear answers. Only the desire to remain in agreement with the prevailing philosophy compels this.

Hilbert embraced a particular philosophy about the infinite, about what has meaning, about what objects can be recognized as existing, and about what objects can be intuited. This part of Hilbert's approach is refuted by the incompleteness theorems. In the Husserlian language quoted earlier, Hilbert's program does not give mathematical essences or abstract concepts their due. It exhibits a certain blindness or prejudice about them. It is an attempt, in effect, to show that directedness toward essences or abstract concepts can be reduced to directedness toward concrete, finite sign-configurations and combinatorial operations on such objects. On the other hand, Hilbert's optimism about mathematical problem solving and his concern for achieving certainty in mathematical proofs were admirable ideals. In an interesting unpublished paper written already in the 1930s (Gödel *193?), Gödel says that the incompleteness theorems can be interpreted as having one of two consequences for Hilbert's optimism about mathematical problem solving: either they mean that (i) it is not the case that every clearly posed mathematical problem can be solved or they show that (ii) something was lost in trans-lating the concept of proof as 'that which provides evidence' into a purely formal or mechanistic concept. Gödel says it is easy to see that (ii) is true, since number-theoretic questions that are not decidable in a given for-malism are always decidable by evident inferences not expressible in the formalism. The new inferences turn out to be exactly as evident as those of the given formalism. Perhaps Gödel is overstating the case here. The claim that number-theoretic questions not decidable in a given formalism are always decidable by evident inferences not expressible in the formal-ism seems to be true of questions such as that of the consistency of the formalism (for reasonable formalisms), but it is not so obvious for arbi-trary number-theoretic questions. Nonetheless, the conclusion we should be able to draw is that Hilbert's optimism about mathematical problem solving remains untouched even though formalization or mechanization of mathematical evidence in the domain of number theory is not pos-sible. The reason Hilbert's optimism remains untouched is that we can hope to make these decisions on the basis of our directedness toward and intuition of the underlying abstract concepts or essences. Gödel does not say this in *193?, but by *1961/? it is clear that this is what he has in mind (see § 6). Some parts of mathematics might be completely formal-ized or mechanized, but, on the whole, it is not possible to mechanize mathematical reasoning.

§ 4   The Incompleteness Theorems and Carnap's Program
(Gödel *1953/59)

Another 'leftward' attempt to accommodate mathematics can be found in Carnap's early work. Gödel (Gödel *1953/9, III) says that around 1930 Carnap, Hahn, and Schlick developed a conception of the nature of math-ematics, under the influence of Wittgenstein, which was a combination of nominalism and conventionalism. Its main objective was to reconcile strict empiricism with the a priori certainty of mathematics. Gödel char-acterizes empiricism as the view that all knowledge is based on external or internal sense perception. It holds that we do not possess an intuition into some realm of abstract mathematical objects or concepts. In light of Gödel 's 1961/? paper, it holds that we do not have an intuition of mathematical essences. Mathematics is thought to have an a priori cer-tainty, but because such a realm cannot be known empirically, it must not be assumed to exist at all. The objective of the program of these logical empiricists (or positivists) can therefore be viewed as one of building up mathematics as a system of sentences which are valid independently of sense experience without using mathematical intuition or referring to any mathematical objects or facts. This is to be done by showing that mathematics can be completely reduced to and in fact is nothing but syntax of language. This would mean that the validity of mathematical theorems consisted solely in their being consequences of certain syntactical conventions about the use of symbols. It would not consist in their describing states of affairs in some realm of objects. As Gödel puts it, mathematics is to be viewed as a system of auxiliary sentences without content or object. The syntactical conventions involved in the program are those by which the use of a symbol is defined by stating rules about the truth or assertibility of the sentences containing the symbol, where the rules refer only to the outward structure of expressions and not to their meaning or to anything outside the expressions. Assertions that conflict with the rules are excluded because of their structure, exactly as are assertions that conflict with rules of grammar.

Gödel lays out some basic conditions that would need to be met for the program to succeed. Among these conditions are the following: *mathematics* should mean classical mathematics, since the syntactical program aims to dispense with mathematical intuition but without impairing the usefulness of the full extent to which mathematics can be used in the empirical sciences. Gödel does say, however, that his argument against the syntactical viewpoint works even if the term *mathematics* is restricted to intuitionistic mathematics. Another condition is that *language* must mean some symbolism that can actually be exhibited and used in the empirical world. Sentences will have to consist of a finite number of symbols, for sentences of infinite length would evidently not exist and could not be produced in the empirical world. For the same reason, the rules of syntax must also be finitary.

Another important condition is that a rule about the truth of sentences can be called 'syntactical' only if one can know that it does not imply the truth or falsehood of any factual or empirical sentence. This follows from the concept of a convention about the use of symbols, but also from the fact that mathematics must lack content if it is supposed to be admissible in spite of strict empiricism. This condition implies that the rules of syntax must be demonstrably consistent since from an inconsistency every proposition follows, including all factual propositions. In the proof of consistency, as well as in the rules of syntax, only syntactical concepts may be used. If mathematical intuition and the assumption of mathematical objects or facts are to be dispensed with by means of syntax, then the use of 'abstract' and 'transfinite' concepts of mathematics will have to be based on considerations about finite combinations of symbols. Gödel notes that this same condition is involved in Hilbert's program. He says that nominalism and conventionalism could be used against realism only if mathematics could be interpreted as syntax in accordance with these conditions. By 1961 the reference to realism here can be understood in terms of a phenomenological realism that recognizes the intuition of essences.

Gödel's central argument against Carnap's syntactical program is developed by way of a philosophically interesting application of the second incompleteness theorem. Suppose that mathematics is syntax of language, as understood according to the conditions mentioned previously. In order to know that this was true we would need to know that the rules of this syntactical system are consistent. To know this, we will, by the second incompleteness theorem, need to use mathematics that is not captured by the rules in question. We will need to use mathematical concepts or principles that are formally independent of and stronger than those captured by the rules in question. The supposition that mathematics is syntax of language is thereby contradicted.

The argument suggests that mathematical sentences are not void of content. On the contrary, they must have their own kind of meaning and reference. They are not 'true by definition'; nor are they empty tautologies (see Gödel *1951, *1953/59). The mathematical essences we intuit could not be linguistic conventions. There are constraints on them that we do not freely invent or create. One might also say that this content or meaning will be 'abstract' relative to the rules of syntax. Mathematical intuition will therefore not be eliminable. In Husserl's language, categorial intuition will not be eliminable. Thus, the meaning of abstract and nonfinitary mathematical concepts is not clarified by explaining them in terms of syntactical rules; rather, abstract and nonfinitary concepts are used to formulate the syntactical rules. Mathematical axioms are not justified by reduction to syntactical rules; instead some of these axioms are required to justify the syntactical rules as consistent.

§ 5   Against the Elimination of Rational Intuition

In papers written prior to 1961, Gödel makes various comments about the nature of mathematical content and the 'rational perception' by which we come to have mathematical knowledge (see also Parsons 1995a). It is remarkable how closely these comments parallel Husserl's view of categorial intuition. In some passages, Gödel likens the conceptual content of sentences to Frege's notion of sense (*Sinn*) (Gödel *1953/59, p. 350). He says the conceptual content of sentences, or their 'sense', is objective and nonpsychological. This meaning is not human-made and does not consist merely of syntactical conventions. He says these concepts form an objective reality of their own, which we cannot create or change, but only perceive and describe (Gödel *1951, p. 320). It is clear that he means to reject psychologism, nominalism, empiricism, and Aristotelian realism about concepts. He says we know particular objects and their properties and relations through ordinary sense perception. With mathematical reason, however, we perceive the most general concepts and their relations (Gödel *1953/59, p. 354). This rational perception is analogous in certain respects to ordinary sensory intuition. Gödel suggests there is an analogy in several respects. First, our perceptions in both cases are constrained or 'forced' in certain respects (see especially Gödel *1951 and 1964). Second, we can be under illusions in each case (see Wang 1974, pp. 85–86). Third, there is a kind of inexhaustibility in each case (Gödel *1951, *1953/59). Gödel says the 'inexhaustibility' of mathematics makes the similarity between rational perception and sense perception even closer because it shows that there is a practically unlimited number of perceptions also in the case of rational perception (Gödel *1953/59, p. 353). This inexhaustibility appears not only through foundational investigations, such as the incompleteness theorems, but also in the actual development of mathematics. It appears, for example, in the unlimited series of axioms of infinity in set theory which Gödel says are analytic and evident in the sense that they explicate the general content of the concept of set.

Similar themes are found in the 1947 version of "What Is Cantor's Continuum Problem?" and Gödel elaborates on the notion of mathematical intuition in the supplement to the 1964 version of the paper. Gödel is already at pains in the early part of the 1947 version of the paper to show that the continuum hypothesis (CH) is meaningful and determinate enough that one can expect an unambiguous answer to be forthcoming in a natural extension of set theory. He does say that the solution of this problem will require an analysis of the meanings of the terms occurring in the axioms of set theory that is more profound than mathematicians are accustomed to providing, but he rules out the meaning analyses proposed by Brouwer, Weyl, and Poincaré on the grounds that they change the original meaning of the problem. The negative attitude toward Cantor's set theory that is found in these writers is the result of a philosophical conception of mathematics that admits objects only to the extent that they are interpretable as our own constructions or can be completely given in mathematical intuition. In effect, these mathematicians are considering categories or essences that are different from the essence of 'set' as it is given in Cantorian set theory. Gödel says that there is a satisfactory foundation of Cantor's set theory provided that one is willing to concede that mathematical objects exist independently of our constructions and of our having an intuition of them individually, and requires only that the general iterative concept of set be sufficiently clear that one can recognize the soundness and the truth of the axioms concerning it. Thus, instead of acquiescing in these reductionist views, we need to find new axioms that only analytically unfold the content of the given (abstract) iterative concept of set. At the time, Gödel suggested the use of large cardinal axioms, but he also said that other unknown axioms may play a role.

In the 1964 supplement to the 1947 paper, Gödel says that despite their remoteness from sense experience we do have "something like a perception of the objects of set theory." This is seen in the fact that the axioms force themselves upon us as being true.

I don't see any reason why we should have less confidence in this kind of perception, i.e., mathematical intuition, than in sense perception, which induces us to build up physical theories and to expect that future sense perceptions will agree with them, and, moreover, to believe that a question not decidable now has meaning and may be decided in the future. (Gödel 1964, p. 268)

Gödel goes on to make some additional remarks about the nature of mathematical intuition.

It is worth noting Gödel's suggestion that in order to make progress in set theory we need not have an individual intuition of the sets of higher set theory. All that is required is that our grasp of the 'general mathematical concepts' be sufficiently clear. In other words, there need not be an intuition of arbitrary individual instances of the essence under analysis in order to obtain some clarification of the essence. We are already forced in some ways by the essence or concept. It is not freely variable. This is why Gödel adds the qualification that we have *something like a perception* of the objects of Cantorian set theory. There arguably are intuitions of some individual instances of the Cantorian concept of set. There may be 'object-like' perceptions, however, in cases in which our thinking is directed by our intentions (meanings) even if we cannot intuit arbitrary individual instances of the essence.

§ 6   From the Incompleteness Theorems to Phenomenology

We have seen how Gödel uses his incompleteness theorems to argue against two of the fundamental twentieth-century schemes in the foundations of mathematics. Gödel says the correct attitude appears to be that the truth lies in the middle, or consists of a combination of the leftward and rightward viewpoints. Hilbert's effort to combine the two directions had been too primitive and too strongly oriented toward the leftward direction. Carnap's program failed for the same reason. We can perhaps hope to obtain a workable combination in a different way. In this case, the certainty of mathematics is not to be guaranteed through proving certain properties by projecting them onto material systems such as physical sign-configurations. Since the second incompleteness theorem suggests that we must reflect on meaning, or on abstract concepts, we can instead try to obtain a workable combination of the two directions through

cultivating (deepening) knowledge of the abstract concepts which themselves lead to the setting up of those mechanical systems, and further, according to the same procedures [for clarifying meaning], seeking to gain insights about the solvability, and the actual methods for the solving of all meaningful mathematical propositions. (Gödel *1961/?, p. 383)

These comments run directly parallel to Husserl's view. (The reader might find it helpful to read again the passages by Husserl quoted earlier.) How is it possible to extend our knowledge of these abstract concepts or essences? How is it possible to make these concepts precise, and to gain comprehensive and secure insight into the fundamental relations that hold among them (i.e., into the axioms that hold for them)? This cannot be done by trying to give explicit definitions for concepts, and proofs for axioms, since we would then need other undefinable abstract concepts with their own axioms. Gödel says that the procedure must consist, at least to a large extent, in a clarification of meaning that does not consist in defining.

It is at this point that Gödel explicitly mentions Husserl's philosophy. With its emphasis on the nonreductionistic, descriptive clarification of meaning, Husserl's view provides a combination of the two directions that can yield a workable foundation for mathematics. (It should be noted that he does not suggest some other philosophical view. Since he would have been aware of a wide range of views at this time, it is worth asking why he does not think that other views would provide a workable combination. I return to this question in § 8.) Gödel says that there "exists today the beginnings of a science which claims to possess a systematic method for such clarification of meaning, and that is the phenomenology founded by Husserl." Continuing, he says,

here clarification of meaning consists in concentrating more intensely on the concepts in question by directing our attention in a certain way, namely, onto our own acts in the use of those concepts, onto our own powers in carrying out those acts, etc. In so doing, one must keep clearly in mind that this phenomenology is not a science in the same sense as the other sciences. Rather it is [or in any case should be] a procedure or technique that should produce in us a new state of consciousness in which we describe in detail the basic concepts we use in our thought, or grasp other, hitherto unknown, basic concepts. (Gödel *1961/?, p. 383)

Gödel argues that the phenomenological approach cannot be dismissed on a priori grounds. Empiricists, in particular, should be the last to suppose there is an a priori argument against the phenomenological approach since a priori arguments about such matters are not available to them and would merely be dogmatic.

Thus, Gödel's path from the incompleteness theorems to Husserl's phenomenology is not surprising at all. Indeed, there is a sense in which the incompleteness theorems support and are supported by a phenomenological view. They cohere with a Husserlian view of meaning and the clarification of meaning, of what kinds of objects exist, and of categorial intuition. They support a phenomenological view in the sense that they suggest that an intuition of mathematical essences (or a grasp of abstract concepts) that cannot be understood reductionistically is required in order to solve certain mathematical problems, to obtain consistency proofs for formal systems, and to facilitate the development of mathematics. They are supported by a phenomenological view in the sense that phenomenology, unlike other recent viewpoints, does give mathematical essences their due. If it is held from the outset that we intuit mathematical essences and that reductionism about intuiting essences is a prejudice, then it is highly unlikely that one would believe that mathematical essences could be completely captured in formal or mechanical systems. There would not be the same kind of blindness to 'ideas' that one finds in the various leftward viewpoints. The 'radical lack of prejudice' needed to let essences have their own particular value might in fact be cultivated. From this perspective, however, logicians such as Hilbert, Carnap, and Skolem appear to display a certain kind of blindness or prejudice. Of course the incompleteness theorems are not derivable from this phenomenological attitude alone. One must also develop a very precise understanding of the essence of formal or mechanical systems in order actually to give a mathematically rigorous proof of incompleteness. The basic point about the irreducibility of mathematical essences comes into even sharper focus, however, with the subsequent clarification of the essence 'mechanically computable function' in Turing's work and with the fact that many different characterizations of this notion were shown to be equivalent.

It might even be suggested that the incompleteness theorems and related results on decidability and consistency are in fact examples of philosophy as rigorous science. They are, that is, supported by a particular philosophy and they show in a rigorous way the limits of a purely formal, syntactical, or mechanical conception of mathematics. They show that the essence 'arithmetical truth' is not exhausted in a purely formal, syntactical, or mechanical (and hence relative) concept of provability. Indeed, the theorems show that virtually none of the mathematical concepts one might like to axiomatize can be exhausted in formal concepts and methods. This does not, however, undermine mathematical rigor. It does not preclude informal rigor or axiomatization. 'Rigorous science' need not be identified with purely formal science.

In the 1961 paper Gödel therefore says that not only is there no reason to reject phenomenology out of hand, but one can even present reasons in its favor. Gödel argues that the incompleteness theorems show how a further development in the rationalistic direction takes place even without the application of a conscious and systematic phenomenological procedure. New axioms of mathematics, which do not follow by formal logic alone from those previously established, become evident to us again and again. One example of this can be found in the unlimited series of new arithmetic axioms, in the form of Gödel sentences, that one could add to the present axioms on the basis of the incompleteness theorems. These axioms become evident again and again and do not follow by formal logic alone from the previous axioms. We can use these axioms to solve problems that were previously undecidable. We are simply unfolding our intuition of an essence. Of course in this particular case the axioms are not mathematically very interesting. The incompleteness theorems for arithmetic have, however, led to interesting results such as those of Paris and Harrington (Paris and Harrington 1977). The Paris/Harrington theorem is a genuinely mathematical statement that refers only to natural numbers but is undecidable in PA. Its proof requires the use of infinite *sets* of natural numbers. It provides a good example of Gödel's idea of the necessity to ascend to stronger, more abstract (in this case, set-theoretic) principles to solve lower-level (number-theoretic) problems.

Another one of Gödel's favorite examples that is mathematically more substantial, but also more controversial, concerns extensions of the axioms of set theory. This example is mentioned in many of his other papers (e.g., Gödel *1951, *1953/59, 1964, 1972a) and is also related to his views on the inexhaustibility of mathematics. Gödel refers to the unlimited series of axioms of infinity in set theory, which are evident in the sense that they only explicate the meaning or content of the general concept of set. He says that such a series may involve a very great and perhaps even an infinite number of actually realizable independent rational perceptions. This is seen in the fact that the axioms concerned are not evident from the beginning, but only become so in the course of the development of mathematics. To understand the first transfinite axiom of infinity, for example, one must first have developed set theory to a considerable extent. One could then rise to a 'higher' state of consciousness at a later stage in which one understood the next axiom of infinity, and so on. This is also an example of meaning clarification, albeit one that did not result directly from an application of the kind of conscious and systematic procedure that Gödel seeks. Gödel hoped that in unfolding our intuition of the essence 'set' for which particular set-theoretic problems could be formulated, we could solve those problems, including the continuum problem. Perhaps we know enough today to say that adding strong axioms of infinity will not induce the kinds of constraints needed to settle the continuum problem in a convincing way. Gödel did, however, suggest that other unknown axioms might play a role.

One might argue that we obtain the unfolding of the concept of set only *if* we are willing to take this concept as given in the first place, and that the concept is very problematic. Perhaps it is even inherently ambiguous or vague. There are of course many more issues about this matter to be discussed than I can go into here. Nonetheless, I shall mention a few things that can be said on the basis of the previous comments. First, a general iterative concept of set *is* given in our experience even if it is not completely understood. There seems to be no a priori reason why it should not be explored. In unfolding our intuition of a general iterative concept of set we have not yet found a contradiction. At the same time, we do not have a precise grasp of the consistent extensions of it that may exist. But how could we be certain at this point that it is inherently ambiguous or vague? What would it mean to *know* this? The fact that the unfolding is even possible suggests that the concept gives some direction to research and has some coherence and stability. Indeed, consider all of the new results and methods to which researchers have been led by exploring aspects of this concept (see now also Hauser, forthcoming).

Gödel goes on to say in *1961/? that it is not excluded by the incompleteness results that every clearly posed mathematical yes-or-no question is nevertheless solvable through cultivating our knowledge of abstract concepts (or through developing our intuition of essences), for it is this activity in which more and more new axioms become evident on the basis of the meaning of the primitive concepts that a machine cannot emulate. Gödel suggests in other writing (Gödel 1934, from the postscript added in 1964; 1972a) that mental procedures may extend beyond mechanical procedures because there may be finite, nonmechanical procedures that make use of the meaning of terms. The intuition of mathematical essences or concepts would be just such a procedure (see also Chapters 7 and 10). Given the incompleteness theorems, there can for most mathematical essences be no consistent machine that solves all of the well-defined yes-or-no questions that are left undecided by the original sets of axioms for those essences. Human reason, however, may be able to achieve such a development by virtue of its ability to reflect on essences or concepts. The mind can constantly develop without diagonalizing out of the mathematical essence it is intuiting. There might be a constant development of machines to capture more of the essence but only by diagonalizing out of each particular machine under consideration. Thus, it is through an adjusted philosophical viewpoint according to which we intuit essences that Gödel seeks to make a place for Hilbert's optimism about mathematical problem solving as well as Hilbert's idea that in mathematical proofs we should strive for certainty.

Gödel says that the intuitive grasp of ever newer axioms that are logically independent of earlier ones is necessary for the solvability of all problems even within a very limited domain. He says that the appeal to this kind of intuition is in principle compatible with the Kantian conception of mathematics. Gödel points out, however, that Kant was wrong to think that for the derivation of elementary geometrical theorems we always need new intuitions, and that a logical derivation of these theorems from a finite number of axioms is therefore impossible. In the case of mathematics in a more general sense, however, Kant's observation is correct. Gödel says that many of Kant's assertions are false if literally understood, but that they contain deeper truths in a more general sense. It is Husserl's (transcendental) phenomenology that for the first time does justice to the core of Kantian thought. It avoids both the "death-defying leap of idealism into a new metaphysics as well as the positivistic rejection

of every metaphysics" (Gödel *1961/?). One could argue that the phenomenological approach is not prey to the excesses and lack of balance
that characterized earlier 'rightward' viewpoints.

§ 7   Why Phenomenology?

Gödel worked mostly on philosophy after 1942. It is interesting to ask
why he would have settled on Husserl's phenomenology and not some
other viewpoint. He certainly knew about extended forms of Hilbert's program, intuitionism, predicativism, and other foundational views. Quine's
ideas had become very influential. He could have appealed to earlier
forms of platonism. There were even developments in post-Husserlian
phenomenology.

The answer to the question is straightforward. First, most of post-
Husserlian phenomenology had itself succumbed to leftward pressures.
Second, consider the views about provability, about what has meaning,
what objects can be recognized, and what can be intuited in (i) modified
forms of Hilbert's program, (ii) traditional intuitionism, and (iii) predicativism. To insist on restrictions such as those found in (i), (ii), or (iii) is
virtually to ensure there will be certain clearly posed mathematical problems that will not be solved, including questions about the consistency of
formal systems. It is, as Husserl might say, to ensure blindness or prejudice. In the case of the limitation to PRA, for example, it is to ensure that
the Gödel sentence for PA is undecidable, or that no consistency proof for
PA would be forthcoming. Or consider whether the following is a clearly
posed mathematical question: is Zermelo-Fraenkel (ZF) + continuum
hypothesis (CH) consistent or not? Adherence to the views (ii) and (iii)
would probably lead (or would have led) one to believe it is not a clearly
posed mathematical question, much less that it has a clear-cut solution.
On the basis of the meaning theories associated with these views, it would
be difficult to see how one could give meaning to the problem. In any
case, one would be blinded to the solution on views (i)–(iii) because the
proof that CH is consistent with the axioms of ZF requires impredicatively
specified sets. It also requires transcending of the intuitionistic ordinals
and acceptance of the classical ordinals as given. For some problems or
consistency proofs there is a need to ascend to higher types, new axioms
of infinity, classical ordinals, and so on, where this emerges naturally in
the course of unfolding the given essence.

Quine's views on meaning, on what kinds of objects can be recognized,
and so forth, are beset with analogous problems (see Chapter 8). It will not

help to be told that one can only recognize as legitimate the mathematics that is needed to round out our theories of nature, much less that open mathematical problems can or should be decided on the basis of whatever mathematics this happens to be.

Most modern philosophical conceptions of mathematics are more skewed in the leftward direction than ever before. Thus, the trend that Gödel saw has not abated but has become stronger than ever. To such philosophers as Husserl and Gödel this trend signals a crisis. Indeed, it is one form of the crisis that Husserl writes about in *The Crisis of the European Sciences and Transcendental Phenomenology* (see also Tieszen 1997b).

A few words should also be said about platonism. Husserl criticizes earlier forms of platonism or realism, which he considers to be naive. A central reason to avoid earlier, naive forms of platonism is that they place essences outside all possible experience. They treat essences, in effect, as abstract things-in-themselves. On the phenomenological view, however, we are clearly directed toward and have access to essences. At least some essences are here in our experience, if only partially. They are not all completely outside or beyond our experience. To deny that essences are here in our experience is simply to deny that we can grasp various categories in our experience without attempting to reduce them to something else. This, in turn, is to deny the undeniable: that consciousness exhibits intentionality.

I conclude by briefly considering a skeptical objection to the so-called phenomenological method. The objection is this: what exactly is the method supposed to be? How does one learn it? What are some examples of the method? What are some of its fruits? My response is to argue that these questions misplace the proper emphasis. For students of Gödel's work, the substantive point about the phenomenological approach lies in its distinctive form of realism and its antireductionism about mathematical concepts and concept analysis. All questions about method should be viewed in this light. Suppose, for example, that the arguments made against psychologism by Frege and Husserl had not been successful. Would thinking of logic and mathematics in terms of empirical psychology have advanced logic and mathematics? Would moving logic and mathematics into the psychology department be an advance? It seems clear that this would hardly be a desirable development. How would the 'methods' of logic and mathematics be conceived in those circumstances? One can ask similar questions about strict formalism, mechanism, nominalism, and other reductionistic and naturalistic viewpoints. In each case, the resulting conception of the methods of logic and mathematics, I would argue, is at least inadequate and possibly even harmful. A correct method would be one that gives mathematical essences their due without necessarily denying the usefulness or importance of formalization. That would be the method of a phenomenological or critical realism in which the epistemological role of categorial intuition is not eliminable.

7

Gödel and the Intuition of Concepts

Gödel has argued that we can cultivate the intuition or 'perception' of abstract concepts in mathematics and logic (see, e.g., Gödel 1944, 1947, 1964, *1953/59, *1961/?, 1964, 1972).[1] Gödel's ideas about the intuition of concepts are not incidental to his later philosophical thinking but are related to many other themes in his work, and especially to his reflections on the incompleteness theorems. I will describe how some of Gödel's claims about the intuition of abstract concepts are related to other themes in his philosophy of mathematics. In most of this chapter, however, I will focus on a central question that has been raised in the literature on Gödel: what kind of account could be given of the intuition of abstract concepts? I sketch an answer to this question that uses some ideas of a philosopher to whom Gödel also turned in this connection: Edmund Husserl. It is not my goal in this chapter to give a complete account of Husserl's own views on the intuition of abstract or ideal objects, and, in any case, it would not be feasible to do so. For more details about Husserl's specific views on categorial intuition and related topics see Tieszen (2004).

Parts of this essay were presented to the spring 2000 Working Group on the History and Philosophy of Logic at UC-Berkeley and to the Spring 2000 Logic Lunch group at Stanford. Thanks for comments are due to the audience members and especially to Solomon Feferman, Grisha Mints, Paolo Mancosu, Tom Ryckman, Joel Friedman, Richard Zach, Ed Zalta, Johan van Bentham, and John Etchemendy. I also thank the *Synthese* referees.

[1] These arguments have been discussed by a number of writers. See especially Wang 1974, 1987, 1996; Parsons 1995a; Tragesser 1977; Tieszen 1994b; and Chapters 4, 5, 6, and 8 of this book.

Gödel's comments on abstract concepts are a product of his philosophical interpretation of his incompleteness theorems. In § 1 of this chapter I present an argument, based on the incompleteness theorems, that is supposed to take us from recognition of the existence of concrete sign-configurations to recognition of the existence of abstract concepts. I then describe the basic features of concepts that Gödel mentions in various papers. We will see why Gödel thinks that clarification of the meaning of basic mathematical concepts is required for deciding undecided mathematical propositions, finding consistency proofs, augmenting mathematics with new axioms, and developing mathematics in general. By 1961 Gödel is arguing that it is Husserl's phenomenology that offers useful ideas about clarification of the meaning of mathematical concepts. In § 2 I locate the place of concepts in a Husserlian view of mathematical cognition; § 3 presents some examples that show how abstract concepts are involved in everyday perception as well as in mathematical experience. On the view I present, we ascend to the awareness of mathematical concepts in the more active and reflective parts of our experience. Central to this view of concepts is the notion of intentionality. I argue that an account of the intuition of abstract concepts depends crucially on the fact that human cognition exhibits intentionality. In § 6 it is argued that the phenomenological ontology of concepts must be understood on this basis. The alleged problem of 'epistemic access' to abstract concepts, discussed in § 8, must also be understood on this basis. In § 9 of the chapter I discuss several views that, according to Gödel, either ignore or attempt to eliminate the intuition of concepts.

Why would anyone think that there is or could be an intuition of abstract concepts? We now proceed to the elements of an answer to this question.

§ 1   From Concrete Signs to Abstract Concepts

The philosophical ideas that lie behind the shift from concrete signs to abstract concepts are part of Hilbert's program, although they are certainly not all Hilbert's inventions. They have a rather deep background in the history of philosophy. Hilbert proposed to put mathematics on secure foundations after the discovery of the set-theoretic paradoxes. The idea would be to show that formalized parts of mathematics were consistent by using *only* a special theory, let us call it C (for 'concrete mathematics'). C could not be just any theory. What would make C special is that it would possess the kinds of properties needed to ensure reliability or security. It should be finitary and not infinitary, for we do not understand the infinite very well and in our thinking about the infinite we may be led into inconsistencies and paradoxes. It should be possible to understand C as a theory involving only concrete and not abstract entities, for abstract entities are mysterious and we should avoid postulating them whenever possible. The concrete entities in this case are finite sign-configurations. C should be concerned with what is real and not with what is ideal. Its sentences and proofs should be surveyable in immediate intuition. Immediate intuition is a nonmysterious form of sensory perception. C should not be a creature of pure thought or pure reason, for we may be led by pure reason into antinomies and hopeless confusion. C represents the part of our mathematical thinking that is contentual and meaningful. The contrast is with parts of mathematical thinking that we may regard as purely formal and 'meaningless' in the sense that we need not consider their purported references to abstract or infinitary objects or concepts. On this kind of view, the only way to understand mathematical rigor is as *formal* rigor.

For formal theories T that contain enough mathematics to make Gödel numbering possible, Gödel's first incompleteness theorem says that if T is consistent, then there is a sentence G, the Gödel sentence for T, such that $\nvdash_T G$ and $\nvdash_T \neg G$. The second theorem says that if T is consistent, then $\nvdash_T \text{CON}(T)$, where 'CON(T)' is the formalized statement that asserts the consistency of T. Theorem 1 tells us that G cannot be decided by T but that it is true if T is consistent. The theorem does not tell us that G is absolutely undecidable. Indeed, Gödel frequently emphasizes how sentences that are undecidable in some theories are in fact decided in certain natural extensions of those theories (e.g., by ascending to higher types). T can be, for example, primitive recursive arithmetic (PRA), Peano arithmetic (PA), or Zermelo-Fraenkel set theory (ZF). The theorems tell us that such formal theories T cannot be finitely axiomatizable, consistent, and complete. Theorem 2 suggests, generally speaking, that if there is a consistency proof for a theory T, then looking for $\vdash_{T'} \text{CON}(T)$, where T is a proper subsystem of T', will be necessary.

A very likely candidate for C is PRA. PA is arguably less suitable, given the distinctions described in the first paragraph of this section. In whatever manner we construe C, however, from theorem 1 it follows that if C is consistent, then the Gödel sentence for C cannot be decided by C even though it is true. It follows from theorem 2 that if C is consistent, then C cannot prove CON(C). Given the way we have characterized C, it follows that deciding the Gödel sentence for C or proving CON(C) must require objects or concepts that *cannot* be completely represented in space-time as finitary, concrete, real, and immediately intuitable. In other words, deciding the Gödel sentence for C or proving that CON(C) must require appeal to the meanings of sign-configurations, to objects or concepts that are in some sense infinitary, ideal or abstract, and not immediately intuitable.

Either there are such 'abstract' entities or there are not. Suppose that there are no such entities or that we can have no insight into such entities. (There are a variety of philosophical views on which such a claim would be made. I return to this point in § 9.) Then it would follow, within the perspective of C, that we must stop or 'become static' with respect to deciding the Gödel sentence for C or obtaining a proof of CON(C). That is, we could not decide some clearly posed mathematical problem. (I mean we could not decide it unless the decision were to be made arbitrarily or on nonmathematical grounds. I return to this point later in order to ward off possible objections.) However, for some T (e.g., PA) we in fact do have proofs of CON(T) and decisions of related problems. Contradiction. Therefore, there are such entities and we must have some insight into them.

It is worth noting that we can obtain a reasonably good understanding of the sense in which the objects or concepts used in the decisions or consistency proofs must be abstract, infinitary, and not completely captured in immediate intuition. In the case of PA, for example, the consistency proof requires the use of transfinite induction on ordinals $<\varepsilon_0$, or it requires primitive recursive functionals of finite type. Perhaps the level of abstraction involved here is not very substantial. It is certainly more substantial, however, in the case of consistency proofs for real analysis or in proofs of the consistency of ZF + the continuum hypothesis (CH) or of ZF + ¬ CH.

It needs to be emphasized again that C cannot be just any formal theory. We cannot keep extending C with new axioms that allow us to decide sentences that were previously undecidable and still expect to have concrete mathematics, for then we might as well have started with something such as ZF in the first place. As a purely formal theory ZF is of course concrete in the same sense as PRA, but the difference is that ZF certainly could not codify concrete, finitary mathematics. To hold that it could is to subvert the philosophical basis of Hilbertian proof theory completely.

The kinds of remarks I have made here on the incompleteness theorems are expressed in many of Gödel's writings. For example, Gödel (Gödel 1972, pp. 271–272) says that

P. Bernays has pointed out on several occasions that, in view of the fact that the consistency of a formal system cannot be proved by any deduction procedures available in the system itself, it is necessary to go beyond the framework of finitary mathematics in Hilbert's sense in order to prove the consistency of classical mathematics or even of classical number theory. Since finitary mathematics is defined as the mathematics of *concrete intuition*, this seems to imply that *abstract concepts* are needed for the proof of consistency of number theory.... [What Hilbert means by *Anschauung* is substantially Kant's space-time intuition confined, however, to configurations of a finite number of discrete objects.] By abstract concepts, in this context, are meant concepts which are essentially of the second or higher level, i.e., which do not have as their content properties or relations of *concrete objects* (such as combinations of symbols), but rather of *thought structures* or *thought contents* (e.g., proofs, meaningful propositions, and so on), where in the proofs of propositions about these mental objects insights are needed which are not derived from a reflection upon the combinatorial (space-time) properties of the symbols representing them, but rather from a reflection upon the *meanings* involved.

The idea that it is necessary to reflect upon meaning is given a central role in the 1961 paper (Gödel *1961/?) in which Gödel invokes Husserl's ideas. The 1961 paper clearly contains the idea that reflection on meaning (or intuition of concepts) is of a 'higher level' than reflection on the combinatorial properties of concrete symbols.

It is well known that these ideas are also part of Gödel's antimechanist views on computation and cognition. Consider, for example, the following remarks in an early paper:

The generalized undecidability results do not establish any bounds for the powers of human reason, but rather for the potentialities of pure formalism in mathematics.... Turing's analysis of mechanically computable functions is independent of the question whether there exist finite *non-mechanical* procedures ... such as involve the use of abstract terms on the basis of their meaning. (Gödel 1934, p. 370)

The idea that (finite) humans intuit abstract concepts or meanings is a central theme in Gödel's comments on the difference between minds and machines.

In his discussion of Husserl's views in his 1961 paper Gödel distinguishes 'leftward', empiricist views of mathematics from 'rightward', rationalist views. Hilbert had attempted a particular kind of reconciliation of these views. The incompleteness theorems, however, show that Hilbert's attempt to reconcile these views is unworkable. Since the theorems suggest that we must reflect on meaning, or on abstract concepts, we can instead try to obtain a workable combination of the two types of view through

cultivating (deepening) knowledge of the abstract concepts which themselves lead to the setting up of those mechanical systems, and further, according to the same procedures [for clarifying meaning], seeking to gain insights about the solvability, and the actual methods for the solving of all meaningful mathematical propositions. (Gödel *1961/?, p. 383)

Gödel says that there "exists today the beginnings of a science which claims to possess a systematic method for such clarification of meaning, and that is the phenomenology founded by Husserl." Continuing, he says,

Here clarification of meaning consists in concentrating more intensely on the concepts in question by directing our attention in a certain way, namely, onto our own acts in the use of those concepts, onto our own powers in carrying out those acts, etc. In so doing, one must keep clearly in mind that this phenomenology is not a science in the same sense as the other sciences. Rather it is [or in any case should be] a procedure or technique that should produce in us a new state of consciousness in which we describe in detail the basic concepts we use in our thought, or grasp other, hitherto unknown, basic concepts. (Gödel *1961/?, p. 383)

Gödel argues that the phenomenological approach cannot be dismissed on a priori grounds. Empiricists, in particular, should be the last to suppose there is an a priori argument against the phenomenological approach since a priori arguments about such matters are not available to them and would merely be dogmatic.

Gödel makes a number of remarks on the nature of concepts in various writings. They are mentioned, for example, in his general views on logic. Gödel often speaks of logic in a manner that calls to mind ideas in Leibniz, Bolzano, Husserl, and other philosophers who associated logic with the idea of science as rational inquiry in a very broad sense. Logic, Gödel says, is about the most general abstract (and 'formal') concepts and the relationships of these concepts to one another (see, e.g., Gödel *1953/59, p. 354). Sense perception is about particular objects and their properties and relations. In some passages, Gödel likens the conceptual content of sentences to Frege's notion of sense (*Sinn*) (Gödel *1953/59, p. 350). Concepts are abstract, objective intensional entities (Gödel 1944, 1958, 1972, *1961/?). They are not merely subjective, and, unlike conscious states, they do not themselves have temporal phases. Concepts are not sets, although one might conjecture that every set is the extension of some concept. Concepts, it seems, are objects sui generis. One could make a list of things with which they are not to be identified or to which they are not to be reduced. Gödel rejects those forms of reductionism that ignore or seek to eliminate the intuition of concepts. He mentions specifically Hilbert's formalism, mechanism, nominalism, conventionalism, positivism, empiricism, psychologism, and Aristotelian realism (Aristotelian universals) (Gödel *1951, *1953/59, *1961/?). In § 9 of this chapter I briefly consider some of the reasons for rejecting each of these positions. Hao Wang recorded a remark from a conversation with Gödel that states the point succinctly, albeit rather roughly (Wang 1996, p. 167):

> Some reductionism is right: reduce to concepts and truths, but not to sense perceptions.... Platonic ideas [what Husserl calls "essences" and Gödel calls "concepts"] are what things are to be reduced to. Phenomenology makes them clear.

§ 2 The Place of Concepts in Husserlian Phenomenology: Intentionality and Objects of Experience

Gödel, as we have seen, says that in phenomenology "clarification of meaning consists in concentrating more intensely on the concepts in question by directing our attention in a certain way, namely, onto our own acts in the use of those concepts, onto our own powers in carrying out those acts, etc." Concepts are abstract, objective intensional entities. We now consider in more detail the place of such concepts in Husserlian phenomenology. This will give us a basis for understanding the idea that we do or can intuit concepts. A key ingredient in the description of the intuition of concepts is the fact that human cognition exhibits *intentionality*. This fact is often overlooked in recent work in the philosophy of logic and the philosophy of mathematics, but understanding the idea of intuiting concepts is probably not possible unless one starts with it. Without this insight the idea of intuiting concepts may in fact be absurd.

The assertion that cognition exhibits intentionality means that our cognitive acts are always *about* or directed toward something. Consciousness is, for many types of acts, always consciousness *of*. One can see this clearly in acts of believing, knowing, willing, desiring, remembering, hoping, imagining, and so on. For example, an act of belief is always directed toward a particular object or state of affairs. Try to deny that cognition exhibits intentionality and you are faced with an absurdity. Imagine for a moment what it would be like for a belief not to be about anything in particular.

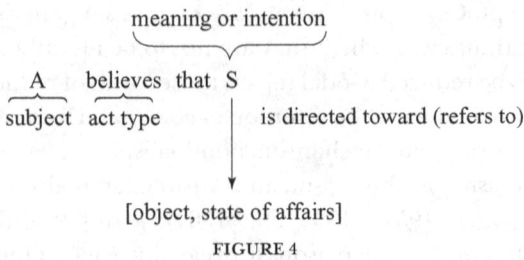

[object, state of affairs]

FIGURE 4

Intentionality means directedness. *For a cognitive act to be directed in a particular way means that it is not directed in other ways.* Cognitive acts are perspectival, and we cannot take all possible perspectives on an object or domain. We do not experience everything all at once. Our beliefs and other cognitive acts at any given time, in other words, always involve certain *categories* of objects. The mind always categorizes in this way. It cannot help doing so, for our experience would have to be very different otherwise. This means that we are always *interpreting* the world is some way or other. As I have been saying, categories or concepts are always at work in our experience, even in everyday sense experience. It is a basic feature of intentionality that directedness is always directedness by way of concepts, even if we do not always reflect on these concepts. In fact, we may not reflect on them at all. There is no such thing, however, as directedness that is not directedness by way of some concept.

The standard form of intentionality can be characterized as in Figure 4. 'S' expresses a sense or meaning, and it is by virtue of this meaning that we are directed toward a particular object or state of affairs. We 'bracket' the existence of the object or state of affairs since it may not always exist. There can be directedness toward an object by virtue of the meaning of a cognitive act even if, as sometimes happens, the object does not exist. In order to impute existence to the object we must have evidence for existence. Evidence takes us beyond the merely possible to the actual. When there is evidence for an object the object stabilizes and takes on a certain identity or invariance in the field of our experience.

Suppose S has the form Pa. As a matter of terminology, I will say that 'P' expresses a *concept*, and that 'a' is an expression for an object. Pa is merely the *form* of an expression that has a content or meaning. The following are examples of informal expressions of mathematical concepts by virtue of which the mind would be directed in different ways: "x is a triangle," "x is a natural number," "x is a function," "x is a mechanically computable function," "x is a set," "x is an infinite set," "x is a measurable cardinal," "x is a weakly compact cardinal." (Some of these are compound concepts.) These differences disappear if we have the pure form Px. Consider how your mind is directed in this latter case.

In a strictly defined formal language we could assign a Gödel number to the expression 'Px'. 'Px' could, for example, be characterized as a certain bit string. As such it would have a particular length and algorithmic complexity. This is to be contrasted with the meaning or concept expressed, for example, by "x is a set." Meanings or concepts do not have a particular length. We cannot measure the complexity of a concept in terms of the shortest program that would generate the bit string we take to express the concept.

It should be noted that in the context of the intentionality scheme presented previously, concepts or categories are intensional entities. Concepts are not, for example, to be confused with sets, and their identity conditions are not determined extensionally. We will say that a concept *applies* to or is true of an object or it fails to apply to or to be true of an object. Similarly, we can say that an object falls under or fails to fall under a concept. Applicability is a relation that should in general be distinguished from the set-theoretic relation of membership '$\in$'.

It is also important to note that, quite generally, meanings (or concepts) are different from mathematical objects (e.g., numbers, sets, functions). A concept or meaning can itself become an object of consciousness when we are directed toward it, but being directed toward the meaning of a mathematical expression is not the same as being directed toward a particular mathematical object. It should not be assumed that intuiting the one is the same as intuiting the other.

§ 3   Some Examples of How Concepts Function in Our Experience

Let us consider an example of how concepts are already present in our everyday perceptual experience. Suppose that I am standing some distance from a wall on which I perceive a round dark dot. As I approach it I see that there are colored surfaces in its interior. Moving closer I can discern some markings in its interior. As I continue in this manner it occurs to me that it is a mandala. I can then make many further determinations about it, noting various designs used in it, and so on. According to the thesis of intentionality, we always experience things by way of concepts. Reflection on this example, therefore, makes it clear that a progression of concepts must be involved in gaining knowledge in ordinary sense perception. Given the description of the experience, one might think of this sequence as beginning with the concept "x is a dark round dot on a wall" and proceeding through "x has colored surfaces in its interior," "x has markings in its interior," "x is a mandala," "x contains thirty-three concentric circles," and so on. These concepts are already present in the (prereflective) experience but they are of course not themselves the objects of the experience. They are individuated and become objects only in reflection on the experience. Once we reflect on the experience we can describe many aspects of the experience in terms of the concepts involved and their relations to one another. The concepts involved at earlier stages, for example, are successively filled in by and harmonized with the later concepts, where the content of the earlier concepts is obviously retained in a certain way at later stages. What was quite indeterminate becomes more and more determinate in the process. The concepts in the progression must be related to one another in certain ways in order for there to be a continuous, harmonious development of the perception. The earlier concepts provide a basis for the later concepts. They are not rejected but are supplemented and mutually adjusted to more determinate concepts. The concepts, taken together, form a consistent set, but some may be more determinate than others.

The determinacy/indeterminacy feature of concepts in our experience simply reflects our epistemic situation. A horizon of possibilities about what may fall under the concept is determined by the concept, against the background of its context, but many things may be left indeterminate. The *horizon* of a concept Px may be defined, for our purposes here, in terms of those concepts that would in future possible experience be consistent with our present experience regarding Px. The possibilities (and probabilities) with respect to Px are always to be considered in the context of and against the background of our previous experience. The concepts in the example all have their own horizons, but clearly the range of possible experience of the object x is narrowed at later phases in the sequence. The horizon of possible experience associated with "x is a dark round dot on a wall" leaves open many possibilities that are closed off by the subsequent determinations. Among other things, this means that some questions about x that could not be decided at earlier stages of the experience are *decidable* at later stages as a function of the new concepts that emerged in the course of the experience.

One might imagine that this little slice of my history resulted from my active interest in the object and in my coming to know more about it. On the other hand, I may not have approached it in any conscious and systematic way but just happened to notice these things. In natural science one takes an active, conscious, systematic approach in obtaining knowledge about natural objects.

Of course the development is not always harmonious. There can be various frustrations, annulments, and so on. For example, I might perceive an object x as a snake at a given stage of my experience and perceive it at a later stage as a coiled rope. In this case we do not have a harmonious progression and filling in of the earlier experience but a misperception. There can be no x in our experience such that x is a snake and x is a coiled rope. These concepts are inconsistent with one another. At later stages I perceive the object under the concept "x is a coiled rope."

Examples of the sort we have considered show how concepts are already at work in our ordinary, engaged prereflective experience and how in reflection we can become aware of these concepts and make them the objects of our experience or intuition. Of course in ordinary everyday experience we do not intuit concepts, and it is not the case that we must intuit concepts in order for our everyday experience to be what it is. Instead, concepts are implicit in our everyday experience and make it possible. They are very much in the background of experience. Since they exist in experience in this manner, however, they can be brought into the foreground and they can become explicit in our experience through reflection. We can in reflection become conscious of the concept itself and its relations to other concepts. The fact that we can reflect on concepts does not at all entail that we will immediately have a perfect grasp of the concept and its relations to other concepts. On the contrary, our grasp of the concepts that are at work in our experience is often quite hazy and imperfect. In reflection we can clarify our grasp of a concept and its relations to other concepts. We can try to perfect our intuition of the concept itself.

Some degree of reflection is required if there is to be anything more to human knowledge and awareness than simple forms of everyday perception. Indeed, all of the 'higher', more theoretical forms of experience, such as those found in logic and mathematics, will require some degree of reflection. These higher, more theoretical forms of experience are founded on sense experience and need not exist. Perhaps there were times in the history of human beings when they did not exist. To the extent that they do exist, however, they show that there must be some intuition of concepts. At a certain point, for example, one comes to know that some things are not natural numbers and some things are. This knowledge presupposes some grasp of the concept of natural number. Reflection is a condition for the possibility of such sciences as mathematics and logic.

All higher forms of awareness involved in science and culture are in this sense built up from reflection on everyday, prereflective, passive forms of awareness. The existence of sciences such as mathematics and logic indicates a more active and systematic cognitive achievement.

Concepts have the same basic function in these more theoretical forms of awareness that they have in everyday awareness. Of course in pure mathematics, unlike in everyday perception, the sensory constraints on experience fall away. There are, however, still logical, conceptual, or meaning-theoretic constraints on the experience. The concepts involved simply direct us toward different kinds of objects in the two cases. Following Husserl, we might also argue that the concepts of logic and mathematics are exact in that they involve idealizations, whereas the concepts involved in everyday perception are inexact or 'morphological' since idealization is not involved in this case (see, e.g., Husserl *Ideas* I, § 74). The points, lines, planes, circles, and spheres of Euclidean geometry, for example, are idealizations of the shapes of objects given to us in everyday sense perception. It is possible to give a detailed account of this cognitive activity of idealization, but there is no space to do so here (see, e.g., Tieszen 2004).

Now certainly we can distinguish particular numbers from the concept of number, particular sets from the concept of set, particular functions from the concept of function, and so on. Imagine a cognitive life, for example, in which particular natural numbers bore no more relation to one another than they did to any other particulars. There could be no systematicity for these objects. Each natural number would be utterly unique and singular and could bear no more relation to another natural number than to, say, a chair. Of course this idea flatly contradicts our experience with natural numbers and it contradicts the fact that we have a science of these objects. The concept of natural number is operative in our experience with natural numbers. We also know, for example, from our experience with natural numbers (as this is shown in mathematical practice) that some concepts are not consistent with "x is a natural number" or that some are simply from the wrong categories to be applicable to the objects that fall under this concept. An example of the latter type would be a concept such as "x has a color." Some of the concepts that are consistent with "x is a natural number" are "x is even," "x is prime," and "x is perfect," Each of these concepts in turn has its own horizon. The fact that the concepts have horizons and are not grasped with perfect clarity is shown by the fact that there are open problems involving combinations of these concepts. There are often various refinements and adjustments over time in our intuition of concepts.

The same points could be made about sets. However incomplete and unfinished our thinking about sets may be, it nonetheless displays the kind of systematicity and organization that indicate the presence of concepts. Our early experience with the concept "x is a set," for example, appears quite indeterminate by later standards. The early experience has been filled in with many concepts with various branchings, refinements, and adjustments. Against this historical background we might single out the concepts associated with Zermelo-Fraenkel set theory and expect new concepts that consistently unfold the existing concepts to emerge. Such new concepts would add further determinations that might allow us to decide questions about these objects that were not decidable at earlier stages. The situation here is analogous to the way that decisions may come about through the progression of concepts involved in the perception of the mandala. Gödel indeed says that the idea of deciding undecidable mathematical propositions by extending mathematical theories with new axioms or by ascending to higher types is best described as developing or unfolding our intuition of the concepts (meanings) that appear in these propositions (Gödel 1947, 1958, 1961/?, 1964, 1972). These decisions will be nonarbitrary, as they are in the preceding example of sense perception. They will be forced or constrained in certain ways. Misperceptions are also possible here.

As another example one might consider the concept of proof. The intuitive concept of proof is clarified by the incompleteness theorems themselves because they show that provability in any given formal system cannot fully capture the intuitive concept (Wang 1974, p. 83). Purely formal proof is always 'relative' to a given formal system. The incompleteness theorems show us that number-theoretic provability is not the same as number-theoretic truth, whereas at earlier points it was not clear whether they could be equated or not.

Our examples show how concepts can become objects in reflection. There is, in these cases, intuition of concepts. If we take intentionality seriously, then we must say that *in being directed toward a concept or category of concrete particulars we are not (primarily) directed toward the particulars themselves.* It is the concept that we grasp. The key idea to remember here is that we are speaking of what the mind can be directed toward. Intuition is to be understood only in these terms. It is a basic fact of consciousness that we can be directed toward many different kinds of things, including, as our examples are meant to show, concepts. If different categories of concepts will be applicable depending upon whether we are directed toward concrete particulars or concepts, then, according to the thesis of intentionality, we must be directed toward different things. The properties and relations of concepts are indeed different from the properties and relations of concrete particulars.

An immediate consequence of our remarks on directedness is that directedness toward concepts is not the same as directedness toward sets. The latter are governed by extensional identity conditions and the former are not. We can of course also be directed toward sets in our thinking, as is the case when we use the concept "x is a set" or some specification of this concept (e.g., "x is a predicatively defined set").

§ 4   Some Relations of Concepts and the Space of Concepts

In following out the consequences of the thesis of intentionality and the earlier examples we have argued that there is directedness toward concepts. We can also be directed toward the relations of concepts to one another. Not all relations are relations of concrete, sensible particulars. Certain relations are naturally construed as being relations of concepts. Some of these relations can be characterized to a first approximation by using the notation of an intensional logic for possibility, '$\Diamond$', and necessity, '$\Box$', where these are relativized to what can occur in our experience.

The inconsistency of concepts P and Q, Inc(P, Q), can be characterized as

$$\neg\Diamond(\exists x)(Px \wedge Qx).$$

Similarly,

Con(P, Q) : $\Diamond(\exists x)(Px \wedge Qx)$ (consistency of concepts P, Q),

Imp(P, Q) : $\Box(\forall x)(Px \rightarrow Qx)$ (concept P implies concept Q),

Equiv(P, Q) : $\Box(\forall x)(Px \leftrightarrow Qx)$ (equivalence of concepts P and Q),

Ind(P, Q) (independence of concepts P and Q),

and so on.

Consider again some of the previous examples. If the x in our experience is a dark round dot on a wall, then is it possible that x is a mandala? Certainly these concepts are consistent with one another, but the second concept is not implied by the first. On the other hand, the concepts "x is a snake" and "x is a coiled rope" are not consistent with one another. Note that this is not a purely formal contradiction, in the sense of $Px \wedge \neg Px$. One must know the origin of the symbols Px and Qx. It is necessary to know what the symbols represent. To shift to some mathematical concepts: if x is a natural number, is it possible that x is prime? Again, these concepts are consistent with one another. If x is a natural number, is it possible that x does not have a successor? Here we find an inconsistency given the standard concept of the natural numbers. Here is another example of inconsistent concepts: x is a triangle and x has four sides. It is worth noting that we can often know that concepts are inconsistent with one another without knowing everything about the concepts. We need not know *exactly* what their applications are in order to know that they are inconsistent.

Critics of realism will sometimes ask the supposedly embarrassing question of *where* logical or mathematical concepts and objects are. They are supposed to be 'out there' in some platonic heaven or third realm. Since the position I am outlining is not a form of naive ontological realism it is easy to respond to this worry. The view I am describing holds that mathematical concepts and at least some mathematical objects are given to us in our experience as transcendent and as existing 'in themselves'. They acquire this meaning of being in our experience. From my point of view the question of where concepts are is based on a category mistake. Concepts are obviously not in some place as physical objects are in a place. Concepts, however, can be thought of as having a 'place' in the network or system of logical and mathematical objects, concepts, and propositions even though we may not have a clear grasp of the network. They are in the (conceptual) space of logic and mathematics. This is, in a sense, a 'world', but it is a cognitive world which, being based on the intentionality of consciousness, is not at all inaccessible to us even though much of it may be implicit at any given stage. These concepts and propositions imply one another, are (in)consistent with one another, independent of one another, equivalent, and so on. To know their 'place' is to know their relations to other concepts. To know the place of a concept Px is to know what it implies, is implied by, is (in)consistent with, and so on. It seems that one could come to know concepts and their places as well as one knows about physical objects and their places. Perhaps only a few have an overview of this mathematical and logical space. To learn mathematics or logic is to have it to some extent if only implicitly.

§ 5   Meaning Clarification

In the preceding discussion of intentionality it was said that we are directed toward objects by way of the 'meanings' associated with our cognitive acts. In reflection on experience we can explore and unfold the meanings of our concepts. Thus, we are in effect describing a kind of meaning clarification. One might seek to clarify, for example, the concept "x is a natural number" or the concept "x is set," along with related concepts. We can take a more or less active role in this meaning clarification. Meaning clarification often takes place in these sciences as a matter of course, albeit not always in a reflective and systematic manner. We may cease to make progress if, under some reductionistic scheme, the meaning of mathematics is lost or distorted. This is what lies behind Gödel's remarks on meaning clarification in phenomenology.

Gödel and Husserl have made remarks about the phenomenological method that have raised great expectations in some philosophers and logicians about applying the method to obtain solutions to open mathematical problems. Gödel no doubt had some hopes of using Husserlian ideas to make more progress in mathematics, but I think one has to be cautious about what can be achieved. No one has a method for solving deep open problems, and one ought not to expect phenomenology to have one either. What the phenomenological view does foster, among other things, is recognition of the existence and importance of meaning and informal concepts in mathematical practice. As I see it, it holds that pure mathematics is an autonomous science with its own concepts and that it is possible to clarify such concepts in a variety of ways (including but not limited to the technique of formalization). It is a view about the nature of mathematics that can be contrasted with other philosophical views. It describes mathematical practice in terms of unfolding and clarifying our grasp of the meaning of mathematical concepts through activities such as abstraction, generalization, specification, use of analogies, idealization, formalization, axiomatization, and free variation in imagination. Some of the activities to which it would appeal in the analysis of *mathematical* cognition are different from the activities to which it would appeal in the analysis of concepts involved in everyday perception. This phenomenological view must have seemed especially appealing to Gödel because it stood in such stark contrast to the ideas of the ax-wielding logical positivists of his time, who held, among other things, that mathematical statements were contentless tautologies. So, in a general sense, it is a view that does give more hope for progress in mathematics than many of the other reductionistic or dismissive 'leftward' views of mathematics that were and still are around. Fortunately, the philosophical attitude toward mathematics does not seem quite as grim now as it did in the heyday of positivism.

What one can do, therefore, is to give examples of informal concepts that have been clarified through the kinds of activities mentioned. Gödel, Kreisel, Wang, and Troelstra have given examples of this type (e.g., the concept of a mechanically computable function, of choice sequence, of continuity, of proof, of points as parts of the continuum). One could do more of this sort of thing with an emphasis on specific ideas in phenomenology.

It might be useful to pause in particular over Husserl's idea of free variation in imagination since it is something he specifically recommends. This is a way in which meaning clarification might be made more active and reflective, given the context of the concepts involved and the appropriate background and expertise. Frege (Frege 1892), for example, has said that comprehensive knowledge of a reference requires us to be able to say immediately whether any given sense belongs to the reference. Similarly, we could say that comprehensive knowledge of a mathematical object requires us to be able to say immediately whether any given concept belongs to the object. One way to try to improve our mathematical knowledge therefore would be (1) to start with an actual or imagined arbitrary instance of the type of mathematical object under consideration; (2) to vary the instance as many ways as possible, and freely or arbitrarily, in imagination, that is, try to imagine it under as many different concepts as possible; and (3) as (2) proceeds to determine under which concepts, if any, the object could no longer be experienced as an instance of the same type of phenomenon. Schematically, for an arbitrary x that could be imagined or perceived to be a P, determine whether x can also be Q, R, S, T,... and so on: if Px, then the tree shown in Figure 5 applies.

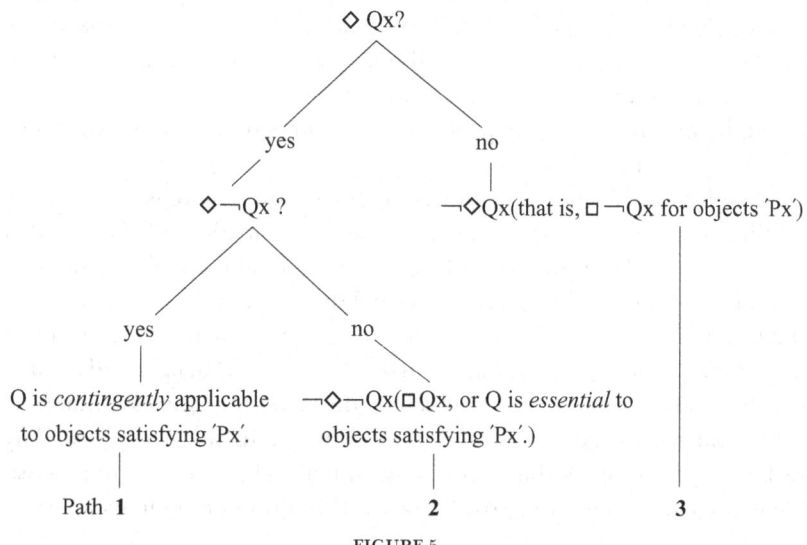

FIGURE 5

Compared to some of the restrictive or constraining conditions of reductionist methodologies, this way of looking at things might help to free the mind. It could help to open up the mind, on the basis of the use of imagination in the service of reason. One generates a bunch of concepts and then looks back through them at what was given. The general formulation just mentioned is of course quite open ended and nondeterministic, but the process is significantly more determinate once we contextualize it against the background and expertise found in our actual practice in various spheres (see, e.g., Chapter 3).

One need not have a great wealth of experience in mathematics, for example, to notice that a set or a natural number cannot be imagined to be just anything. Our imagination here is not absolutely free (compare with Gödel's remarks on free creation and constraints, Gödel *1951, p. 314, and elsewhere). It is bounded or limited in certain respects. This is not a matter of convention. Social or linguistic conventions are susceptible to variation and could be imagined differently. This is the nature of conventions. The limits under which we may vary our thinking of a mathematical object, however, are determined by something that is invariant: the concept of the object.

It should also be noted that there are two types of variation or change that we might observe in this process: essential change and alteration or 'accidental' change. In the case in which we determine that $\Box Qx$ given the concept Px we are dealing with the kind of information found in axioms or theorems, the denial of which would amount to essential change in the objects; in the other case we arrive at "accidental" properties or relations concerning the objects. In the latter case we can vary the concepts under which an object of a given type is thought without changing the object, whereas in the former case we cannot do this. We see this necessity, for example, in thinking of the procedure in connection with the axioms of Peano arithmetic.

Meaning clarification is premised on the fact that there is much that is implicit in the intuition of a concept that could be made explicit. The idea is that we extend our understanding of the possibilities involving the phenomenon and that necessary or essential properties will appear through the multiplicity of variations. By generating a multiplicity of variations in imagination one consciously and purposely creates a background against which necessities may emerge. It is only the stability of the essential concepts that allows us to vary the other concepts in the first place. Any variation presupposes that something which makes the variation possible is constant. Imagining possibilities in this way might help us to arrive at a new, higher state of consciousness concerning the phenomenon. Through the process of free variation one might be able to uncover a "rule" governing the horizon of the concept which would not have been seen prior to the process of variation. The rule, even if only partially understood or not fully determinate, would suggest further possible insights. In some cases we would be forced outside a given concept. But we are not necessarily forced into a void. We may have stumbled onto a new concept that can be consistently developed.

As noted, Gödel says that a nonreductionistic clarification of meaning, or this kind of informal rigor, can be expected to play a role in the nonmechanical decidability of open mathematical problems. In his paper on Cantor's continuum problem (Gödel 1947, *1951, 1964), for example, Gödel says that a proof of the undecidability of the generalized continuum hypothesis (GCH) from the present axioms of set theory does not solve the problem of GCH. Its undecidability from the axioms being assumed today only means that these axioms do not contain a complete description of the well-determined reality in which GCH must be either true or false. By a proof of undecidability a problem loses its meaning only if the system of axioms under consideration is interpreted as a hypotheticodeductive system, that is, if the meanings of the primitive terms are left undetermined. If the meanings of the primitive terms are not left undetermined, then we can develop and sharpen our intuition of the concepts involved.

On the basis of his own work on (un)decidability, Gödel also notes that sometimes we might be expected to ascend to higher levels of abstraction and meaning clarification to decide problems at lower levels (see Gödel 1947, *1951, 1964). It can be shown that the axioms involved in extensions of transfinite set theory, for example, have consequences outside the domain of very great transfinite numbers, which is their immediate subject matter. Each of the axioms of infinity, under the assumption of consistency, can be shown to increase the number of decidable propositions even in the field of Diophantine equations.

§ 6   The Phenomenological Ontology of Concepts

I considered some simple examples in § 2 in which there is directedness toward concrete sensible particulars in space-time in order to show how there is a natural shift in some of our cognitive activities to directedness toward concepts that are applicable to concrete particulars. There can also be a shift of directedness toward concepts that are *not* directly applicable to concrete, sensible particulars. It is in the nature of pure mathematical thinking that it is not directed toward concrete, sensible particulars in space-time. Even in applications of mathematics we come to interpret the natural world in a special way: we mathematize it. The mathematical concepts and structures used in applications shape and organize the way we see the natural world into more idealized and abstract forms.

In mathematical thinking there is of course directedness, but it is now directedness toward 'abstract' or ideal objects such as numbers, sets, and functions. Pure mathematical concepts (e.g., "x is a triangle," "x is a natural number," "x is prime") have abstract or ideal objects as instances. Husserl distinguishes, for example, 'real' individuals and universals from 'ideal' individuals and universals (Husserl *LI*, Investigation II, § 2). The directedness toward ideal objects as instances appears at higher, founded levels of awareness of the type we see in mathematics. Our awareness moves from sensory individuals up to the more theoretical, conceptual forms of directedness. Of course mathematical concepts are not themselves typically the objects of mathematical thinking. The objects that fall under such concepts are what mathematics is about. One could, however, always reflect on and analyze mathematical concepts. This may usually be more of a philosophical, logical, or meaning-theoretic task.

In all of this I am speaking about what the mind is directed toward, and even if there is always a more or less vague background of sensory awareness in all of our cognitive activities, it is not the case that the mind is always directed toward objects of sense experience. If the mind were directed only toward objects of sense experience, there would be no mathematics and logic.

I wish to take mathematics and logic (in a broad sense) as sciences that are given. They are or can be part of our conscious life, and they provide data that may be substituted for S in the intentionality schema described earlier. Once we accept the idea that consciousness exhibits intentionality we can see why we must think of mathematical concepts and objects as abstract or ideal. Indeed, the putative 'ontology' of concepts should be arrived at by way of the application of the notion of intentionality to what is given in mathematics and logic. One consequence is that in being directed toward a concept that applies to particulars we are not directed toward the particulars themselves. The concepts or relations that apply to concrete particulars are different from the concepts or relations that apply to concepts. Guided by the idea of directedness, we see that there are many types of concepts and relations that do not apply to concepts themselves.

The phenomenological ontology of concepts results from the simple observation that certain categories of concepts and relations are not applicable to concepts. For example, categories involving spatial relations are not applicable to concepts. Concepts do not have spatial properties and do not stand in spatial relations. Concepts do not have temporal properties and do not stand in temporal relations. By way of contrast, the mental process of conceptualizing does have temporal duration. Concepts also do not stand in causal relations to one another or to anything else (e.g., to us). Because they do not have temporal properties they are distinct from subjective phases of consciousness. Phases of consciousness do have temporal duration and vary through time. Concepts are in this sense objective. They are invariants in the stream of consciousness. They are independent of phases of consciousness in the sense that they are not variable, 'real' elements in the stream of consciousness, but they are not mind independent in the sense that they are not in the stream of consciousness at all. At least some concepts are in the stream of consciousness as 'abstract' or 'ideal' and invariant. Thus, they are not mind independent in a manner that precludes our being directed toward them (or by them). Of course some concepts do apply to concepts and concepts do stand in certain kinds of relations to one another. Not all relations are empirical, for instance, spatial, temporal, or causal. Some relations are meaning theoretic, logical, or formal.

The form of rationalism that starts with the notion of intentionality and takes mathematics and logic as given yields a kind of phenomenological realism about mathematics and logic. It can be contrasted with classical metaphysical realism because it is a realism that starts from inside the domain of our conscious life and, given the intentionality of human reason, finds that the mind is directed, prima facie, toward different kinds of things in its different activities. Concepts are among the things toward which it may be directed in reflection. It is not an untenable realism that places concepts outside all possible experience. The concepts it recognizes are only those that are objects of possible experience.

To speak of *phenomenological* ontology is to hold that all we have to go on are the invariants in our experience, even if we are under some massive illusion about these for many years. We can distinguish real invariants from apparent invariants relative to our experience thus far, but we are never in a position to do this once and for all. This is just the nature of phenomenology and is part of its way of avoiding a naive and insupportable metaphysical realism.

§ 7   Concepts Are Not Subjective Ideas

We have been saying that concepts are not subjective. They should be distinguished from the individual changing ideas of each human subject. Since subjects are supposed to be directed toward objects by way of concepts it might be objected that consciousness could not be directed by abstract, objective entities but only by individual subjective ideas. This objection gives expression to a very basic misunderstanding of my view of concepts or senses. Thus, it is important to respond to it.

First, in Fregean/Husserlian fashion, we can say that many different subjective ideas may be associated with the same abstract concept. 'Ideas' in this sense occur at particular times in particular subjects. They are unique to each subject. One person's idea is not that of another. The same concept need not always be connected with the same idea, even in the same person. The concept, however, can be shared by many different people at different times and places. Indeed, we see how a common store of concepts is invariant from generation to generation, especially in the sciences of mathematics and logic. A good start on an account of this phenomenon can be found, for example, in Husserl's essay "The Origins of Geometry." Thus, various subjective ideas may be associated with the same, invariant concept. The relationship is another instance of the variation/invariance structure of consciousness. Epistemically speaking, concepts and subjective ideas depend on and emerge in relation to one another. The existence of a concept is required for the existence of a subjective idea, and the existence of a subjective idea is required for knowledge of a concept.

It is true that if there were no subjective ideas, then a subject's acts would not be directed. A subjective idea must, in a sense, be present if there is to be directedness toward an object. The point, however, is to see that directedness would not be possible without concepts. The presence of a subjective idea is a necessary condition for directedness, but it is not sufficient. The reason it is not sufficient is that without concepts there would be no invariance across different subjective experiences within the same subject, to say nothing of invariance across different subjects at different times and places. Directedness toward objects (= identities in experience) requires such invariance. Otherwise there would be nothing that these diverse ideas or experiences would have in common. The concept or sense is precisely that which diverse ideas can have in common. Thus, the directedness of consciousness requires the notion of concept or sense. Moreover, as we have been saying, it is not possible to deny that consciousness is directed.

Frege liked to illustrate some of these ideas with his telescope analogy (Frege 1892). Consider the observation of the Moon though a telescope. The Moon is the object or reference (invariant), mediated by the real image projected by the glass in the interior of the telescope, and by the retinal image of the observer. The real image is like the sense (another invariant), and the retinal image is like the subjective idea or experience. The optical image in the telescope is one-sided and dependent upon the standpoint of the observation, but it is still objective and invariant inasmuch as it can be used by different observers or even the same observer on different occasions. Each different observer, however, would have his or her own retinal image as a result of the diverse shapes of the observers' eyes, different physiological features, and so forth.

§ 8   Conceptual Intuition and the Alleged Problem of 'Epistemic Contact'

On the view of conceptual or rational intuition that we have been developing, concepts are not abstract 'things-in-themselves' that are forever cut off from consciousness. Rather, they are given as invariants in our experience toward which we can be directed. Directedness is not a spatial, temporal, or causal relation. As our examples show, the absence of a causal relation makes not the slightest difference to the fact that we can be directed toward concepts. Its absence does not somehow block cognitive directedness toward such objects. The real mystery is why anyone would insist that if directedness is not a causal relation, then it is an illusion. This insistence is probably just a misapplication of the desire to be 'scientific' (i.e., empirical). One need not accept the assumption, however, that all science is empirical science. One can perfectly well remain scientific without such misplaced zeal. Directedness is not a causal relation, but it is a fundamental and undeniable feature of consciousness. To hold that we cannot be directed toward such different objects is to lose track of the facts of conscious life. Once we understand intuition in terms of intentionality and directedness, and not in terms of causal relations, we can develop an account of the intuition of concepts. Such an account is a natural consequence of adopting the intentional stance on our experience, as opposed to the physical stance or the design stance (see, e.g., Dennett 1971).

Of course in being directed toward ordinary sensory objects causal relations do play a role, but even here it can be argued that causal attributions are parasitic on invariance in the field of consciousness. Since we can misperceive and be under illusions about objects, the 'object' to which we take ourselves to be causally related will be that which stabilizes and becomes invariant in our experience. If I perceive a house at one point and then see at a later stage that it is actually only a house façade (as in a movie prop), then at the later stage I should not be able to say that I was causally related to a house at any point in the experience. What I take myself to be causally related to will be a function of what remains invariant in my experience. In a phenomenological account of knowledge we can speak of what motivates knowledge in various domains without having to suppose that motivation is always the same as a causal relation to a sensory object.

The more fundamental structure, even in sense perception, is therefore directedness toward invariants. What this means is that our directedness model applies uniformly across the domains of mathematics and sense perception and that we simply particularize it for each of these domains. Virtually every empiricist account of knowledge and intuition I know of starts with sense perception and never gets beyond it. Some philosophers who have developed these accounts admit outright that their models do not apply to mathematics and that they do not know how to account for mathematics. Since empiricist accounts do not apply uniformly across domains we are presented with dilemmas about how we could have mathematical knowledge. The locus classicus of this in recent times is Benacerraf's essay "Mathematical Truth" (Benacerraf 1973). It should be obvious that I challenge Benacerraf's claim that our best account of knowledge will be some kind of causal account. Benacerraf's paper ignores the fact that in "knowledge" we have an example par excellence of intentionality. If we have knowledge anywhere, it is in the sciences of mathematics and logic. This is arguably our most secure and highest grade of knowledge. Why downplay it and insist on empirical knowledge as the paradigm? The way to account for mathematical knowledge is in terms of directedness toward invariants. The fact that intentionality in mathematics has not been investigated has led to a huge gap in current thinking about mathematical knowledge, which contributes to the feeling that Benacerraf's dilemma is genuine.

§ 9   Some Positions That Ignore or Seek to Eliminate the Intuition of Abstract Concepts

Gödel mentions a number of philosophical positions that either ignore or seek to eliminate the intuition of abstract concepts. It is useful to consider these positions in order to obtain a sharper view of the place of concepts in mathematics.

Hilbert's formalism would provide a way to ignore or eliminate the intuition of abstract concepts. If Hilbert's original plan for proof theory had succeeded, one could appeal only to the intuition of concrete, finite sign-configurations and to finitary, purely formal, or mechanical rules for manipulating these sign-configurations in order to establish the consistency and completeness of mathematical theories. Gödel's incompleteness theorems, however, show that Hilbert's original plan for proof theory will not succeed. Mathematics cannot be completely represented in finitely axiomatizable, purely formal systems if it is consistent. The intuition of abstract concepts cannot be ignored or eliminated, because the incompleteness theorems show that the appeal to only concrete, finite particulars and mechanical operations on these will not suffice for obtaining decisions about formally undecidable sentences, for obtaining consistency proofs for formal systems, or for the general development of mathematics. There must be some informal meaning that has not been captured in the finitary formal system. This notion of meaning, and the kind of informal rigor called for in dealing with it, goes far beyond the narrow 'contentual' meaning that Hilbert assigned to finitist, concrete mathematics. What will be required is the intuition of abstract concepts and meaning clarification. The finitary restriction of Hilbert's program is replaced with the realization that there is an inexhaustibility to many mathematical concepts. These concepts exceed or transcend our intuition at any given stage. The intuition of concepts is partial. All we can do is to work in a more or less systematic and reflective manner to unfold more and more of the meaning of these concepts.

Given the technical relationships that exist between finitary formal systems and the mathematical concept of mechanical computation, Gödel applies many of these same remarks to mechanism (see, e.g., Gödel 1934, *193?, 1972a). The intuition of abstract concepts could be ignored or eliminated if mathematics could be completely mechanized. The incompleteness theorems, however, cast doubt on the complete mechanization of mathematics. The intuition of abstract concepts cannot be completely mechanized. What cannot be captured in a machine is the reflection on abstract informal meanings, where these meanings transcend our grasp but are nonetheless still partially available to us. The finitary restrictions on machines are contrasted with the fact that there is an inexhaustibility to many mathematical concepts. It is this kind of view that underlies the late note on a "philosophical error in Turing's work" (1972a) that

Gödel appended to his 1972 paper. Gödel says that at one point Turing presents an argument to show that mental procedures cannot go beyond mechanical procedures. The problem is that there may be finite, nonmechanical procedures that make use of the meaning of terms. Turing does not recognize that

> mind, in its use, is not static, but constantly developing, i.e., that we understand abstract terms more and more precisely as we go on using them, and that more and more abstract terms enter the sphere of our understanding. (Gödel 1972a, p. 306)

> Therefore, although at each stage the number and precision of the abstract terms at our disposal may be *finite*, both (and therefore, also Turing's number of *distinguishable states of mind*) may *converge toward infinity* in the course of the application of the procedure. (Gödel 1972a, p. 306)

Human minds are not machines because human minds intuit abstract concepts (see also Tieszen 1994b). Moreover, intuition of concepts in my description is to be understood in terms of the intentionality of consciousness, and intentionality is just what machines lack.

Nominalism is the view that only concrete particulars exist. Universals do not exist, or, in the language I have used, abstract concepts/objects do not exist. Nominalism is usually part of an empiricist epistemology and ontology. Nominalists would certainly seek to eliminate the idea of intuiting abstract concepts. According to the arguments in this paper, however, we can say that concepts must exist in our experience in order for it to be the way that it is. Either nominalism is a naive metaphysical view that ignores the nature of our experience, or, if it does not ignore the nature of our experience, then it is untenable. It is not the case that only concrete particulars exist in our experience (see §§ 1–3). Gödel notes that Hilbert's formalism may be viewed as a kind of nominalism. In his 1953/59 papers he also shows how the incompleteness theorems can be applied in particular to Carnap's positivism, which can be viewed as a combination of nominalism and conventionalism.

One of Gödel's main objections (*1951, *1953/59) to conventionalism is that it portrays mathematics as our own free creation or invention in a way that does not square with the facts. As we said, not just anything falls under a given concept. Not just anything could be, for example, a natural number. We are constrained or forced in certain ways by the meanings of concepts and we cannot change these meanings at will. There may be a certain amount of freedom in unfolding a concept, depending on how indeterminate the concept is. Conventionalism, however, misrepresents this. It treats mathematics and logic as 'human made' and variable at will.

Conventions can be varied and imagined differently. Concepts, however, are not variable in this manner and are not human made.

Gödel's remarks about positivism are related to his remarks about conventionalism and nominalism. A central problem with positivism lies in the way that it distinguishes statements of logic and mathematics from empirical statements. The latter, being based on sense experience, are supposed to have content and to be meaningful. The statements of logic and mathematics, on the other hand, are supposed to be tautological or void of content. They do not have their own content or meaning; nor do they refer to distinctively mathematical objects and facts. They are true 'by definition', where this is a matter of linguistic convention. They are not considered to be true on the basis of the abstract concepts they contain. The positivist view thus leaves no room for intuition of abstract concepts and meaning clarification in mathematics. Apart from its conventionalism, Gödel thinks, positivism does not square with facts about how, in light of the incompleteness theorems, we will need to decide undecidable but meaningful mathematical statements and obtain consistency proofs. It does not do justice to mathematics.

Psychologism is a species of empiricism. According to psychologism, the objects logic and mathematics are mental objects and the truths of logic and mathematics are truths about our factual psychological makeup. They are empirical, a posteriori invariants that could be different from the way they now are. As is the case with many of the views that Gödel rejects, this introduces an untenable relativism into mathematics and logic. We can see why it would be set aside. The properties of the mental objects and invariants found by empirical science are not the properties of concepts.

The general 'leftward' worldview (see Gödel *1961/?) that underlies nominalism, conventionalism, positivism, and psychologism is empiricism. It may also be seen as underwriting Hilbertian formalism and mechanism. Empiricism is the view that all knowledge is derived from sense experience. Thus, there cannot be anything like an experience of abstract concepts. Gödel thinks that attempts to reconcile mathematics and logic with empiricism have failed, and the incompleteness theorems and related results can be used to show how they have failed. What is needed is a more balanced view that includes elements of rationalism that accord with our experience but that excludes earlier excesses or mistakes associated with naive rationalism. I think that the view I have outlined indicates some of the major empiricist sins of omission. Chief among these is its blindness about certain facts of experience, including that consciousness exhibits intentionality and that the mind can be directed in many different ways. The mind is not locked onto sense experience and only sense experience, even if some elements of sense experience are always in the background of consciousness. In particular, concepts make our experience possible, and the mind can be directed toward these concepts.

Finally, Aristotelian realism is not necessarily a view that eliminates intuition of abstract concepts. The problem is that it is just too limited to do justice to mathematics and logic. It allows only those concepts or universals that have physical, concrete instances. No other concepts or universals exist. This is clearly not acceptable if we are to do justice to pure mathematics and pure logic.

In light of more recent work on theories of concepts one could add that concepts are not mereological sums, sets of actual or possible particulars, tropes, and so on. It might be possible to devise a formal theory of concepts (see, e.g., Wang's suggestions about this in Wang 1987, pp. 309–311) but, as we have seen, Gödel makes various remarks about concepts and meaning clarification that suggest the necessity to develop insights into the content as well as the form of concepts.

About the process of the intuition Gödel says, in summary, that our intuitions of concepts (1) are constrained or 'forced' in certain respects (e.g., not just anything falls under the concept 'natural number'); (2) are fallible; (3) are more or less clear and distinct, precise; and (4) are for many mathematical concepts inexhaustible (*1951, 1947, 1964, and citations in Wang).

§ 10   Conclusion

Why would anyone think that there is or could be an intuition of abstract concepts? I have now provided some of the elements of an answer to this question. It is an answer that preserves some of the rationalistic, 'rightward' features (Gödel *1961/?) that have always been associated with logic and mathematics while avoiding the metaphysical excesses of naive rationalism and naive realism. Our experience in mathematics is not random or arbitrary. It displays some degree of systematicity, which is due to the concepts at work in experience. We should be able to uncover additional relations among concepts. Viewed aright, the idea that we can intuit concepts should hardly be the great mystery its detractors have made it out to be. This of course is not to say that more work is not needed.

8

Gödel and Quine on Meaning and Mathematics

Charles Parsons (1995b, p. 309) has noted that Gödel never discussed the deeper issues about meaning that are addressed by Quine. Parsons says that the only place where Gödel even begins to approach these issues is in the essay "The Modern Development of the Foundations of Mathematics in the Light of Philosophy" (Gödel *1961/?). In it Gödel argues that a foundational view that would allow us to cultivate and deepen our knowledge of the abstract concepts that underlie formal or 'mechanical' systems of mathematics is needed. It should be a viewpoint that is favorable to the idea of clarifying and making precise our understanding of these concepts and the relations that hold among them. Gödel says that phenomenology offers such a method for clarification of the meaning of basic mathematical concepts. The method does not consist in giving explicit definitions but "in focusing more sharply on the concepts concerned by directing our attention in a certain way, namely, onto our own acts in the use of those concepts, onto our powers in carrying out our acts, etc." It is through such a methodological view that we might hope to

A version of this essay was presented at the Berkeley Logic Colloquium, April 1996, and at the Stanford Philosophy Colloquium, May 1996. I thank members of both audiences for comments, and especially Charles Chihara, Sol Feferman, Dagfinn Føllesdal, Thomas Hofweber, David Stump, and Ed Zalta. I also thank Michael Resnik for comments and for a spirited defense of some of Quine's views. I doubt that he will be fully satisfied with my responses.

In preparing this chapter, I have especially had in mind Charles Parsons' writings on Gödel and Quine, and some of his remarks on Husserl and Kant (Parsons 1980, 1983a, 1983b, 1990, 1995a, 1995b). Indeed, my essay can be read as a response to the comments that Charles makes at the end of "Quine and Gödel on Analyticity" (1995b).

facilitate the development of mathematics and to gain insights into the solvability of meaningful mathematical problems.

The current philosophical climate is perhaps not favorable to this idea. Gödel wrote a number of other papers (e.g., *1951, *1953/59, 1947, 1964, 1972a) in which, in effect, he criticized views of mathematics that were not favorable to it. His criticisms were based on his ideas about the incompleteness theorems, consistency proofs, and the solvability of mathematical problems, and he applied these ideas to the views of Carnap, Hilbert, and others. As far as I know, Gödel never wrote about Quine's philosophical views. In this chapter I want to extend some of Gödel's arguments to Quine's view of mathematics. Quine's view has been very influential, and it stands as a major rival position to Gödel's. I shall argue that, as in the work of Carnap and other empiricists, Quine has no place for the kind of nonreductive meaning clarification that is needed to facilitate the development of mathematics and to gain insights into the solvability of meaningful mathematical problems. Quine's view is like the other views criticized by Gödel in failing to reconcile mathematics with empiricism. It fails to achieve the kind of balance between empiricist and rationalist views that Gödel argues for in the 1961 paper. These are the central claims I want to argue for in this chapter. I do not attempt to give detailed support to all of the ideas that are part of Gödel's alternative view of mathematics (e.g., on rational intuition), although I do think it is possible to support some of them. Furthermore, I do not discuss the kinds of extrinsic or a posteriori grounds for developing mathematics that Gödel mentions in some of his papers. Gödel does not mention these in the 1961 paper. My focus will be on the intrinsic, meaning-theoretic grounds that are part of Gödel's view of conceptual intuition.

I agree with Parsons that it is only in the 1961 paper that Gödel begins to enter the circle of ideas in which Quine's discussion of meaning moves. Accordingly, I will start with a few observations on the 1961 paper and on the papers from the fifties (*1951, *1953/59) that are part of its immediate background. These observations are intended to set the stage for the argument that follows.

§ 1   The Call for Meaning Clarification (1961)

In the essays from the fifties we see that Gödel thinks mathematical expressions have their own content and that they refer to objects and facts in a way that is analogous to perceptual reference. Gödel thinks there is a distinction between empirical science and mathematics, but he holds that there is an analogy between our experiences in these two domains. Carnap makes a sharp distinction between empirical science and mathematics and holds that the two are not analogous. Quine of course thinks it is not possible to make such a sharp distinction between the two. Empirical science is continuous with mathematics.

According to Gödel, we learn about mathematical concepts, objects, and facts by a kind of rational perception or intuition that is analogous to sensory perception. Gödel thinks the analogy holds on several counts: in each case our intuitions are forced or constrained in certain respects; there is a kind of inexhaustibility in each case; and we can be under illusions in each case. The notion of rational perception is construed so broadly that it may include the perception of concepts themselves (e.g., the general concept of set, or the concept of the natural numbers) and, in some cases, the objects (e.g., sets, natural numbers) or facts to which mathematical sentences may refer. There can evidently be an awareness of the concepts even if we do not or cannot intuit individual objects falling under the concepts (see, e.g., Gödel 1964, p. 258). We can, as it were, grasp the meanings of mathematical terms that express systems of concepts independently of knowing whether there are individual objects falling under the concepts.

Gödel argues, against Carnap, Hilbert, and others, that the second incompleteness theorem suggests that we must reflect on the meanings or 'thought contents' of mathematical expressions. We must try to cultivate and deepen our knowledge of the abstract concepts that underlie formal or 'mechanical' systems of mathematics. It is through such a methodological view that we might hope to facilitate the development of mathematics and to gain insights into the solvability of meaningful mathematical problems. The incompleteness theorems show that we cannot adequately capture mathematical concepts in formal systems. Instead, we refer through a formal system to what is not presently enclosed in that system, which only shows that our mathematical intentions or concepts were not adequately captured in the first place. As was noted earlier, Gödel states that this "clarification of meaning consists in concentrating more intensely on the concepts in question by directing our attention in a certain way, namely, onto our own acts in the use of those concepts, onto our own powers in carrying out those acts, etc." The phenomenological view to which Gödel is appealing holds that it is by virtue of the contents or 'meanings' of our conscious acts that we are referred to objects and facts, whether in mathematics or in other domains of experience (see Chapters 4, 6, and 7 earlier). Meaning or content in mathematics determines extension (if there is one). Generally speaking, the extension of a mathematical expression underdetermines its meaning or content. The same object or fact can be given from a number of different perspectives, or under a number of different contents. I will also say that objects can be given under different 'concepts', and I will henceforth think of concepts as that part of the intentional content expressed by predicates. The viewpoint thus recognizes both intensional and extensional aspects of mathematics and logic.

It is usually possible to identify at least some of the concepts that are at work in a given domain of experience. This much is not usually an epistemological mystery, even if we do not understand everything about the concepts we are using. Simply consider the contents of the 'that'-clauses we use to express our experience in the domain. Although an actual or imagined object can be given under different concepts, it cannot be given under just any concept. Concepts under which an object might be given can, for example, contradict or be consistent with one another, where the only way to determine this is to reflect on the meaning of the terms that express the concepts. Concepts can be of different categories or types, thus making category mistakes possible. There are bounds on what can fall under a concept, although there may be a wide range of variation. This is all part of the phenomenological view of intentionality, and, according to this view, the logic of concepts is not to be identified with purely formal logic. If reference in mathematics is a function of meaning or content and reference can be indeterminate in various respects, then there is a need for meaning clarification. We do not have a grasp of the precise boundaries of all of the basic concepts we use in mathematical thinking. Thus, we should try to determine the boundaries of our basic mathematical concepts, what is compatible with them and what is not, and what will fall under a given concept and what will not. It should be noted that a purely *formal* theory of concepts could not be sufficient for this epistemological task. One has just to plunge in and acquaint oneself with the concepts in a given domain. Thus, I think that it is not an objection that Gödel's view makes sense only if we have or can devise an 'adequate' formal theory of concepts. This is not, however, to say that we do not need a better account of concepts.

According to the thesis of intentionality, consciousness is always consciousness of something or other. The fact that there is directedness implies that there is categorization in our experience. Mathematical reason, by virtue of exhibiting intentionality, is responsible at any given stage for the nonarbitrary categorization and identification of mathematical objects and facts. There can be different meaning categories or categories of concepts and corresponding regional ontologies within mathematics itself. To see this categorization we need only to look to the existing science of mathematics, and to the different domains of mathematical practice. It follows, for example, that there need not be only one true concept of set. There could be different concepts of set (e.g., predicative, intuitionistic, maximal iterative), and then we simply develop our intuitions with respect to these different concepts. Mathematical propositions are true or false relative to different concepts. Indeed, there are different sets of axioms for different concepts.

The content of mathematical acts is therefore a condition for the possibility of the science of mathematics. This content is seen in the different areas of mathematics. It is also seen in differences in the intended meanings of formal mathematical theories. Sometimes we attempt to clarify or make precise the intended meanings of mathematical theories. When we do not discern the intended meaning of a formal system we often attempt, quite automatically, to supply one.

I might also note that a phenomenological view allows for more immediate and more theoretical aspects of experience in both mathematical and physical experience. Propositions that are more or less obvious can be found in both cases. Empirical theories are built up through reflection, generalization, and abstraction, as are the more theoretical parts of mathematics. In mathematics, however, there is even greater generalization. In works such as *Ideas I* Husserl also argues that mathematics depends upon a type of 'formal generalization' that should not be confused with empirical generalization. I return to this point later.

§ 2   Meaning Clarification and Reductionism

One of the most important consequences of these ideas is that we should be critical of programs that are reductionistic about mathematics. I think this is a point that Gödel makes in many of his philosophical papers (see especially *1951, *1953/59, *1961/?). The point is made succinctly in a remark of Gödel's recorded by Hao Wang (Wang 1996, p. 167):

Some reductionism is right: reduce to concepts and truths, but not to sense perceptions. . . . Platonic ideas [what Husserl calls "essences" and Gödel calls "concepts"] are what things are to be reduced to. Phenomenology makes them clear.

Meaning clarification is supposed to amount to the (more or less systematic and conscious) analytic unfolding of the content of mathematical concepts. But the methods of 'meaning clarification' offered by Carnap's program, Hilbert's formalism, nominalism, and psychologism misconstrue mathematical content. The point applies to a wide range of 'isms': empiricism, naturalism, fictionalism, instrumentalism, pragmatism, and even mechanism and logicism (see Chapters 4, 6, and 7). The methods offered by these philosophical views are simply not adequate to the tasks of facilitating the development of mathematics and helping us gain insights into the solution of meaningful mathematical problems. All of these viewpoints try in one way or another to eliminate or modify the kind of act/content/object structure involved in the intentionality of mathematical experience. In particular, they try to substitute other kinds of content for the given mathematical content. In some of his early work Carnap tried to substitute syntax for mathematical content. I think Gödel (*1953/59) has given us good reason, however, to believe that this effort fails. Hilbert's original program, which also fails, contains a variation on this theme: substitute finitary, concrete content, which may be understood in terms of syntax, for abstract, infinitary mathematical content. The lines are drawn distinctly in Hilbert's method. We are to start with the part of mathematics that is finitary, concrete, real, contentual, meaningful, and available to immediate intuition, and this can be set off from what is infinitary, abstract, ideal, merely formal, and 'meaningless', and a product of thought without intuition. The incompleteness theorems show that the proposed substitution is unworkable.

Generally speaking, these programs insist on a substitute for the given mathematical content, where the substitute involves notions of possibility, necessity, and generality that are more restricted than or different in type from the given or intended content. It is as if all mathematical content *must* somehow be conservative over the favored substitute and if it turns out that it cannot be so construed, then it is to be neutralized or downplayed. The incompleteness theorems tell us, however, that in mathematics we are sometimes faced with genuine extensions of content. Even intuitionism substitutes a different (albeit mathematical) content for the sentences of higher set theory. We can see this, for example, in the fact that the continuum hypothesis splits up into many different statements in intuitionism (which may be quite interesting in their own right).

Gödel's point is that we cannot smuggle in such substitutions unless we are willing to solve mathematical problems in terms different from those in which the problems are put (see especially the comments in the early part of Gödel 1947, 1964). Philosophers of these persuasions do not, in effect, make the phenomenological reduction when they come to concepts in sciences such as mathematics. They do not set aside their ideological prejudices (see Husserl 1911, 1913). This surely explains Gödel's remark (Wang 1987, p. 193) that we might be able to see concepts more clearly if we practiced the phenomenological epoché. This is an important part of what it means to practice the epoché. It is not the silly, quasi-mystical undertaking that some commentators have made it out to be. I will consider later how Quine's philosophy distorts mathematical content.

§ 3   The Contrast with Quine

It is clear from the 1961 paper that Gödel means to reject a wide range of viewpoints about mathematics, including the views of Carnap and Hilbert. By Gödel's sights, Quine's view of mathematics would also not strike the appropriate balance between empiricist and rationalist views. Quine is an empiricist about mathematics, but he wants an empiricism without the two dogmas of logical positivism. If we use the classification of worldviews given in the 1961 paper, then Quine's empiricism can be seen in his skepticism about concepts, meaning, intensions, intentionality, and even the parts of mathematics that are not applied. Indeed, Quine is skeptical of a notion of reason that would depend on these ideas. In his critique of the analytic-synthetic distinction we see that Quine 'reconciles' the a priori nature of mathematics with empiricism by attempting to undercut this division with a pragmatic holism.

Quine's holism, as is well known, is quite broad in scope. It encompasses the fields of empirical science, mathematics, and logic. Of course it could be even broader were it to encompass additional fields such as ethics. Quine does not say much about ethics, but there are holists who would also want to include ethics in the mix (see Føllesdal's discussion of these matters in Føllesdal 1988). Others have gone even further. Richard Rorty, for example, criticizes Quine for marking off the whole of science from the whole of culture. Rorty's determination to avoid such a division issues in a rather forlorn characterization of science as solidarity. In the other direction, one might be a holist within the fields of empirical science, mathematics, and logic separately, without attempting to combine them. Thus, it seems that in principle one could be a more localized holist or one could try to extend the view to an unbounded holism.

Quine's characterization of his holism toward the end of "Two Dogmas of Empiricism" is still worth quoting:

Total science is like a field of force whose boundary conditions are experience. A conflict with experience at the periphery occasions readjustments in the interior of the field. Truth values have to be redistributed over some of our statements. . . . Having reevaluated one statement we must reevaluate some others, which may be statements logically connected with the first or may be statements of logical connections themselves. . . . No particular experiences are linked with any particular statements in the interior of the field, except indirectly through considerations of equilibrium affecting the field as a whole. (Quine 1951, pp. 42–43)

Quine says that if his extended form of holism is correct, then it is folly to seek a boundary between synthetic statements, which hold contingently on sense experience, and analytic statements, which hold come what may. Any statement can be held true come what may if we make drastic enough adjustments elsewhere in the system. Conversely, no statement is immune to revision, including statements of mathematics and logic.

The view that there is no interesting philosophical distinction between mathematical and empirical truths reappears in many of Quine's writings (see, e.g., Quine 1960, 1966a, 1970, 1974, 1992). There are only gradations of abstraction and remove from the particularities of sense experience in these truths but no sharp boundaries and no qualitative differences or differences in type. Granted, mathematical content cannot be understood along the lines of earlier, cruder forms of empiricism (e.g., Mill). It cannot, for example, be understood in terms of empirical induction. Instead, mathematical content will have to be more like the content of the theoretical hypotheses of natural science. It will be more centrally located in the holistic web of belief, less likely to be revised in the face of recalcitrant experience, but in principle revisable. In this scheme, there is no clear demarcation point of the the the analytic, the a priori, or the 'necessary'. Notions of mathematical necessity, possibility, and generality are assimilated to natural necessity, possibility, and generality. Similarly, there is no clear demarcation point of 'certainty', and the alleged certainty of mathematics will have to be understood accordingly. (I leave aside discussion of the certainty of mathematics in this essay since I think the issue is complicated by a number of factors.) In this context it will also be difficult to understand the other rationalist element of mathematics that Gödel mentions in 1961, which is the idea that a kind of 'meaning clarification' based on conceptual intuition might play an important role in facilitating the development of mathematics and in helping us to solve open mathematical problems. In Quine's work there is certainly no notion of meaning clarification based on conceptual intuition. The very idea of such a type of meaning clarification would be met with skepticism.

In contrasting Quine's pragmatic holism with Gödel's view I will argue that it is the breadth of Quine's holism that leads to problems. A holism that attempts to encompass the fields of empirical science, mathematics, and logic does not do justice to mathematical meaning or mathematical content. Quine's deflationary remarks on analyticity depend upon this overarching holism. On a more localized holism within mathematics, or within mathematics and logic, the deflation may not be possible. One would not be forced to hold that mathematical or logical statements are revisable on the basis of what happens in empirical science (e.g., to simplify empirical theory), but one could hold that they may be revisable on the basis of what happens inside mathematics itself. That is, they may be revisable on the basis of uniquely mathematical evidence.

§ 4 Analyticity

It will be useful to establish first a few simple reference points about the notion of analyticity. We know that Gödel distinguishes a narrow from a wide notion of analyticity and rejects the claim that mathematics is analytic in the narrow sense. Gödel (1944, p. 139) says that 'analyticity' may have the purely formal sense that terms can be defined (either explicitly or by rules for eliminating them from sentences in which they are contained) in such a way that axioms and theorems become special cases of the law of identity and disprovable propositions become negations of this law. In a second sense, he says, a proposition is called 'analytic' if it holds on account of the meaning of the terms occurring in it, where meaning is perhaps undefinable (i.e., irreducible to anything more fundamental). Gödel makes a similar distinction in a later paper (Gödel *1951, p. 321), except that in this context he is thinking more specifically of the notion of analyticity in positivism and conventionalism. In this later paper he says that to hold that mathematics is analytic in a broad sense does not mean that mathematical propositions are "true owing to our definitions." Rather, it means that they are "true owing to the nature of the concepts occurring therein." This notion of analyticity is so far from meaning 'void of content' that an analytic proposition might possibly be undecidable. Gödel claims that we know about propositions that are analytic in the broad sense through rational intuition of concepts. Analyticity is thus linked to a kind of concept description and analysis.

Quine's arguments in "Two Dogmas of Empiricism" and related works are focused on a much narrower notion of analyticity than Gödel's preferred notion. Quine looks to the views of Carnap and the logical positivists, and to the tradition that stems from accepting Hume's distinction between matters of fact and relations of ideas. Gödel could perhaps even agree with these arguments. Indeed, we should keep in mind Gödel's comments on what the incompleteness theorems show about the analyticity of mathematics, namely, that mathematics could not be analytic in the narrow senses that he indicates. If mathematics is analytic in the wide sense, however, then relations between mathematical concepts must be of a rather substantial nature. Determining relations between concepts, as we described it earlier, must be different from determining purely formal relations, relations of synonymy, explicit definition, convention, and 'semantical rules' of the type that Quine considers in "Two Dogmas." At the same time, it must not be based on sense experience and it is not simply a function of rounding out our theories of sensory objects.

It is very important to keep in mind Gödel's examples of propositions that are analytic in the wide sense. In particular, I think we must start with examples in mathematics (and possibly logic) before we begin to worry about whether there are wide analytic propositions containing terms drawn from other domains of our experience. Gödel says in various works that there exist unexplored series of axioms that are analytic in the sense that they only explicate the content of the concepts they contain. An example from foundations is provided by the same phenomenon Gödel uses to refute Carnap's idea that mathematics is syntax and is void of content: the incompleteness theorems. On the basis of the incompleteness theorems, an unlimited series of new arithmetic 'axioms', in the form of Gödel sentences, could be added to the present axioms. These 'axioms' become evident again and again and do not follow by formal logic alone from the previous axioms. Here we might say that by way of a series of independent rational perceptions we are only explicating the content of the concept of the natural numbers. The Gödel sentences we obtain are compatible with this concept and do not overstep its bounds. In the procedure for forming Gödel sentences, we do not diagonalize out of this concept, even though we do step outside the given formal system. Moreover, new propositions or axioms may help us solve problems that are presently unsolvable or undecidable. One can already look at the undecidable Gödel sentence G for a formal system F in this way. We see that F will prove neither G nor $\neg G$. But metamathematical reasoning shows us that G is true if F is consistent, and we can thus 'decide' G on these grounds. By adding this sentence to F we can create a new formal system that will solve (albeit trivially in this case) a problem that was previously unsolvable. This idea is related to Gödel's comments on finding a viewpoint that is conducive to solving meaningful mathematical problems through a clarification of concepts in which we ascend to higher forms of awareness.

In the case of the construction provided by the incompleteness proof it is not clear that we reach a 'higher' state of consciousness in a significant way, for these 'axioms' are not mathematically interesting. However, the incompleteness results for arithmetic do open up the possibility of finding interesting results such as those of Paris and Harrington (Paris and Harrington 1977). The Paris/Harrington theorem is a genuinely mathematical statement (a strengthening of the finite Ramsey theorem) and is undecidable in Peano arithmetic (PA). The finite Ramsey theorem itself is provable in PA. The Paris/Harrington theorem refers only to natural numbers, but its proof requires the use of infinite sets of natural numbers. This is a good example of Gödel's idea of having to ascend to stronger, more abstract (in this case, set-theoretic) principles to solve lower level (number-theoretic) problems.

Gödel has the same model in mind for other parts of mathematics. He notes how we can solve problems that were previously unsolvable by ascending to higher types, and how we can also obtain various kinds of 'speed-up' results in this way (see especially the trenchant conclusion that Boolos reaches in Boolos 1987). An example that Gödel presents in many papers is based on axioms of infinity in set theory. These axioms assert the existence of sets of greater and greater cardinality, or of higher and higher transfinite types, and Gödel (e.g., in Gödel 1972a, p. 306) says they only explicate the content or meaning of a general iterative concept of set. In other words, these axioms do not overstep the bounds of this concept but are in fact compatible with it. So here we have axioms that are different from one another, but that appear to explicate the same concept(s). We have a kind of unity through these different axioms. Such a series may involve a very great and perhaps even an infinite number of actually realizable independent rational perceptions. Gödel says this can be seen in the fact that the axioms concerned are not evident from the beginning but only become so as the mathematics develops. To understand the first transfinite axiom of infinity, for example, one must first have developed set theory to a considerable extent. Once again, we are able to solve previously unsolvable problems with these new axioms. Gödel even thought we might find an axiom that would allow us to decide the continuum hypothesis (CH).

One might think of CH as an example in the following way. CH is known to be independent of first-order ZF (ZF^1). One can argue, however, that ZF^1 is not really adequate to the intended interpretation. ZF^1

does not come close to the goal of describing the cumulative hierarchy with its membership structure since ZF^1 has many nonstandard models. For the intended interpretation we do better to look to second-order ZF (ZF^2) (compare, e.g., Kreisel 1967, 1971). (The possibility of viewing set theory in this manner may be closed to Quine, given his strictures about higher-order logic.) Although ZF^2 is not itself categorical, its models are known to be isomorphic to an inaccessible rank. Now on the basis of the first incompleteness theorem one might suppose that the concept of set-theoretic truth is richer than the concept of set-theoretic provability. There will be truths of ZF^2, for example, that are not provable in ZF^2. In particular, CH should have a truth value in the universe of ZF^2, even if we do not presently know what it is. The truth or falsity of CH depends on the breadth of the set-theoretic hierarchy and not on its height. The relation of $\aleph_1$ and $2^{\aleph_0}$ is determined by the internal structure of the stages of the hierarchy. One can therefore argue that the truth value of CH is fixed by the contents of an initial segment of the hierarchy. By stage ω, sets of cardinality $\aleph_0$ appear. By stage $\omega + 1$, sets of cardinality $2^{\aleph_0}$ appear. And by stage $\omega + 3$, the pairing functions necessary for the truth of CH will have appeared. In other words, there is some reason to believe that CH should have a truth value under the *intended interpretation* of the axioms of this theory. The intended interpretation is to be understood in terms of the comments made earlier about intentionality. That is, it is to be understood in terms of the idea that we are directed toward a domain or universe by virtue of the meanings or contents of our acts, and we can then further explore this domain. In this directedness we have an example of what Gödel calls 'rational intuition'. As we said, rational intuition need not always be fully determinate. That is precisely why meaning clarification is needed. We need not have and usually do not have a fully determinate understanding of a domain from the outset. Moreover, there are a variety of ways in which the intended interpretation might be corrupted. We must be careful not to substitute some other content for the given or intended content, even though this is what reductionist views (such as empiricism and naturalism) would have us do.

§ 5   Rational Intuition and Analyticity

Gödel gives examples in which there is a conceptual or meaning-theoretic link between some given axioms and new axioms that are logically or formally independent of the given axioms. The link cannot therefore be an analytic link in the sense of being formally analytic. Thus, it cannot be a link that excludes intuition in the way that formal logic is supposed to exclude intuition. Something outside the given logical formalism must be involved. But we also do not learn about the link on the basis of sense experience. This is why Gödel says we learn about it through 'rational intuition'. There must be a (partial) intuition of a concept (intention) whereby the earlier axioms are related to the new axioms. There must be a grasp of the common concept or concepts (intentions) under which the axioms are unified. We can then begin to explore additional concepts that consistently extend a given concept.

The notion of intuition of concepts is not mysterious if one is prepared to recognize the fact that mathematical awareness exhibits intentionality. As we said, the fact that there is directedness implies that there is categorization in our experience. For an act to be directed in a particular way means that it is not directed in other ways. Our beliefs are always about certain *types* or *categories* of objects, and it is just these categories that we are referring to as 'concepts'. It is safe, for example, to say that we *know* that certain things are not instances of the concept 'natural number' and that other things are instances of this concept. Thus, we must have some grasp of this concept even if our grasp is not fully precise and complete. Instead of saying that we have a partial 'grasp' of a concept such as this we might as well say that we 'intuit' the concept. The term *intuition* is used because the concept is immediately given as a datum in our mathematical experience once we adopt a reflective attitude toward this experience. It is given prior to further analysis of the concept and to the consideration of its relation to other concepts. The objections raised to Quine's views in this paper seem to me to point toward such a notion of rational intuition. If it is possible to show that there are serious flaws in Quine's view of mathematics and if the exposure of these flaws seems to presuppose the notion of rational intuition, then we have all the more reason to take the notion of rational intuition seriously.

The view of analyticity just sketched is far from the tight little circle of preserving or obtaining truths by synonym substitution or by the semantical rules that Quine considers in "Two Dogmas." The relation of one axiom to another cannot be one of synonymy, or of 'truth by semantical rules'. Indeed, I shall suggest later that wide analytic truths share some (but not all) features with the theoretical hypotheses of natural science to which Quine generally wishes to assimilate mathematics.

Recognizing wide analytic truths is, in a sense, just a way of making room for notions of meaning and the a priori that are needed to account for mathematical developments that have taken place since the late nineteenth century. It is a way of making room for the meaning of modern mathematics. When Quine says there is no difference in principle between empirical science and mathematics we must keep in mind that this claim derives what plausibility it possesses from a purely extensionalist or truth-functional view of the sameness and difference of the meaning of predicates and sentences. Quine's 'meaning holism' is a thoroughly extensionalist viewpoint. On such a viewpoint it is possible to say that sets of sentences cannot be self-contained or autonomous in terms of their meaning or content because the concepts associated with predicates are ignored and only extensions of predicates or truth values of sentences are considered. If we consider the concepts, then we will find various compatibilities and incompatibilities, concepts with their own horizons and boundaries, categories of concepts, and so on.

§ 6   Mathematical Content and Theoretical Hypotheses of Natural Science

Gödel's notion of wide analyticity yields something more like what Quine obtains by assimilating mathematical content to theoretical hypotheses about nature. (Quine has acknowledged this in response to Parsons' "Quine and Gödel on Analyticity"; see Leonardi and Santambrogio 1995, p. 352.) Mathematical content is like the content of theoretical hypotheses in that it is more general or universal. The content of theoretical hypotheses is further removed from the particularities of sense experience. Neither mathematics nor natural science is content neutral. Also, Gödel would want to say that neither type of content is arbitrary. We are forced or constrained in some respects. Mathematical sentences, like theoretical hypotheses, are not true by definition, linguistic convention, or synonymy. In both cases, we go far beyond a narrow notion of the analyticity of mathematics. Both have real content.

There are, however, some important disanalogies. Charles Parsons has pointed out some of the most important differences. Parsons (Parsons 1983b, p. 195) states that theoretical and experimental physics are about the same subject matter, that experiments are carried out to verify or falsify the theories of theoretical physics, and so on. There is no similar unity of subject matter or of purpose of mathematics and physics. Mathematical truth does not depend on the tribunal of sense experience, whereas theoretical hypotheses do, even if only indirectly. Propositions of mathematics are not falsified when physical theories are abandoned or modified. If the mathematics changes in applications, it changes in the sense that a particular structure from the mathematician's inventory is replaced by another. There do not seem to be competing theoretical hypotheses in core parts of mathematics that are replaced with the passage of time.

Parsons' remark that theoretical and experimental physics are *about* the same subject matter should, in my view, be read in terms of the concept of intentionality. When he says that there is no similar unity of subject matter or purpose for mathematics and physics, this assertion signifies that the mind is directed in different ways in these fields and toward different goals. Different sets of properties and relations are applicable to the objects or concepts in each case.

Parsons notes that, according to Quine, there is no higher necessity than physical or natural necessity. Set theory is supposed to be on par with physics in this respect. The problem is that there is tension in Quine's own view of this. It conflicts with his view of mathematical existence. Not only does Quine treat mathematical existence and truth as independent of the possibilities of construction and verification, but he also treats them as independent of the possibilities of representation in the concrete (Parsons 1983b, p. 186). Quine is a platonist about set theory. The notion of object in set theory, however, and the structures whose possibility it postulates, are much more general than the notion of physical object and spatiotemporally or physically representable structure. But then how can Quine maintain that these possibilities are 'natural' and that the necessity of logic and mathematics is not 'higher'?

Although it is sensible to hold that neither mathematical propositions nor the theoretical hypotheses of natural science are topic neutral, mathematics is nonetheless closely connected with logic in that its potential field of application is just as wide (Parsons 1980, p. 152). Quine does mention the applicability of mathematics to other sciences. Parsons argues that this applicability indicates that mathematics has greater generality, and he notes that we might think of the generality of mathematics and logic as different in kind from that of laws in other domains of knowledge. He (Parsons 1983a, p. 18) cites with approval Husserl's idea that there is a difference in kind between the 'formal generalization' involved in mathematics and logic and the generality of the laws of particular regions of being, such as the physical world.

This is related to another difference noted by Parsons: elementary mathematical truths do not seem to be even more rarefied and theoretical than the theoretical hypotheses of natural science. On the contrary, they seem quite obvious. Quine's view cannot explain the obviousness of elementary mathematics and parts of logic (Parsons 1980, p. 151). In fact, there seem to be very general principles that are universally regarded as obvious, whereas on an empiricist view one would expect them to be bold hypotheses about which a prudent scientist would maintain reserve. For an empiricist such as Quine, obviousness should not typically accompany general hypotheses about nature. Think of how these hypotheses have to be built up over time, and of how they are refined and adjusted in various ways in the process. The generalizations of physical theory are not typically elementary and obvious, whereas formal generalization may be elementary and obvious. Formal generalization need not be involved only in highly theoretical assertions.

Finally, Parsons, Dummett, and many others have suggested that differences such as those between classical and intuitionistic mathematics are naturally explained as differences about *meaning* or content. The truth of mathematical propositions in these different areas is based, broadly speaking, on different notions of meaning.

§ 7   Application of Ideas on Incompleteness, Consistency, and Solvability

All of this suggests that mathematical content differs in significant ways from the content of theoretical hypotheses about nature. Mathematical content has different dimensions in terms of the type or degree of generality, obviousness, possibility, and necessity involved. As was mentioned earlier, the second incompleteness theorem indicates that if we are interested in consistency proofs, then, generally speaking, we cannot substitute content involving narrower notions of possibility and generality for content involving wider notions of possibility and generality. For example, there is a clear sense in which mathematical induction in PA involves a narrower notion of mathematical possibility than is involved in transfinite induction on ordinals $<\varepsilon_0$. There is a clear sense in which the introduction of primitive recursive functionals of finite type involves a notion of possibility or generality that goes beyond what we find in PA.

The point is that we cannot substitute less general for more general content, contents involving different types of generalization, or contents involving a 'lower' necessity for those involving a 'higher' necessity. Gödel would say we cannot eliminate rational intuition. Rational intuition is needed to find consistency proofs, and it is generally involved in the cognition of consistency. Appeals to more general principles, and wider conceptions of possibility are needed to justify the consistency of principles that involve less general, narrower notions of possibility. Of course Quine could say that he is not interested in proof-theoretic consistency proofs anyway, since all we really need to worry about is what works in or is indispensable to natural science. There are a number of difficulties with this response. Here I mention only the fact that consistency problems in the practice of mathematics and logic are perfectly legitimate and interesting. Quine also wants to help himself to the reduction of mathematics to set theory. Which set theory should we chose? Quine's NF or ML? Or ZF? Considerations about the likelihood of the consistency of these theories should presumably have some bearing on how we answer this question. It is not clear how appeals to natural science could help.

In any case, Gödel's work has shown us how these considerations about incompleteness and consistency are related to the matter of finding new axioms, solutions to mathematical problems, and speed-up results, and to the development of mathematics itself. On Quine's naturalistic holism we always want to stay within the narrowest set of possibilities and generalities needed to round off the mathematics required by natural science. This practice conflicts with the need to go to wider notions of generality and mathematical possibility to obtain solutions to *mathematical* problems (or to obtain consistency proofs). If we want a solution to a mathematical problem in the form 'P or not P' that is not now solved, we will not find it a good idea to insist that we substitute narrower or different ideas (or even ideas at the same level) of generality and possibility for the mathematical content given in the sentences that form the background of the problem. (See the examples of the Paris/Harrington theorem and CH in § 4, and of V = L in § 9.) What we need is a new principle that is independent of the existing principles but consistent with them, one that we see (by rational intuition) to be true, given that we are willing to take the other principles to be true. A kind of informal rigor is involved here. If in particular we substitute something that is supposed to be assimilable to *theoretical hypotheses about nature* for the given mathematical content of P, then we are not going to see that P is true (or, e.g., that a particular theory is consistent) and thus find a solution to 'P or not P'. We would simply not recognize the possibility of a sentence that could be true but that lies outside these narrower generalities or possibilities, because the latter are officially the criteria of truth. It is conceptual intuition that accounts for the fact that we are not restricted by these bounds. This is similar to the problem of recognizing the possibility of a true but unprovable formula for a formal system given that one is supposed to remain in the narrower sphere of immediate concrete intuition of the type that is supposed to accompany finitary mathematics (see Tieszen 1994b). The difference is that in Quine's case it is natural science, not finitist or constructive mathematics, that is supposed to provide the most secure and reliable basis of knowledge. Natural science is epistemically privileged. The problem is that it is privileged to the extent that it can blind us to the discovery of consistency proofs, new axioms, and solutions to open problems, and to the general development of mathematics.

To avoid these objections, Quine could try to revert to his set-theoretic platonism. The problem with this maneuver, as we have seen, is that he is then in a bind with his own views on natural necessity. He does in fact make some effort to inactivate unapplied parts of set theory, as we will see later. This effort suggests that he wants to backtrack on his platonism, but without embracing constructivism. Another problem with the maneuver is that Quine's view of set theory does not embody a notion of content that would allow us to avoid the basic objection.

§ 8   Mathematical Content and Quine's Conception of Set Theory

It is telling that, in practice, Quine actually assimilates set theory more to logic than he does to theoretical hypotheses about nature. In this guise, Quine thinks of sets as extensions of predicates. The axioms of set theory attribute extensions to certain predicates (see, e.g., Quine 1937, 1940, 1969, 1974, 1992). As Parsons has noted (Parsons 1983b, p. 198), this logic-like conception of sets is rather Fregean. This makes Quine's views 'deviant' relative to most contemporary thought on the subject. We need to keep in mind how the ideas associated with the iterative or 'mathematical' conception of set differ from Quine's views.

Now mathematical content could not be like the content of theoretical hypotheses about nature and at the same time be assimilable to logic or Quinean set theory. But mathematical content differs from what we find in logic, and it differs from what we find in set theory as construed by Quine. It differs from logic because logic is supposed to be content neutral whereas mathematics is not. Mathematical content also cannot be assimilated to Quinean set theory. If we suppose otherwise, we are saddled with a very impoverished notion of mathematical content. Mathematicians do not actually do mathematics from within NF or ML (or even from within other forms of set theory). To hold otherwise is to ignore mathematical practice. Indeed, what is the intended interpretation of NF or ML? These are syntactically motivated theories. It is not clear that there is a concept behind them. Quine's view of set-theoretic content does not yield the proper directedness and regulation.

Quine has been very slow to warm to the idea of the cumulative hierarchy as an interpretation of the axioms of ZF. His view has not included the idea of advancing to higher sets/types or to new axioms of infinity as a natural and even intrinsic extension of set theory. Hence, the ideas about new axioms and problem solving that are described by Gödel are simply missing in Quine's work. Gödel's examples of the analytic unfolding of the content of the concept of set are always built around the general, 'mathematical' concept of set. In Quine's conception of set theory we would have nothing like the idea of solving problems by advancing to higher types or to new axioms of infinity. We would simply not be able to obtain this idea from Quine's conception. Given the concept Gödel works with, however, such ascension suggests itself. Why not look into it? It is in the horizon of the concept, so to speak, and it has been fruitful and led to many new development and results which would be overlooked or downplayed if Quine's views were taken seriously.

These reflections on set theory are related to another problem for Quine's view: the problem of accounting for unapplied parts of mathematics.

§ 9   Mathematical Content and Unapplied Parts of Mathematics

As we noted, one of the problems with Quine's view is that the elementary, more obvious parts of mathematics must be assimilated to highly theoretical hypotheses about nature. On the other hand, Quine has difficulties accounting for the theoretical parts of mathematics that do not have applications in natural science. This is the fate of Quine's view. Pragmatic holism allows us to do justice neither to elementary, obvious parts of mathematics nor to advanced, theoretical parts of mathematics. That is, it does not allow us to do justice to mathematics. The view I favor runs orthogonal to this. On a more localized holism we could have the distinction between general and specific, or between the theoretical and the applied, in empirical science as well as in mathematics. Unapplied parts of mathematics often do seem to have content or meaning, and we have already indicated how this is possible. On Quine's view, this is a serious problem.

Many of Quine's most explicit remarks on the problem have been responses to Parsons' promptings. Here are two examples:

So much of mathematics as is wanted for use in empirical science is for me on a par with the rest of science. Transfinite ramifications are on the same footing insofar as they come of a simplificatory rounding out, but anything further is on a par with uninterpreted systems. (Quine 1984, p. 788)

I recognize indenumerable infinities only because they are forced on me by the simplest known systematizations of more welcome matters. Magnitudes in excess of such demands, e.g., $\beth_\omega$ [the cardinal number of V_ω (N) and of $V_{\omega+\omega}$] or inaccessible numbers, I look upon only as mathematical recreation and without ontological rights. (Quine 1986, p. 400)

In one of his later books, Quine (Quine 1992) reexamines objections to his views on set theory. He says that sentences such as CH and the axiom of choice, although not justified by their applications in natural science, can still be submitted to considerations of simplicity, economy, and naturalness that contribute to molding scientific theories generally. Such considerations, he says (Quine 1992, p. 95), support Gödel's axiom of constructibility, 'V = L'. This axiom inactivates the more gratuitous flights of higher set theory. On the basis of the argument in §§ 1, 2, 4, 5, and 7, however, we should have the strong sense that the point has been missed. Should these really be the reasons, or the only reasons, for *deciding* that V = L? It is certainly not clear that we should close off investigation for these reasons. To do so would be to ignore what is in the horizon of the general mathematical or iterative concept of set.

It seems to me that we are at least owed an account of the meaningfulness of higher set theory, even if there is reason to be skittish about its ontology or its radical platonism about set-theoretic objects and facts. On this matter, the problems with Quine's view are manifold. First, what constitutes a 'simplificatory rounding out' of the type cited in the first passage quoted? Perhaps we should just accept all of first-order ZF. But the other remarks cited run counter to the idea that this is the simplest known systematization of more welcome matters. In any case, if a theory such as ZF turns out to provide the simplest systematization, then Quine's stipulations about simplicity could conflict with his stipulations about natural necessity. Given the ontological commitments of ZF, the bind with a pragmatic holism that assimilates mathematics to natural science emerges once again. It is also questionable whether research in higher set theory should be discouraged on the grounds of what are at a given time the simplest *known* systematizations of more welcome matters.

The simplest known systematization today may not be the simplest known systematization tomorrow.

It is also not clear how the idea of a simplificatory rounding off could be compatible with advance to the more robust notions of mathematical possibility and necessity that are required in the face of the incompleteness theorems. Advancing to higher types or sets is a natural and intrinsic extension of Gödel's concept of set, but whether it answers to Quine's concerns for simplicity and economy with respect to natural science is not clear. In short, it is not clear what the appeal to simplicity and economy could amount to in the presence of the incompleteness phenomena and, more generally, of the *inexhaustibility of mathematics.* The kind of ascent needed to solve open mathematical problems takes us further from natural science and its problems and has nothing to do with rounding out the mathematics needed for the natural sciences. Empiricist accounts of mathematics tie mathematical concepts more closely to acts of sense perception in one way or another and thus place constraints on mathematical thinking instead of fostering its expansion. On the other hand, acts in which there is a free *imagination* of possibilities, when placed in the service of reason, lead to a far less constrained view of mathematical possibility.

One has to ask how to proceed with research about open mathematical problems. For Quine, the 'simplificatory rounding out' is always a rounding out of what is needed for natural science. What cannot be handled in this way is treated as on a par with uninterpreted systems or mathematical recreation and, as we see, is regarded as gratuitous. But what is gratuitous or simplest with respect to the objective of solving *mathematical* problems? One could argue that with such an objective in mind, deciding in favor of V = L would be gratuitous and perhaps even contrary to canons of simplicity.

In a sense, the problem goes beyond Quine's view of higher set theory. The problem is that Quine has no place at all for the intended meanings of mathematical theories. Quine has on occasion said that a serious divergence over logic or set theory is just a confrontation of rival formal stipulations or postulates. He has said that we make deliberate choices and set them forth unaccompanied by any attempt at justification other than appeals to elegance or convenience. Although this may be true of Quine's attitude toward mathematics and set theory, it does not gibe with the usual attitude toward these subjects. It is not accurate as a report on mathematical practice.

It is worthwhile to consider the meaning and motivation of Quine's set theories ML and NF in light of these comments on rival formal stipulations. In point of fact, ZF has enjoyed far more popularity in mathematics than Quine's ML or NF. This popularity is most likely due to the fact that there is some sense of what the intended interpretation of ZF is, even if it is not perfectly clear. This yields more confidence in the consistency of ZF than in that of ML or NF. Given the choice between ZF, NF, and ML as the set theory to which mathematics is to be reduced, clearly ZF is preferable. Considerations of this type show that we do and must worry about the consistency of higher set theory. This concern cannot be overridden by or collapsed into a pragmatic or instrumentalist criterion that focuses only on what works in or is indispensable to natural science.

As Wang (Wang 1986, p. 162) characterizes it, Quine tries to combine an emphasis on formal precision with a gradualism that tends to blur distinctions and emphasize relativity or difference in degree. The drive for precision gives preference to reference over meaning, extensional objects over intensional objects, language over thoughts and concepts. But science generally possesses far less formal precision than classical predicate logic or Quine's ML or NF. In contrast, I am urging that much of science depends on informal rigor, that is, on rigorous but informal analyses of concepts. Mathematical content has the function of determining research and making it meaningful even if this research is not applied to or 'justified' by what is needed in natural science. Mathematical content is not exclusively a function of finding the simplest systematizations of the mathematics required by the natural sciences. It is underdetermined (and possibly even corrupted) by such a process. The meaningfulness of the propositions involved in stating an open mathematical problem, for example, must be autonomous in this sense. There is evidence that is unique to mathematics because there is content or meaning that is unique to mathematics. It is very difficult to see how, on Quine's view, we should explain problem solving in those parts of mathematics that have no applications in natural science, for we are not provided with a way to understand the meaningfulness of the propositions involved in these problems.

I am arguing that these mathematical intentions or contents are indispensable to research in mathematics. They are conditions for the possibility of the science of mathematics. It is possible, although not recommendable, for theoretical mathematicians to turn their backs on natural science and to get along quite well. Natural science is dispensable to the work of a theoretical mathematician, as it is to the work of a logician. However, mathematics is indispensable to natural science. Quine's indispensability arguments do not speak to the question of what is required for the practice of mathematics itself.

§ 10 Mathematics and (Propositional) Attitude Adjustment

It is important to note how these problems about mathematical content are related to the unstable place that the concept of intentionality occupies in Quine's philosophy. Quine generally wants to be a behaviorist about mental phenomena. In his flight from intentionality and intension he does not wish to recognize concepts or other intensions as abstract entities; nor does he wish to recognize mental entities. At the same time, he wants somehow to recognize intentionality and its irreducibility, but to downplay it (see, e.g., Quine 1960, pp. 219–221). However, he has more recently recommended Dennett's work on these matters (Quine 1992, p. 73), and so he would perhaps accept a pragmatic or instrumentalist view of intentionality and of the need for intentional content. The situation is rather unclear.

It is a generally accepted fact about intentionality that the content or meaning of intentional states determines what those states are about. The notion of meaning or content is thus inserted into an account of our thinking in various domains. I argue that an appeal to concepts is needed to explain the kinds of directedness, categorization, and regulation that are involved in research in mathematics, and that such an appeal is not available in a behaviorist account of mental phenomena. As Gödel suggests, what we really need to do is to focus "more sharply on the concepts concerned by directing our attention in a certain way, namely, onto our own acts in the use of those concepts, onto our powers in carrying out our acts, etc." Content or meaning determines what our research is about in various domains of mathematics, and it can do so independently of applications in natural science. The content required for this function is simply not the kind of content that Quine's view prescribes. If Quine were to accept this claim, on Dennettian grounds, then he could do more justice to mathematics. He would then have to change his thinking about mathematics. Quine's present position on mathematics indicates that he does not want to take intentional content seriously and that he wants to be a behaviorist. Thus, there is tension between Quine's views on intentionality and his view on mathematics.

§ 11 Conclusion

If the arguments in this essay are correct, then Quine fails to reconcile mathematics with his pragmatic brand of empiricism. The attempt to extend holism across the fields of empirical science, mathematics, and logic does not do justice to mathematical content. Thus, he fails to arrive at a position with the appropriate balance of the empiricist and rationalist views that Gödel advocates in various papers. In surveying Quine's work it appears that we must think of mathematical content in terms of either (i) the content of theoretical hypotheses of natural science; or (ii) extensions of predicates, as in Quine's view of set theory; or (iii) formal stipulation that is governed by considerations of elegance and convenience. Quine invariably substitutes one or another of these for mathematical content, depending on the issue at hand. It is clear that mathematical content could not be like the content of theoretical hypotheses about nature and at the same time be assimilable to logic or Quinean set theory. As I have indicated, mathematical content cannot be like the content of theoretical hypotheses in the relevant ways. It is also not exhausted by logic or Quinean set theory. Finally, mathematics cannot just be formal stipulation governed by considerations of elegance and convenience. Reflection on the incompleteness theorems and related ideas on consistency and solvability helps us to establish these facts.

In attempting to reconcile mathematics with empiricism, Quine's philosophy distorts mathematical content. The Gödelian views I have discussed in this chapter provide a better account of the meaning of mathematics.

9

Maddy on Realism in Mathematics

Realism in Mathematics (RM) by Penelope Maddy (1990) is a delightful, thought-provoking book which contains interesting ideas on almost every page. Maddy attempts to develop and defend a novel form of "naturalized" set-theoretic realism, which she portrays as "compromise platonism." The compromise is supposed to be between Quine/Putnam platonism, on the one hand, and Gödelian platonism on the other, and the focus of the book, to make the project manageable, is on set theory in particular. In this chapter I shall discuss the arguments of her book in some detail.

Chapter 1 of *RM*, entitled "Realism," leads the reader through a variety of positions in the philosophy of mathematics on the way to a characterization of compromise platonism. A number of antirealist positions are briefly evaluated – intuitionism, formalism, if-thenism, the logicism of the logical positivists, and conventionalism – and some issues concerning different theories of truth are canvassed. Traditional platonism about mathematics is then characterized as the view that mathematical entities are abstract, outside physical space, eternal, unchanging, and acausal. Knowledge of such entities is supposed to be a priori and certain, and mathematical truths are supposed to be necessary truths. Like traditional platonism, Gödelian platonism holds that mathematical entities are abstract, and it takes its lead from the actual experience of doing mathematics. Maddy says that, unlike Quine/Putnam platonism, it recognizes a form of evidence intrinsic to mathematics. It holds that there is a kind of intuition of abstract entities and that this mathematical intuition plays a

I thank Michael Resnik and Charles Chihara for comments on this essay.

role in mathematics analogous to the role that sense perception plays in the physical sciences. So, for example, axioms of mathematical theories force themselves upon us as explanations of the intuitive data in mathematics as much as the assumption of medium-sized physical objects forces itself upon us as an explanation of our sensory data.

Gödel's platonistic epistemology is, as Maddy sees it, two tiered: some parts of mathematics are intuitive and some parts are more theoretical. Just as there are facts about physical objects that are not perceivable, so there are facts about mathematical objects that are not intuitable. She argues that, according to Gödel, simpler concepts and axioms are justified intrinsically by their intuitiveness, and more theoretical hypotheses (or unobservable facts) are justified extrinsically, by their consequences, their role in our theories, their explanatory power, their predictive success, their fruitful interconnections with other well-confirmed theories, and so on. On Maddy's construal, the second tier of Gödelian epistemology leads to departures from traditional platonism. Extrinsically justified hypotheses are not certain and, given that Gödel allows for justification by fruitfulness, are not a priori either. On these latter points Gödel's platonism would be similar to Quine/Putnam platonism.

Representing Gödel's view as two tiered in this way, however, is probably not accurate since Gödel seems to present his 'inductive' criterion of the truth of axioms, based on their 'success', as simply an alternative to the view of the intrinsic necessity of axioms that would be provided by mathematical intuition. Nor does it fit well with his view that axioms of mathematical theories force themselves upon us as explanantions of the *intuitive* data in mathematics much as the assumption of medium-sized physical objects forces itself upon us as an explanation of our sensory data. Maddy notes later (*RM*, footnote 91, p. 76) that Gödel's "intuitions" do cover more esoteric cases than are covered by her own account of intuition, cases involving what she would consider to be theoretical evidence. I think this divergence from Gödel's view is related to some problems with Maddy's compromise platonism indicated later.

Quine/Putnam platonism, on the other hand, arises from the fact that mathematical entities – in particular, sets – appear to be required for natural science. We are committed to the existence of such mathematical objects because they are indispensable to our best scientific theories. And by what better canons could we hope to judge our claims about such objects than by scientific ones? The claim that we can do no better than to appeal to science to determine the answers to such epistemological and ontological questions lies at the center of the Quinean effort to "naturalize" philosophy. From the naturalized perspective there is no point of view prior to or superior to that of natural science. And if science can get by with extensional entities such as sets, as it evidently can, then there is no reason to recognize the existence of other abstract entities, such as universals (which also happen to be intensional entities). In this version of platonism, mathematical knowledge cannot be considered to be a priori because it is justified only by the role it plays in our empirically supported scientific theory. It is also not certain, since our empirical theories are subject to revision. Mathematical truths are perhaps also not necessary truths on a Quinean view.

Maddy raises several serious objections to Quine/Putnam platonism. First, unapplied mathematics is completely without justification on this view, whereas Gödel provides for it by recognizing a uniquely mathematical form of evidence. The Quine/Putnam view ignores the actual justificatory practices of mathematicians. Second, mathematics is assimilated to the most theoretical part of science on the Quine/Putnam view, but, as Charles Parsons has pointed out, a statement such as '7 + 5 = 12' hardly seems to be a highly theoretical principle. Quine/Putnamism in unable to account for the obviousness of elementary mathematics. Gödelian platonism accounts for this with its notion of mathematical intuition. The central problem with Gödel's view, as Maddy sees it, is that his account of mathematical intuition is unpersuasive, and how it could be naturalized is not clear. What is needed is a scientifically feasible account of mathematical intuition. In order to provide such an account something has to give somewhere and so Maddy parts company with Gödel very significantly on the nature of the objects of mathematical intuition. Gödel insists on a traditional platonist account of mathematical objects, whereas the novelty of Maddy's account derives from the bold claim that (some) sets are part of the physical world and can literally be perceived. Maddy thus proposes a compromise platonism which she believes preserves the best of the other versions but avoids their problems.

Compromise platonism is intended to be a naturalistic view which recognizes the Quine/Putnam indispensability arguments as supports for the (approximate) truth of classical mathematics but which also recognizes, as does Gödel, purely mathematical forms of evidence, a mathematics/science analogy, and a place for mathematical intuition in a two-tiered analysis of mathematical justification. The objections to Quine/Putnam platonism are thus to be blocked, and objections to Gödelian platonism will be bypassed by offering a naturalized account of the perception of mathematical objects as part of the first tier of mathematical epistemology. Obviously, the fate of Maddy's compromise platonism depends heavily on the success of this latter undertaking.

In Chapter 2, "Perception and Intuition," Maddy takes up the challenge of providing a naturalized account of mathematical perception. The problem of mathematical knowledge is framed in terms of a dilemma posed by Paul Benacerraf. If traditional platonism is true, then mathematical entities are abstract. They do not inhabit the physical universe and do not take part in any causal interactions. But if we start with the part of epistemology that we understand best, some kind of causal interaction between persons and objects appears to be required for basic forms of noninferential knowledge. Thus, if platonism is true, we can have no knowledge of mathematical objects. If we assume that we do have such knowledge, platonism must be false. Maddy chronicles developments in epistemology since the time that Benacerraf raised the problem in this form and concludes that on newer reliabilist accounts of knowledge we still need an explanation of the reliability of mathematical beliefs that are not inferred. The processes by which we come to believe claims about x's must ultimately be responsive in some appropriate way to actual x's if they are to be dependable. Maddy's response to the problem, as has been mentioned, is to reject the traditional platonist's characterization of mathematical objects. Sets of physical objects are, on her view, part of the physical world and have spatiotemporal location where the physical stuff that makes up their members is located. Even an extremely complicated set would have spatiotemporal location as long as it has physical things in its transitive closure. Such sets are objects that can be directly perceived. Of course something will have to be said about 'pure' sets (such as the empty set), about all those sets whose members cannot be construed as physical objects, about unit sets, and about sets such as the set of real numbers or other large sets whose existence is routinely recognized in such systems as Zermelo-Frankel set theory. These matters are taken up at later points in the book. Real numbers and sets of reals are discussed in parts of Chapter 3, "Numbers," and in the opening pages of Chapter 4, "Axioms," and the other cases are discussed mainly in Chapter 5, "Monism and Beyond."

Maddy's effort to suggest what a scientifically respectable account of the perception of sets of physical objects might look like is based on an influential neuroscientific model of the lower levels of perception due to Donald Hebb. Hebbian theory postulates that the connectivity of the brain changes as an organism learns different functional tasks and that different cell-assemblies are created by such changes. Repeated activation of one neuron by another through a particular connection (i.e., synapse) increases its conductance so that groups of weakly connected cells, if synchronously activated, tend to organize into more strongly connected assemblies. In this way we develop complex, neural object-detectors through repeated stimulation by external stimuli. An object-detector is partly the result of the structure of the brain at birth and partly the result of early childhood experiences with objects. According to Maddy, it is these cell-assemblies that bridge the gap between what is interacted with and what is perceived. This Hebbian view is certainly plausible for some forms of cognition. Maddy wishes to extend it. She hypothesizes that we develop a set-detector, on the analogy with the development of other neural object-detectors. Several grounds for skepticism about such an extension of the Hebbian view are discussed later.

Now Maddy says that an initial baptism of a set, given this background, might go as follows. We are to imagine a set-dubber declaring, for example, "These three things – the paperweight, the globe, and the inkwell – taken together, regardless of order, form a set." By such a process the baptist picks out samples of a kind, and the word *set* refers to the kind of which samples such as this are members. Maddy argues that the person in this circumstance literally perceives a set. The appropriate kind of causal interaction between a person and a set exists, for some aspect of the set stimulates a phase sequence of appropriate cell-assemblies.

Maddy immediately considers several objections that could be raised to this account. Perhaps the central objection is associated with the ontology of traditional platonism: sets simply do not have location in space or time. Maddy argues, however, that there is no real obstacle to saying that a set of physical objects comes into and goes out of existence when the physical objects do, and that the set is spatially and temporally located exactly where the physical objects are. The same applies to higher-order sets. Moreover, any number of different sets could be located in the same place. She argues that none of this is any more surprising than saying that fifty-two cards can be located in the same place as a deck. If this means that sets no longer count as 'abstract', so be it. Maddy says she attaches no special importance to the term.

Another objection that she considers is meant to challenge the analogy between the perception of physical objects and the perception of sets that lies at the core of her epistemological account. The objection is that a person who dubs a sample of natural stuff, such as gold, causally interacts with her samples, whereas the set-dubber interacts only with the members of her samples. This objection, as I see it, is related to the ontological objection since the analogy should break down if sets are not located in space and time. Thus, I will consider the objections together.

First, concerning the ontological point, it is probably safe to say that no argument is needed for the claim that a deck of cards is physical. At the same time, we do not have a science made up of theorems about decks of cards. One of the most striking characteristics of mathematics is that its axioms and theorems, once established, seem to have a remarkable stability over time. That is, investigations of the sort that lead to axioms and theorems are taken by mathematicians to be about objects that are identical across times, places, and persons. Different mathematicians at different times and places believe they are dealing with the same object (e.g., the set of prime numbers or the number π) in their investigations, and this accounts for why the science of mathematics is possible at all. And doing justice to mathematical practice requires taking such basic beliefs of mathematicians into account. If mathematical objects were not (believed to be) characterized by such invariance, mathematics would be constantly shifting, with different objects and different theorems for different times, persons, and places. We would see a kind of relativism in mathematics that is alien to the way that this science is actually given. (Naturalism, it is worth noting, has always been accompanied by some form of relativism.) Now Maddy's perceived 'sets' simply do not have this kind of stability or invariance. They could not be such invariants in mathematical experience because, as noted previously, they change in many ways. The origin of our awareness of sets may have something to do with seeing groups of physical objects, but one could argue that only when one has grasped this kind of invariant does one have a set as it is understood in mathematics. So what is needed is an account of the knowledge of these invariants. And Maddy's views on generalization, presented later in the book (Chapter 4), do not help (see later discussion). If these comments are correct, then set theory could not itself be about physical things that shift locations, blink in and out of existence, and are constantly changing. It may have applications to these things, but even its applications involve a certain degree of idealization.

This point is directly related to the objection that challenges the analogy between the perception of physical objects and the perception of sets, for if sets are not objects in the environment to be detected by neural cell-assemblies, one would expect the analogy to break down. Maddy's initial response to this objection is to point out that the gold-dubber has in fact causally interacted only with the front surface of a time slice (an aspect) of the sample and that, by analogy, the set-dubber also interacts with 'aspects' of sets, their members. On closer analysis, however, the analogy becomes strained. The set-dubber actually has only aspects of the members of her sample sets within her causal grasp. But then the interaction with the set must be unlike the interaction with gold. It must be more complicated. Maddy allows that it must be a composition of the aspect/object relation and the element/set relation. This portends trouble with her epistemological claim. It seems that the element/set relation plays no role in the case of any other kind of perceptual object, but that the aspect/object relation suffices. This makes the role of the element/set relation look ad hoc. Or consider, for example, the question of how the aspects are taken to be aspects of the same object. For objects whose status as perceptual objects is not in question the aspects seem to be just automatically combined by the mind, as it were. Maddy's example of the set-dubbing, however, is quite unlike this, for we evidently must actively or reflectively 'form' or 'take together' the aspects (or aspects of members) of an object and be concerned about the order in which aspects are taken together, in order to perceive one object. The identity of the object is not immediately given in our experience but is something we have to constitute, as it were. In the case of sets, only mathematical, as opposed to perceptual, considerations will lead us to recognize something in addition to the members.

Under normal circumstances in ordinary perception we cannot help but see one object, even if we always see it from some perspective. We also cannot will it to disappear or to change, even though not seeing the paperweight, globe, and inkwell (or aspects of them) as aspects of one object seems to be rather easy. In fact, we could imagine a group of people who do not have perceptual handicaps or short attention spans and who perceive things such as inkwells, pens, and chairs but who just do not possess the concepts that would enable them to see these things as members of a set. In other words, in order to make Maddy's 'set' disappear or change, I only have to do something mental, such as change my concept. Change concepts appropriately, it seems, and the 'set' disappears. But this is not at all analogous to the case of a genuine perceptual object such as an inkwell or, say, three dots with a circle drawn around them on a blackboard. Light does reflect from these latter objects, and changing my concept or not attending to them will not make them disappear, although some other appropriate causal interaction with them may. And it is precisely this kind of resistance to or constraint on our awareness that has long been taken to be a condition for distinguishing the actual perception or intuition of an object from the mere conceptualizaton of an object. On Maddy's account the degree of plasticity associated with the number of sets we can take to be in one location – and she thinks that there are many – suggests that we are on the conceptualization side of this distinction. All of this just bears out the view that Maddy's analogy between the perception of ordinary physical objects and the perception of sets fails, just as one would expect if sets are not objects in space and time.

Maddy contends that a numerical belief such as "There are three eggs in the carton" is also perceptual and that this perceptual belief is about a set. She points out, following Frege, that the belief could not be about the (non-set-theoretical) physical stuff in the carton because that has no determinate number property. There are three eggs, but many more molecules, even more atoms, and so forth, and there is no predetermined way it must be divided up. But then what is the subject of the number property? There are several candidates to be the bearer of the number property: a set, an aggregate, a concept, and a class, among other things. She argues that the properties that separate these kinds of objects are not perceptual but are theoretical, for example, extensionality. The same applies to the properties that we use to distinguish physical objects from one another. So nailing down the number bearer's more esoteric properties is a theoretical matter. To decide the case between these possible bearers, therefore, we need to look not to our perceptual experiences but to our overall theory of the world, and to ask which of these candidates is best suited to playing the role of the most fundamental mathematical entity. Her answer is that sets are best suited, on the grounds of the indispensability arguments mentioned. Thus, the perceptual, numerical belief here is a belief about a set, not one of the other things. She notes that this conclusion is contingent on the fact that our best scientific theory seems to demand that sets exist. This is another place in the argument in which one senses a kind of relativism that one might not expect in mathematics. And one might wonder how the corresponding epistemological account would have to change should new or alternative mathematical objects be required by natural science. Maddy says more about the claim that the numerical belief is perceptual in her next chapter.

Chapter 3, "Numbers," opens with the statement of a basic ontological challenge to platonism, again formulated by Benacerraf. The challenge emerges against the background of the effort, starting with Frege, to define natural numbers in terms of sets. The challenge is put this way: if numbers are sets, they must be some particular sets, but any choice of particular sets will exhibit properties that go beyond what ordinary arithmetic tells us about the numbers. There are no arguments that one of these choices is the correct one. Therefore, numbers are not sets. In fact, numbers are not objects at all, because any objects we identify them with will have additional properties which are superfluous to the object's numerical functioning. Again, there are no arguments that one of these choices is the correct one. Therefore, numbers are not objects at all. But then what is the nature of the ontological relationship between numbers and sets?

Maddy accepts the conclusions that numbers are not sets and are not objects. To continue her effort to naturalize mathematics, however, she wishes to locate numbers in space-time. She holds that numbers are properties of sets, analogous to physical properties, in particular, to physical quantities. Just as the perception of physical objects includes the perception of their properties, so the perception of sets does. She argues that knowledge of numbers is knowledge of sets because numbers are properties of sets. Moreover, knowledge of sets presupposes knowledge of number since, for example, Piaget's work indicates that subset relations cannot be properly perceived before number properties. Arithmetic, from this perspective, is part of the theory of hereditarily finite sets. The problem of multiple reductions of numbers to sets in set theory is now to be viewed in the following way: consider the von Neumann ordinals. They are nothing more than a measuring rod against which sets are compared for numerical size. So the choice between the von Neumann ordinals and the Zermelo ordinals, for example, is no more than the choice between two different rulers that both measure in meters, and asking whether a number is one of these kinds of sets is like asking whether an inch is wooden or metal.

Maddy discusses objections to the identification of natural numbers with properties of sets raised by Frege and Benacerraf and argues that the effort to construct a neo-Benacerrafian dilemma for property theories – which properties are really the numbers? – fails. Her argument, which I do not find convincing, rests on the view that numbers are to be understood in a particular way as scientific properties, not as sets or predicates. She leaves open the question whether properties (as universals) should be included in the set-theoretic realist's ontology, pointing out that the question of the ontology of properties is just as pressing for physical science as it is for mathematics. Further discussion of a Fregean proposal for what numbers are leads to some interesting comments on proper classes and the nature of the distinction between sets and classes.

At the beginning of Chapter 4, "Axioms," Maddy notes that her view of the perception of sets gives us only the barest beginning of an account of set-theoretic knowledge. How do we get from knowledge of particular facts about particular sets of physical objects to knowledge of the simplest axioms of set theory, and from there to knowledge of more complex axioms? How do we come to know, for example, that *any* two objects can be collected into a set with exactly those two members, as is proclaimed in the Pairing axiom? Such *general* beliefs underlie even the simplest axioms, such as Pairing and Union. Maddy's view is that particular observations provide support for such general beliefs in a way that is quite different from the way in which general empirical claims are supported by enumerative induction. The particular observations here correspond to primitive general beliefs that we have about physical objects that are not supported by simple enumerative induction, e.g., the child's beliefs that physical objects exist independently of the human viewer, and that objects are independent of their state of motion. Maddy thinks that the Hebbian model can accommodate such general beliefs in the case of physical objects and so the same should presumably apply to sets. This could account for at least our most primitive general beliefs. Maddy calls such beliefs 'intuitive beliefs'. She holds that only the very simplest axioms, such as Pairing and Union, are directly supported by such intuitive beliefs. So, in contrast with Gödel's view, only these axioms evidently force themselves upon us as explanations of the intuitive data.

This is another place in Maddy's argument where skepticism might set in. It is not clear that the Hebbian model explains these kinds of primitive general beliefs about physical objects, much less about sets. The Hebbian model offers an account of generalization and of learning in which the connections between neurons required for the development of cell-assemblies are strengthened as a function of the frequency with which the neurons are synchronously activated. Thus, it offers what is basically an inductive, frequency-relative account of generalization and learning. For a large range of developmental tasks, including those wanted by Maddy, it is really just a new form of associationism. But there has always been a gap between what such frequency-relative models can account for and the kind of productivity, systematicity, compositionality, and inferential coherence that we find in higher forms of cognition such as those involved in set theory. It has been argued by many philosophers that the only way to explain productivity, systematicity, and so forth, and hence to have an account of generalization and learning that is not frequency-relative, but in which certain kinds of intrinsic relations obtain, is to hold that there are structured mental contents and structure-sensitive mental processes. Maddy has nothing like this in her account, however, and does not consider the arguments that seem to force such a conclusion. In the same vein, she is silent on the matter of the intentionality of perception and belief and on the difficult problems, discussed by Quine, Putnam, and others, that stand in the way of naturalizing intentionality.

Maddy holds that whereas axioms like Pairing and Union are supported intuitively, the other axioms of set theory must be treated differently. She focuses on the history of the axiom of Choice in particular and analyzes it in some detail, arguing that it is supported in part by intuitive evidence and in part by extrinsic means. It is not clear, however, how the account of the intuitive support for Choice described in this chapter jibes with the details of her own view of intuition discussed earlier, for here she simply cites Zermelo's remarks on the way the axiom was used by mathematicians before it was explicitly formulated. She takes the axiom of Infinity ('there is an infinite set') to be a bold and revolutionary hypothesis. There is nothing intuitive about it, but it launched modern mathematics, and the success and fruitfulness of that endeavor provides its purely theoretical justification. (But was there not some sort of axiom of infinity in ancient Greek mathematics, for example, in the idea that a line is infinitely divisible?)

Maddy finds that a variety of extrinsic supports are in fact offered by set theorists: appeals to 'verifiable' consequences, to powerful new methods for solving preexisting open problems, to simplification and systematization of theory, to implication of previous conjectures, to the implication of 'natural' results, to strong intertheoretic connections, and to provision of new insight into old theorems. She also discusses how open problems in set theory, such as the continuum hypothesis, should be understood from this perspective and finds, interestingly, that in the second tier there are some arguments for new axioms that do not depend on consequences, but also cannot be based on intuition. The ideas they rest on, which she calls 'rules of thumb', cannot be easily classified as either intuitive or extrinsic.

The line of argument involving the status of the empty set, unit sets, and all those sets whose members cannot be construed as physical objects is picked up in Chapter 5, "Monism and Beyond." In this final chapter Maddy responds to a number of objections that have been raised to her views by Charles Chihara and tries to show that set-theoretic realism is consistent with a physicalistic ontology. Here she says that anyone unafflicted by physicalistic scruples is free to hold that we gain knowledge of pure sets by theoretical inference from our elementary perceptual and intuitive knowledge of impure sets. Why or how this should be so she does not say. In any case, physicalists will complain that this does not solve the problem of spatiotemporal location, and so Maddy argues that pure sets are not really needed. She says that the set-theoretic realist who wishes to embrace physicalism can locate all the sets she needs in space and time. All pure sets can be eliminated. Moreover, she argues that physical objects can be identified with their singletons, so problems about unit sets evaporate. Maddy says that not every set-theoretic realist in her sense must be a physicalist and that nonphysicalists may prefer to retain the 'standard' version of set-theoretic realism with its pure sets. But the identification of physical objects with their singletons and the elimination of pure sets yield a version of metaphysical monism in place of the customary dualism of the mathematical and the physical. On this monistic view, every physical thing is already mathematical and every mathematical thing is based in the physical. This view is offered as an option to those for whom it holds some appeal.

In the final sections of Chapter 5 Maddy compares the monistic (not the standard) position to Hartry Field's nominalism and finds monism to be, on the whole, a better position. Both forms of set-theoretic realism are also compared to and contrasted with the structuralist views of Michael Resnik and Stewart Shapiro.

So we see in this final chapter that in order to solve the problems at hand there is a splitting of set-theoretic realism. The split, however, has the effect of straining if not unraveling the compromise platonism that Maddy wants, for now there appear to be dualistic and monistic forms of compromise Platonism. The monistic version is supposed to solve problems about pure sets and unit sets, but in so doing it is subject to some of the same objections that compromise platonism was meant to avoid. For example, it is not clear that *eliminating* pure sets, or *requiring* that physicial objects be identified with their singletons, is consistent with taking set theory at face value, as compromise platonism was supposed to do. If we do take set theory at face value, then there seem to be some very simple and obvious truths into which only pure sets figure. How is the obviousness of these simple truths to be explained by the monistic version? This is a variant of the obviousness problem raised earlier for the Quinean view. Moreover, it is not clear how the monistic version recognizes a uniquely mathematical form of evidence for (unapplied parts of) set theory.

The dualistic version, on Maddy's own terms, is apparently not consistent with physicalism. Otherwise, why offer the alternative? Perhaps this is not a serious problem, but then how the dualistic version can handle problems about pure sets and unit sets is still not clear. Maddy says very little about how we could gain knowledge of pure sets by theoretical inference from our elementary perceptual and intuitive knowledge of impure sets. And it again appears that there will be an 'obviousness' problem for simple truths involving pure sets since Maddy will now have to sort these truths into the theoretical (versus intuitive) side of her epistemological scheme.

Another problem for the compromise results from the two-tiered epistemological structure. How the two tiers are related is not clear. The first tier is supposed to have perception of sets. The second tier does not have perception of sets but, rather, extrinsic justification of the Quinean type. The second tier is supposed to handle all of the sets that are routinely taken to exist in set theory but are not perceivable (that is, the theoretical part of mathematics). (This suggests, incidentally, that the meaning of the existential quantifier in mathematics is not univocal but must be understood differently, depending on whether it occurs in a statement that is intuitively supported or extrinsically supported.) Now it appears that even if we allow that there is sensory perception of sets, this perception could not be a necessary condition for mathematical knowledge in the second tier. To see this consider, for example, Maddy's account of the difference between the axioms of Pairing and Infinity. The analysis of the epistemological basis of these axioms mentioned earlier suggests that the two tiers are in fact logically independent of one another. It could be argued that there is really a very radical division here. That is why Gödel focuses on intuitive, intrinsically mathematical evidence involving abstract objects even at higher levels of set theory and introduces extrinsic criteria as an *alternative*. On the other hand, if you recognize extrinsic criteria, then, as in Quine's view, why not apply them uniformly to all of the axioms? What could mathematical intuition possibly add? In short, it is not clear that 'compromise' platonism is a coherent position in the philosophy of mathematics.

Apart from these issues, *Realism in Mathematics* prompts a number of general questions about why and even whether we would want mathematics to be naturalized. The claim that there is no point of view prior to or superior to that of natural science is no doubt persuasive as part of a general cultural perspective, that is, as a result of the great success of natural science and of the technologies based on it. Nonetheless, there are some well-known problems with the claim. Consider, for example, the following questions: how we are to conceive of science? and how we are to view the relationships between the various sciences (e.g., neuroscience and psychology)? It is not clear that these are questions that could or should be answered by natural science. To attempt to naturalize mathematics is perhaps inevitably to be a kind of reductionist about mathematics, and this can have a number of negative consequences. It could actually hinder *mathematical* progress at the expense of some philosophical theory. I think that Frege and Husserl had this concern in mind when they criticized the prevalent form of naturalism of their time, psychologism, and that Gödel also had it clearly in mind when he criticized various forms of empiricism and positivism about mathematics. (See also his remarks on why the completeness theorem for first-order logic was not discovered earlier.) Focusing too heavily on inductive criteria or extrinsic justification for the truth of mathematical statements, for example, might blind us to interesting or important results that could be established intrinsically (e.g., on the basis of mathematical intuition, as in Gödel's view). Or trying to naturalize mathematics could lead to an exclusionist attitude about forms of reasoning in mathematics that appear to resist such treatment. There may be ways to naturalize mathematics, however, that are not subject to these problems.

10

Penrose on Minds and Machines

In *Shadows of the Mind: A Search for the Missing Science of Consciousness*
(*SM*) (Penrose 1994), Roger Penrose continues to develop some central
themes of his earlier book, *The Emperor's New Mind* (*ENM*) (Penrose 1989),
but he also strikes off in some entirely new directions. Penrose argues in
ENM that human cognition cannot in principle be fully understood in
terms of computation. The argument, following an earlier argument of
J. R. Lucas (Lucas 1961), was based on Gödel's incompleteness theorem.
Penrose returns to this material in Part I of *SM* and develops it at great
length in response to the many criticisms of the argument in *ENM*. In
Part II of *SM*, Penrose opens a new line of inquiry which is motivated
by the Conclusion of Part I. Since he thinks consciousness is a function
of the brain but cannot be fully understood in terms of computation,
he asks how the brain can perform the needed noncomputational actions. He is not willing to forgo a scientific account of consciousness,
and to develop such an account we must look to neuroscience. But how
might noncomputational actions arise within scientifically comprehensi-
ble physical laws? What physical principles might the brain use? Penrose
is driven to the conclusion that these principles must be subtle and largely
unknown. He suggests in Part II that neuron signals in the brain may be-
have as classically determinate events, but synaptic connections between
neurons are controlled at a deeper level where there is physical activity at the quantum-classical borderline. Specifically, he proposes that in
consciousness some kind of global quantum state must take place across
large areas of the brain, and that these collective quantum effects are most
likely to reside within the microtubules in the cytoskeletons of neurons.

The central ideas in each part of the book are quite interesting, but they are also highly controversial. I will focus on the material on Gödel's theorems in Part I, but I will also comment on some of the assumptions that lead to Part II. There are alternatives to Penrose's attempt to reconcile a neuroscientific account of consciousness with an antimechanistic view that are simply not explored in *SM*.

In Part I, Penrose sets out four main positions on computation and cognition (p. 12): (A) All thinking is computation; in particular, feelings of conscious awareness are evoked merely by carrying out the appropriate computations; (B) awareness is a feature of the brain's physical action, and whereas any physical action can be simulated computationally, computational simulation cannot by itself evoke awareness; (C) appropriate physical action of the brain evokes awareness, but this physical action cannot be properly simulated computationally; (D) awareness cannot be explained in physical, computational, or any other scientific terms. Penrose does not want to abandon science, and so he rejects D. D must involve some kind of antiscientific mysticism. He thinks that C is closest to the truth. Not all physical action can be properly simulated computationally. Penrose argues that A must be rejected on the basis of Gödel's theorem.

The argument from Gödel's theorem takes up much of Part I. The form in which Penrose presents Gödel's theorem bears a strong resemblance to a generalized form of the incompleteness theorem in Kleene's *Introduction to Metamathematics* (Kleene 1952, which is cited by Penrose at various points in *SM*). Kleene's theorem XIII (p. 302) uses a very general notion of a formal system F, where the main condition on F is that its set of 'provable formulas' be effectively enumerable. Suppose that F contains effectively given 'formulas' $\phi(\mathbf{q}, \mathbf{n})$ which are supposed to express the predicate $P(q, n)$ which holds just in case $C_q(n)$ does not halt. F is said to be *sound* for P if whenever F proves $\phi(\mathbf{q}, \mathbf{n})$ then $P(q, n)$ holds. F is *complete* for P if the converse is true. Kleene's theorem (slightly weakened) is that if F is a formal system which is sound for the predicate P, then it is not complete for it. Specifically, there is a k such that $C_k(k)$ does not halt but F does not prove $\phi(\mathbf{k}, \mathbf{k})$.

Penrose states his version of the theorem directly in terms of Turing machine computations (pp. 74–75). For the Penrose-Turing version, suppose that A is a Turing machine which is such that whenever A halts on an input (q, n) then $C_q(n)$ does not halt. Then for some k, $C_k(k)$ does not halt, but A does not halt on (k, k). As Penrose puts it, the soundness of A implies the incompleteness of A. The proof of the theorem in *SM* is a variant of the standard diagonal argument for showing that the halting problem for Turing machines is not effectively decidable. Assuming Church's thesis, the Penrose-Turing theorem and the Kleene theorem are interderivable.

There are a number of technical flaws in Penrose's subsequent discussion of the relation of the Penrose-Turing theorem to the usual presentation of the incompleteness theorems. These flaws have already been clearly described in some detail in Solomon Feferman's review of *SM* (Feferman 1996), and I will not discuss them here. It will be useful, however, to say a few things about the relationship in order to set the stage for some of the critical remarks later.

In response to a key objection (Q10, pp. 95–97), Penrose notes the special role that Π_1 sentences play in connection with Gödel's theorems. Π_1 sentences are of the form $(\forall y)Ry$, where R expresses an effectively decidable ($=$ recursive) property of the natural numbers and the intended range of 'y' is the set of natural numbers. We can also consider Π_1 sentences with free variables, for example, $(\forall x)Rxy$. Gödel constructs a Π_1 sentence $G(F)$ equivalent (in PA) to $(\forall x)\neg \operatorname{Proof}_F(x, \mathbf{g})$, where g is the Gödel number of $G(F)$. $G(F)$ thus provably expresses of itself that it is not provable in F. If h is the Gödel number of the formula $0 = 1$, the Π_1 formula $Con(F): = (\forall x)\neg \operatorname{Proof}_F(x, \mathbf{h})$ expresses that F is consistent. Then Gödel's first theorem says that if F is consistent, then $G(F)$ is not provable in F. It is possible to show that if F is sufficiently strong, then F is consistent iff F is sound for Π_1 sentences. It follows that the hypothesis in the first theorem is equivalent to the soundness of F for Π_1 sentences. Under this hypothesis $G(F)$ is true. We therefore obtain a version of the first theorem in the form of the Penrose-Turing theorem: if F is sound for Π_1 sentences, then F is not complete for them. Gödel's second theorem says that if F is consistent, then F does not prove $Con(F)$: that is, it does not prove its own consistency. One can get from the first to the second theorem by showing that F proves $Con(F) \leftrightarrow G(F)$.

Penrose bases his antimechanist argument on the Penrose-Turing version of the first incompleteness theorem. As does Penrose, I count myself among those who think the incompleteness theorems are not irrelevant to mechanistic views of the mind. The problem is to state precisely how they are relevant. Penrose, as does Lucas, thinks they are relevant in the following way: according to the Penrose-Turing theorem, A is incapable of ascertaining that $C_k(k)$ does not stop even though it does not. From the knowledge of A and its soundness we can actually construct a computation $C_k(k)$ that *we* can see does not ever stop. Thus, A cannot be a formalization of the procedures available to mathematicians for ascertaining that computations do not stop, no matter what A is (p. 75). *We know something* that A is unable to ascertain, and A cannot encapsulate our understanding. The upshot is expressed by Penrose in his thesis **G** (p. 76): human mathematicians are not using a knowably sound algorithm to ascertain mathematical truth.

Penrose presents and responds to twenty objections to this argument. He also shows an awareness of and is responsive to a wide variety of other issues. Among the objections considered are the following: why could not the algorithm A be one that continually changes, as is the case in learning (Q2)? Why is a human being able to do better than any A, not just some particular A (Q5)? Why could not a computer be programmed to follow through precisely the argument Penrose has given (Q6)? Is it not the case that nonterminating computations are idealized mathematical constructions and, as such, are not relevant to real computers or brains (Q8)? Why could we not repeatedly adjoin the Gödel sentence G(F) to whatever system F we currently accept and allow this procedure to carry on indefinitely (Q19)?

The quality of Penrose's responses to the objections is uneven. Some responses are adequate, but others are deficient on technical or philosophical grounds. Some of the responses simply do not cohere with one another or with other claims in the book. One has the sense that Penrose is not arguing from some principled alternative position. Let me give an example of what I mean.

One of the important objections concerning the notion of truth in **G** is stated as follows (Q10, p. 95; see also Q11): "Is mathematical truth an absolute matter?" "Can we trust arguments that depend upon having some vague concept of 'mathematical truth', as opposed to a clearly defined concept of formal 'truth'?" Penrose replies that such seeming uncertainties about the absolute nature of mathematical truth do not affect his argument because we are concerned with a class of mathematical problems of a much more limited nature than those which, as does the axiom of choice or the continuum problem, relate to nonconstructively infinite sets. We are only concerned with statements of the form 'computation x does not ever terminate', that is, with Π_1 sentences. He holds that there is little reason to doubt that the true/false nature of any Π_1 sentence is an absolute matter. This concept of mathematical truth is actually no less well defined than the concepts of 'truth', 'falsity', and 'undecidability' for any formal system F. Penrose claims that only this limited concept of truth is needed for his argument.

When read in the context of other claims in *SM*, however, this response is puzzling. Penrose professes in some passages to be a platonist about mathematical truths and objects, but Π_1 truth does not need to be construed as platonistic. A Π_1 truth predicate is definable in intuitionistic arithmetic. In any case, the appeal to the truth of Π_1 sentences here does not cohere with what Penrose says elsewhere. On p. 91 he says that if a formal system is sound, then it is certainly ω-consistent. This is a different notion of soundness from that discussed earlier, since ω-consistency is stronger than consistency for Π_1 sentences. At this point he seems to have in mind soundness for all arithmetical sentences, as one might expect in a platonistic antimechanist argument based on Gödel's theorems. In the discussion of Q18 (p. 112), Penrose allows that "we cannot properly encapsulate 'soundness' or 'truth' within any formal system – as follows from a famous theorem of Tarski." It therefore appears that soundness is to be taken as the truth of all sentences. Or is he thinking of all arithmetical sentences? By the time the reader reaches the discussion of platonism toward the end of the book Penrose seems to be claiming that mathematical truth *is* an absolute matter. He says (p. 418) that Gödel's proof of the incompleteness theorems argues "powerfully for the very existence of the Platonic mathematical world. Mathematical truth is not determined arbitrarily by the rules of some 'man-made' formal system, but has an absolute nature, and lies beyond any such system of specifiable rules." Thus, what exactly is the response to Q10? And how should we understand **G**?

One might expect Penrose, as a platonist, consistently to emphasize a global notion of truth in his antimechanist argument. Otherwise, why not try to provide an antimechanist argument on intuitionistic or constructivist grounds? The problem is that it is never very clear how we should understand Penrose's platonism and its role in the antimechanist argument. This may be a problem that Penrose inherits from the approach of Lucas, for platonism also played no role in Lucas' antimechanist argument.

It is instructive to compare the Penrose-Lucas argument briefly with Gödel's antimechanistic remarks. Gödel's remarks are typically made in connection with the second incompleteness theorem and with the question of what will be needed for consistency proofs for mathematics. It is odd that neither Lucas nor Penrose places much emphasis on the second theorem in conjunction with the existence of consistency proofs for PA and for other more extensive systems. Penrose does consider (Q15, p. 107) what has become a well-known objection about the role that consistency plays in antimechanist arguments derived from Gödel's theorems. The objection, as he puts it, is this: the formal system F that we choose to use might not actually be consistent. At least we might not be certain that it is consistent. By what right can we therefore assert that G(F) is 'obviously' true? A response to this objection, which neither Lucas nor Penrose offers, is that we in fact do have (relative) consistency proofs for PA and other systems. These results have been accepted into the body of mathematics and they are holistically connected to it. It can be argued that even if we do not have a consistency proof for any choice of F, the existence of the consistency proofs we do have constitutes evidence (which is not to be equated with 'certainty') for the claim that we have insight into objects, concepts, or truths that are abstract and nonfinitary relative to formalism or mechanism. This appears to be one element of Gödel's view.

The argument is simply that it follows from the second theorem, applied to PA, that the concepts or truths needed for a consistency proof for PA are not available in PA. One must step outside PA. To suppose that the concepts or truths that can be adequately represented in PA are concrete, finitary, immediately available to sensory intuition, and so on, is already rather liberal. There is reason to believe that we should say this only for some subsystem of PA. It follows that the concepts or truths involved in the consistency proofs for PA will not possess these properties. As Gödel states the argument, we will need 'higher level' concepts, that is, concepts that are not about concrete objects in space-time, such as finite combinations of symbols. We will need concepts that are about abstract 'thought structures' or 'meanings'. It will be necessary to go beyond reflection on concrete signs and their properties and include reflection on meanings or intentions. From this point of view, which could be either platonistic or intuitionistic, the problem for mechanism is how we could be using a *mechanical* procedure to grasp these *abstract* and nonfinitary concepts or truths. This formulation should be compared with thesis G. I will refer to this line of reasoning as 'the argument from abstract objects'. It can be elaborated in many directions.

The fact that every theorem-generating machine can be recast as a formal system, and vice versa, helps to make this philosophical point about mechanism clearer. In Hilbert's original program, formal systems were supposed to have a foundational advantage by virtue of the fact that they involved only concrete, finitary objects in space-time that were available to immediate perception. Our activities with them would be directly verifiable in sense perception, much as the activities of participants in the Turing test would be directly verifiable in sense perception. This is what is 'real'. We would not need to invoke intended interpretations or 'meanings' of the symbols of mathematical theories. Nor would we need to invoke objects that were 'abstract' or nonfinitary but nonetheless constructively admissible. Interestingly, it is these same features that recent mechanists still highlight. Mechanists still think in many ways as pre-Gödelian practitioners of proof theory. They want to show, for example, that our talk about 'intended interpretations' or 'meanings' is really just talk in a high-level programming language that is reducible to and adequately captured in a concrete, low-level machine language. The language of 'intended meanings', 'abstract objects', and so on, should, in effect, be conservative over the language of concrete particulars. But Gödel's second theorem, when coupled with some other observations, suggests that we cannot ignore intended interpretations of theories by way of this latter-day incarnation of the formalist program. It is notable that there have been developments in philosophical psychology (e.g., in the recent dual-aspect theories mentioned later) that are independent of reflections on Gödel's theorems and that also suggest we should not expect such tight language-level reductions.

If mathematical concepts or truths are abstract or nonfinitary relative to mechanism, then, by the very conception of what a mechanical procedure is, they simply cannot be input for a mechanical procedure. There is a gap, however, between this claim and the claim that they are not available to a finite human procedure that exhibits intentionality. It is somewhat surprising that Penrose does not note or use the fact that consciousness exhibits intentionality. (This is also true of Lucas.) As I read Gödel, he claims that what a machine cannot imitate is the activity in which more and more new axioms become evident to us on the basis of the abstract meanings (or intentions) of primitive mathematical terms. If we start with PA, for example, then we can think of G(PA) as an axiom that could be added to PA. More and more 'axioms' of this type could be added to the successive formal systems obtained, and one could argue that these axioms are evident to us on the basis of our understanding of the concept of the natural numbers. That is, we are *directed* through this sequence on the basis of our understanding of this concept. We could not proceed strictly on the basis of the given formal system because the Gödel sentence obtained for that system is formally independent of the system. In effect, we refer through PA to what is not presently enclosed in PA, and that means that PA did not adequately capture our arithmetic intentions in the first place. The sentences obtained are nonetheless all arithmetical truths if PA is consistent. We develop a consistent extension of our understanding of the natural number concept through these sentences. Although this particular development is not especially fruitful for mathematics itself, Gödel has in mind a similar model of how new axioms of infinity in set theory become evident to us on the basis of our (admittedly imperfect) understanding of the concept of set.

In response to Q15, we could say it is true that we may not be certain of the consistency of PA. But why should this be a problem for the claim that we have some insight into abstract and nonfinitary objects? This epistemic situation in the domain of mathematical experience is arguably analogous to the role of sense perception in sense experience. We cannot actually attain certainty in either case except perhaps in special circumstances. There are degrees of evidence and clarity, and nothing about the situation precludes a form of fallibilism. We nonetheless *expect* to have a consistent insight into domains of objects of different types. This in fact seems to be a condition of scientific rationality. If it were otherwise, science in general and mathematics in particular would not be possible. In the domain of mathematics, however, the objects simply cannot be understood as concrete.

It is possible to be extremely skeptical about mathematical knowledge for fear of inconsistencies. This will lead one to believe that very few mathematical propositions, if any, are true. Skepticism of this sort is disabling and is often applied in a lopsided manner to mathematics alone. There are of course known antinomies or inconsistencies in the foundations of mathematics. It seems that Gödel is correct in saying that one ought not to use the antinomies as a pretext for an extreme empiricist skepticism about mathematical knowledge. It can be argued that although these known antinomies do mark limits on mathematical reason, there is still a gap between these limits and the limits of formal systems or machines. In other words, we cannot identify the limits of formalism-mechanism with the limits of human reason. The science of mathematics falls between the limits marked by the antinomies and a disabling empiricist skepticism about mathematics. Within this range we can grant that it is less clear or obvious that ZF is consistent, and that is not, for example, to say that the propositions of ZF are meaningless. But the consistency of PA seems to be as 'obvious' as much of what one finds in mathematics, if not more so.

It is often claimed that the Lucas-Penrose argument depends on the assumption that we are mathematically omniscient, in the sense that we are able to see the truth of any mathematically true sentence. The argument from abstract objects can avoid the unwanted implications of this claim.

Suppose that mathematical reason exhibits intentionality, is finite, is fallible, and is directed by its contents or 'meanings' to abstract objects which it grasps only partially at a stage of research. It follows that we are not mathematically omniscient at that stage. But this same view stands as an alternative to a mechanistic conception of reason, as was indicated previously. In the presence of the incompleteness theorems we see that mathematical reason refers through a formal system or machine for mathematics to what is not presently enclosed in the formal system or machine, even if this referring is indeterminate in some respects. Gödel calls for a kind of nonreductive, descriptive clarification of the meaning of primitive mathematical terms because reference is a function of meaning and reference may be indeterminate in some respects.

One might argue, as Gödel evidently does, that human reason is able to relate concrete particulars to concepts, and concepts to concepts, in a finitary, meaningful, but partial manner in which not everything needs to be determinate or perfectly clear at a given stage of research. If we cannot collapse the appropriate distinction between concepts and particulars, then the process of subsuming a particular under a concept, of abstracting a concept from a particular, or of relating concepts to one another cannot be mechanical. Machines, by way of contrast, relate concrete particulars to concrete particulars in a finite, mechanical way. Mathematics may in fact be inexhaustible, but that one should identify mechanical undecidability with undecidability by reason would not follow. Furthermore, to hold that there are absolutely undecidable but clearly posed mathematical problems, on the basis of this identification, would then appear as an unwarranted form of pessimism about problem solving in mathematics.

The argument from abstract and nonfinitary objects does seem to leave us with the problem of how the human mind could know about such objects. Antimechanism, however, may not require the abstract objects or truths of classical or naive platonism. It is known, for example, that objects such as intuitionistic proofs are 'abstract' in the sense that they cannot be encoded in the natural numbers. One option that has not been much explored is that the objects in question might be understood as intentional.

Penrose launches Part II of *SM* on the assumption that the noncomputational activity of consciousness requires noncomputational brain activity and, therefore, that some kind of noncomputational physics is required. Although they are highly speculative, Penrose's suggestions about this activity may be of some use in entertaining the possibility of noncomputational processes in neuroscience. After all, it is very difficult to predict what our best physical and biological theories will look like in one hundred or two hundred years. One might emphasize just how primitive our scientific understanding of consciousness is at this time. Some have argued that it is comparable to Neanderthal man's understanding of the theory of relativity. However that may be, it seems that Penrose's suggestions could not possibly help with his own alleged platonism. A conception of physics according to which the brain grasps the abstract, acausal objects recognized to exist in platonism must indeed be *extremely* sublime. In fact, are we not faced with a contradiction here? For Penrose, we need a radically new conception of physics to see how the brain could know about abstract objects. On the other hand, there are philosophical views on which we require a radically new conception of mathematics to see how this is possible. Both views seem to me to be wrong.

Gödel sometimes appears to be attracted to a third possible position. The mind, but not the brain, could have access to abstract objects that are not located in the space-time causal order. Objects that can affect our senses and that are located in the causal order are accessible to the brain. The brain functions basically as a digital computer. Thus, the mind must be distinct from the brain. This view suggests something like a Cartesian or 'substance' dualism. It is a position fraught with many problems. Or it may be that Gödel thinks there is a kind of Leibnizian preestablished harmony between these different domains.

Yet another possibility can be found in dual-aspect theories of mental phenomena. Earlier theories of this type are found in Spinoza and Kant. Recent theorists include Donald Davidson, Thomas Nagel, and P. F. Strawson. Davidson, for example, argues for anomalous monism, which he distinguishes from nomological monism (e.g., materialism, Penrose?), nomological dualism (e.g., Leibniz), and anomalous dualism (e.g., Descartes). Recent dual-aspect theories often hold that consciousness exhibits intentionality and that the language of intentionality or of reason cannot be reduced to the language of neuroscience or physics, even if there is only one kind of thing in the universe. Dual-aspect theorists deny the existence of psychophysical laws. They claim that there are irreducibly mental ways of grouping phenomena. One must presumably also apply this claim to our mental activities in mathematics. A dual-aspect theory might provide a way to maintain that some activities of mathematical consciousness are noncomputational without having to invoke a subtle physics that is supposed to explain how the brain could grasp abstract objects.

PART III

CONSTRUCTIVISM, FULFILLABLE INTENTIONS, AND ORIGINS

11

Intuitionism, Meaning Theory, and Cognition

§ 1 Introduction

In intuitionism there is a fertile confluence of ideas on mathematics, meaning theory, and cognition. Philosophers interested in any one of these latter areas of research would profit from studying intuitionism. In this chapter I want to focus on several connections between intuitionism and some recent, post-Wittgensteinian views in the philosophy of mind, meaning, and language. The views I shall focus on are associated with the claim that human cognition exhibits intentionality and with related ideas in philosophical psychology. This tradition in the philosophy of mind, meaning, and language differs significantly from the tradition in which Michael Dummett has interpreted and expounded upon intuitionism.[1] Dummett has said that he is not attempting to portray accurately the views of Brouwer and Heyting, and more than one commentator has noted that Dummett's view of intuitionism diverges widely in some respects from 'traditional' intuitionism. The manner in which it diverges seems to be primarily in its view of human cognition and, specifically, in its view of

Work on this essay was supported by N.W.O. (Dutch National Science Foundation) research grant 22–266. I thank Dirk van Dalen for sponsoring the grant and for many helpful discussions about intuitionism. I am grateful to Per Martin-Löf and Dag Prawitz for making it possible for me to visit them at the University of Stockholm and for discussing with me some of the ideas in the chapter. Thanks are also due to Anne Troelstra for discussing some of the material and for making available some of Heyting's unpublished correspondence. It should not be assumed that any of these people would endorse the arguments expressed in this essay. Finally, I thank an anonymous referee for *History and Philosophy of Logic* for some helpful comments.

[1] See especially Michael Dummett 1977, 1978, 1991, and 1993.

the key intuitionistic idea that mathematical constructions are mental processes or objects. Dummett has been influenced by Wittgensteinian views on language, meaning, and cognition, whereas I shall be arguing against the application of some of these views to intuitionism and pointing to some shortcomings in Dummett's approach. I do not claim to have the final word on what intuitionistic mental constructions are, but I think I arrive at a view on this matter that is closer than Dummett's to the common intentions of Brouwer and Heyting. In any case, it is a view that should be of interest in its own right.

I begin with a brief description of some central themes of intuitionism. I describe the themes in a way that is neutral with respect to the issues about meaning and cognition I shall discuss later in the chapter.

§ 2   Seven Theses of Intuitionism

Intuitionism is a form of constructivism about mathematics and logic that has its origins in the work of L. E. J. Brouwer and A. Heyting.[2] Its name is derived from the fact that Brouwer held that an a priori form of temporal intuition was the foundation of mathematics. Brouwer says this view is influenced by Kant. Intuitionism is of course different from a platonistic or classical viewpoint. Also important is distinguishing intuitionism both mathematically and philosophically from other forms of constructivism, such as finitism, Bishop's constructive mathematics, constructive recursive mathematics (CRM) in the style of the Markov school, and predicativism. Other types of constructivism have not, for example, made a point of the claim that constructions must be understood as *mental* processes or objects and that an a priori temporal form of intuition is the foundation of mathematics. The fact that intuitionism has made a point of these ideas is reflected in the resulting logic and mathematics. For intuitionists, constructions are not, for example, finite configurations of signs on a piece of paper. They are also not to be identified with Turing computations. Instead, they are in the first instance forms of consciousness or possible experience. I shall say more about this later.

Among the distinctive philosophical theses that have continued to play an important role in the development of intuitionism are the following:

(i)   Mathematics is the most exact part of our thinking. It is concerned with the mental constructions (= proofs) of the mathematician,

---

[2] Readers who are not well acquainted with traditional intuitionism will find central sources of information in Brouwer 1975 and Heyting 1931 and 1971. An extensive survey of results that contains some philosophical discussion is Troelstra and van Dalen 1988.

who should be thought of as 'idealized' in the sense that he or she has an unbounded memory and perfect recall.

(ii) The constructive meaning of a mathematical statement is to be explained by giving conditions under which one has a proof of the statement. This appears very clearly in the Brouwer-Heyting-Kolmogorov (BHK) interpretation of the logical constants. For example, a proof of $A \to B$ is a construction which permits us to tranform any proof of A into a proof of B, and a proof of $(\exists x)Ax$ is given by constructing a $b \in D$ and a proof of Ab.

(iii) Platonism or realism about mathematics is unfounded. There is reason to be skeptical about classical mathematics. Human beings cannot construct just anything, for constructions are carried out in time. There is no reason to expect that a human mathematician, idealized appropriately, could construct some of the infinite totalities or structures that are taken to exist in parts of classical mathematics. The principle of the excluded middle, $A \vee \neg A$, can therefore not necessarily be expected to hold if A is a statement about such a totality. A proof of $A \vee \neg A$ would in fact have to yield a general method for the solution of all mathematical problems. It is a tenet of intuitionism that no objects or truths exist independently of us or of our knowledge.

(iv) What we can do is to generate lawlike and choice sequences of constructible objects, where this process is not supposed to be completable in all cases. Natural numbers, for example, are generated by a lawlike process. The ideal mathematician may also construct longer and longer initial segments $\alpha(0), \ldots, \alpha(n)$ of an infinite sequence of natural numbers α, where α is not determined in advance by some fixed process of producing values. α is an example of a choice sequence, and the development of this notion led Brouwer to his continuity principle and to bar induction, yielding a distinctive form of real analysis. It is a basic theorem of intuitionistic real analysis, for example, that all real-valued functions are continuous. This theorem can easily be accommodated in modern topos theory.[3] It has been argued that intuitionism provides a better perspective on the (intuitive) continuum than we find in classical set theory.

(v) Mathematics does not depend on logic. Rather, logic is part of mathematics. Logic is an applied subject. This point can be directed

[3] See, e.g., MacLane and Moerdijk 1995, pp. 267–346.

against logicism and formalism, but it also has other interesting implications.

(vi) Because it is concerned with mental constructions, intuitionism preserves some intensional aspects of mathematics. Functions as rules are, for example, distinguished from functions as graphs.

(vii) Mental constructions cannot be completely captured in formal systems. Insofar as mechanical computation can be characterized in terms of elementary formal systems, mental constructions cannot be completely captured by mechanical computation. They are not to be understood in terms of formal or mechanical manipulations of symbols. Intuitionism is antiformalistic and antimechanistic in this sense.

Since I want to highlight connections between intuitionism and the philosophy of mind, it should be noted, regarding (vi) and (vii), that intuitionists have raised questions about the truth of Church's Thesis (CT). What is questioned is whether all effectively calculable functions are Turing computable. This point is relevant to the philosophy of mind because, for intuitionists, functions are just mental processes or entities. Intuitionists raise doubts about the claim that all intuitionistic number-theoretic functions are Turing computable, as well as about the more general claim that all constructive functions are Turing computable. It is known by Gödel's incompleteness theorems that not all number-theoretic functions can be captured in any one consistent formal system of number theory. Now in the intuitionistic arithmetical formulation of CT the universal quantification over functions cannot be understood in terms of any one formal system of arithmetic, since it is the intuitionistic functions that are at issue. But once a particular function is represented in a particular formal theory, once there is a closed term, it is possible to obtain the recursive function. As long as the expression is open or absolute, however, obtaining the recursive function is not possible. What is involved in the intuitionistic version of CT is a meaningful reference to functions (i.e., mental processes or entities) that, relative to a formal system or machine, are 'abstract' and are not completely captured in a formal system. Dummett has also noted that the assumption that we can effectively recognize a proof of a given statement of a mathematical theory such as elementary number theory lies at the basis of intuitionistic mathematics, but to hold that there is a recursive procedure for recogizing proofs of arithmetical statements runs afoul of the incompleteness theorems (Dummett 1977, p. 264). This is not to maintain that the set of arithmetical statements provable by some intuitionistically correct means is not recursively enumerable, but only to deny that the totality of intuitionistically correct proofs of such statements can be represented by a formal system with a recursive proof predicate.

The matter of accepting CT is an issue on which intuitionism differs from CRM. CRM is, in effect, based on the arithmetic version of CT. According to CRM, the potentially realizable infinite object that is a number-theoretic function must allow representation by an index. In addition, Markov's principle (MP) is rejected in intuitionism but is accepted in CRM. MP says, in effect, that if it is not the case that a Turing machine does not halt on a given input, then it does halt. It has the form $\neg\neg(\exists x)Ax \rightarrow (\exists x)Ax$, where A is an algorithmically decidable predicate. Intuitionists will not allow that we are actually aware of an x such that Ax given only that we know that it is impossible that there is no such x.

§ 3 Dummett's View of Intuitionism and Some Objections to It

Philosophical research on some of these claims, especially (ii) and (iii), has been influenced heavily by the work of Michael Dummett. Dummett has significantly changed the way many people think about intuitionism as a philosophy. Before Dummett there were Brouwer and Heyting. Intuitionistic philosophy after Dummett has been different. Dummett's approach is very much motivated by a form of the 'linguistic turn' that dominated philosophy in the middle part of the twentieth century. Dummett offers a meaning-theoretic argument in favor of intuitionism and against classical or platonistic viewpoints. This meaning-theoretic and linguistic emphasis was virtually nonexistent in intuitionism before Dummett, just as an emphasis on the notion of intuition is now absent in many philosophical discussions of intuitionism.

Dummett points out that on the classical truth-conditional view of meaning, truth is understood in such a way that the principle of bivalence holds, and truth is divorced from the ability to recognize truth. This view is supported by a platonistic or realistic metaphysics according to which all statements have definite truth values by virtue of some mind-independent reality. Dummett, however, wishes to proceed from general conditions about language use and the ability to learn and communicate meaning, rather than from metaphysical views about whether or not there is a mind-independent mathematical reality. He argues that a notion of truth on which the principle of bivalence holds cannot be justified.

The core of his argument is that the meaning of a statement must be construed in terms of what it is for a person to know the meaning. Knowledge of meaning must ultimately be implicit. That is, if we define or elucidate the meaning of an expression by using other expressions, then there must be knowledge of the meaning of these other expressions. In order to avoid an infinite regress, we must finally arrive at expressions for which meaning is implicit. Implicit knowledge can be ascribed to a person only if is it fully manifestable in his or her behavior or practice. This is Dummett's version of Wittgenstein's idea that meaning is determined by use. Intuitionism supposedly satisfies the condition that the user of a language must be able to manifest fully his or her knowledge of meaning because it explains meaning in terms of the constructions (= proofs) that we actually possess. In particular, the BHK or 'proof' interpretation satisfies this condition. (To be more specific, one should think of the meaning of the statements as given by the introduction and elimination rules of an intuitionistic natural deduction system in the style of Gentzen and Prawitz. The introduction rules are said to give 'canonical' proofs. A noncanonical proof would be a method for obtaining a canonical proof.)

The classical or platonistic view does not explain knowledge of meaning in terms of the constructions we possess. It explains knowledge of meaning in terms of knowledge of truth conditions. But then the knowledge of truth conditions of undecidable statements (e.g., involving quantification over infinite domains) in particular cannot be fully manifestable if truth is understood platonistically or according to the classical view of logic and mathematics. In other words, the truth-conditional view of meaning, if used to support platonism or classical mathematics, yields a notion of meaning that is not recognizable by us, or that transcends our knowledge or understanding. In his more polemical moments, Dummett has portrayed the classical or platonistic view of mathematics as 'incoherent', 'illegitimate', or 'unintelligble'. Human practice is simply limited, and there is no extension of it, by analogy, that will give us an understanding of the capacity to run through an infinite totality. Meaning must be derived from our capacities, not from a hypothetical conception of capacities we do not have.

Dummett has argued that if meaning were not fully determined by use then language would not be learnable and communication would not be possible. If there were some kind of meaning that transcended the use made of an expression, then one would have to say that someone might have learned a language and behaved in every way as if he had learned it, and yet he does not understand the language. Such a view would make meaning private and ineffable.

Dummett has also argued that meaning holism in the style of Quine must be rejected if this argument is to lead to the revision of mathematics and logic, because the question of justifying deductive practices cannot arise for a meaning holist. Dummett has argued for a molecular meaning theory.

Various aspects of these views have been widely discussed and criticized. Any alternative to Dummett's view should include responses to the following problems. First, some critics (see especially Prawitz 1977) have claimed that Dummett's requirement of 'full' manifestability is too strong and should be weakened. Another objection is that a finitist could apply Dummett's argument to intuitionism itself to show that its idealizations of human practice, like those of the platonist, are unfounded (Wright 1982; George 1988). In that case, Dummett's arguments against the platonist are not compelling. It has been argued that, in fact, Dummett's argument for intuitionism is circular (George 1993). Troelstra and van Dalen have suggested some problems that arise from the fact that Dummett's approach is focused too narrowly on logic and arithmetic (Troelstra and van Dalen 1988, p. 851). There has also been some concern about how the notion of intuition has disappeared from intuitionism as a consequence of the focus on meaning-theoretic issues. It has been argued, on the basis of the Kantian background of Brouwer's work, that intuitionism needs a nonsolipsistic notion of intuition (Parsons 1986). In a related vein, it has been suggested that Dummett's view omits or downplays the conscious, experiential character of mental construction (Tieszen 1994a). Moreover, this concern about intuition does not have to preclude a theory of meaning.

§ 4   Replacing the Requirement of Full Manifestability

In order to motivate the alternative to Dummett's view of intuitionism, I will focus on some facets of the requirement of full manifestability. It should be noted that, according to Dummett's argument, it is with our constructions that we supposedly fully manifest our knowledge of meaning in our linguistic behavior or practice. In particular, it is with *intuitionistic* constructions (as opposed to other types of constructions) that we supposedly do this. The problem is that it is questionable whether intuitionistic constructions fill the bill. As noted earlier, there are other forms of constructivism. If full manifestability is what is at stake, why settle on intuitionism? Why not, for example, choose finitism or CRM? A central problem, emphasized by Wright and George, is that finitists or actualists can argue that the reasoning Dummett applies to platonism can be applied to Dummett's own position. It is not the case that we fully manifest our knowledge of meaning in our linguistic behavior or practice with intuitionistic constructions, for intuitionistic constructions involve unjustified idealizations of and abstractions from what we can actually do. A better approximation to full manifestability in linguistic behavior can be had with a finitist conception of construction. Otherwise, it is necessary to admit recognition-transcendent forms of meaning, evidence, and so forth, and then it is not clear how communication would be possible, how it would be possible to learn languages, and so on.

One could take this to be an argument in favor of adopting strict finitism or actualism. The argument does not, however, force such a conclusion. Indeed, it can be argued that although some version of finitism might be interesting to pursue for computational or other mathematical reasons, it is sterile if it imposes restrictions on mathematics for philosophical reasons. If it is philosophical skepticism that motivates the adoption of finitism, then this skepticism can be pressed to the point of leaving us with a barren view of mathematics. An alternative is to hold that Dummett's argument against platonism is not compelling. One might then try to develop a different perspective on intuitionism and platonism. As a way of motivating such a different view, I suggest that the requirement of full manifestability be dropped. The issues it raises can thereby be avoided. I shall argue later that our knowledge of meaning is never really fully manifested in our linguistic practice anyway. I think it is clearly better to opt for this alternative.

In particular, the idea that *intuitionistic* constructions are completely captured in our linguistic behavior should be avoided. Of course Dummett does say, as do Brouwer and Heyting, that intuitionistic constructions cannot be adequately captured in formal systems. He will not allow the same claim to be made about linguistic behavior in general, for he seems to be concerned that the construction would then have to exist in some strange medium. It should be noted, however, that most recent theories in cognitive science and philosophical psychology could avoid worries about a strange medium without adopting Dummett's strictures about the manifestation of knowledge of meaning in linguistic practice.

By linking knowledge of meaning with the full manifestability of that knowledge in linguistic practice, Dummett's view of intuitionism is quite different from Brouwer's view. Brouwer held not only that mental constructions could not be adequately captured in formal systems, but that they were fundamentally languageless.[4] But if one claims that intuitionistic constructions are not completely captured in our linguistic behavior, it does not follow, as Brouwer sometimes suggests, that they are altogether languageless. To claim that intuitionistic constructions either are languageless or must be completely captured in our linguistic practice is to pose a false dilemma. In what follows, I shall think of intuitionistic constructions in terms of group intentionality, by which I mean the capacity for groups of people to share the same intentions and to see objects or states of affairs under the same meanings. Scientific theories are just sets of propositions that are believed to be true by groups of people. Groups of people come to see problems under the same meanings or contents and pursue their research accordingly. I argue for a view like this in the sections that follow.

Philosophers as different from one another as Quine, Searle, and Jackson agree that observable linguistic behavior underdetermines what we know, understand, or believe. It underdetermines the meaning of expressions, as Quine has pointed out with his famous 'gavagai' example. If we understand meaning in terms of constructions, then linguistic behavior also underdetermines constructions. Once this is recognized, philosophers will have different things to say about the matter. I think the most defensible line to take in intuitionism is this: because observable practice in using sentences underdetermines our knowledge or understanding and does not suffice to explain it, one must fill in the explanation by making inferences to unobservable, inner processes or structures that may be shared by different subjects. This need not involve introspection. It is a pattern of reasoning that goes back at least as far as Kant and that is now used widely in linguistics, cognitive science, and artificial intelligence studies.

Alexander George, using some ideas from linguistics and philosophy of mind, has already discussed some of the relevant points in relation

---

[4] For a more detailed discussion of what this claim amounts to, see Dirk van Dalen 1991. One of the interesting points made by van Dalen is that in Brouwer's philosophy 'causal sequences' can have higher or lower degrees of 'egoicity' or dependence on the individual subject, ranging from lawless to lawlike sequences. Lawlike sequences are most independent of the individual subject and are therefore most objective. Van Dalen suggests that Dummett's view can be associated with the part of intuitionism that concerns logic and lawlike sequences. Lawless sequences, however, are highly egoic. My comments in this chapter will be confined to what can be said about lawless sequences insofar as there can be a *science* of these objects.

to Dummett's views (George 1988). George reminds us that a prevalent view in modern linguistics is that a fixed, innate mental structure shrinks the totality of learnable languages and so makes possible the leap children uniformly take from interlocking and limited data to a language containing strings whose interpretations transcend what could be given in observable linguistic practice. An ideal speaker-hearer's linguistic understanding reaches beyond all the uses of expressions that a learner could actually observe. On this view, one cannot issue decrees, as both Dummett and finitists have, about what information can be extracted from elucidations, and about what extrapolations can be effected from observations.

This view of language learning, which emerged in a forceful way after Wittgenstein's work, clearly does not commit us to solipsism about meaning or constructions. The view is compatible with the idea of group intentionality and is in fact often linked with it in recent work in philosophical psychology. If cognition and meaning cannot be adequately understood in terms of observable linguistic practice, it does not follow that meaning is private, unlearnable, and uncommunicable. It is a perfectly sensible hypothesis that human beings are so constituted as to have at least some isomorphic cognitive structure, and that this structure is what makes language learnability and communication possible. On this basis, language learnability and communication would in principle be possible for both intuitionistic and classical mathematics.

It seems to me that the possibility of learnability and communicability is established by the actual practice of mathematics, where both intuitionistic and classical mathematics exist as scientific, social enterprises to which different people make contributions at different times and places. If mathematics, including intuitionistic mathematics, is approached as a given scientific, social enterprise, then there is no room for the solipsistic view that there are meanings or proofs that are in principle understandable to only one person. If constructions are not completely languageless, it also does not follow that they must be completely manifested in our linguistic practice. The ideas I have described neutralize worries about our supposed abilities to make infallible reports on essentially private mental processes, for there are no *essentially* private mental processes and one does not arrive at constructions on the basis of (infallible) introspective reports. One arrives at constructions in the context of a scientific community whose members are trying to solve problems posed by meaningful mathematical propositions. It is a matter of group intentionality.

These ideas enable us to sidestep the purported problems Dummett raises about learning and communicating meaning. With this background in mind, I would now like to turn directly to the matter of how to characterize intuitionistic constructions.

§ 5   What Are Intuitionistic Constructions?

What are intuitionistic constructions? It is clear, as noted earlier, that they are not finite configurations of signs on a piece of paper, and that they are not to be identified with Turing computations. It is not clear how Dummett would answer this question, apart from simply saying that they are mental processes and that solipsism about the mental is untenable. I claim that it is necessary to appeal to the notions of intentionality, consciousness, and inner time to understand intuitionistic ideas about meaning and mental constructions properly. It is necessary to speak directly and explicitly about our mathematical intentions, that is, the meanings or contents associated with our mathematical acts. Intentions, mathematical or otherwise, are not fully manifested in linguistic behavior. Mathematical intentions must be considered in intuitionism because it is a basic fact that awareness, including mathematical awareness, exhibits intentionality.[5] Moreover, it is important to note that it was Heyting (in conjunction with Becker) who explicitly appealed to this fact in the case of intuitionistic mathematics.[6] What are intuitionistic constructions? Heyting held that they are fulfilled or fulfillable mathematical intentions.

It should be recalled that in his famous 1931 comments on the intuitionistic foundations of mathematics Heyting insists that it is necessary to distinguish between propositions and assertions. An assertion is the affirmation of a proposition. Heyting says he uses the word *proposition* for the intention which is linguistically expressed by the proposition. The intention refers not only to a state of affairs thought to exist independently of us but also to an experience thought to be possible. The affirmation of a proposition is the fulfillment of an intention. A mathematical proposition thus expresses a certain expectation. For example, the proposition

---

[5] For a useful general introduction to the notion of intentionality see Searle 1983. Most writers on the subject, including Searle, do not consider the concept in the context of mathematics. I discuss the concept in relation to constructive mathematics in Tieszen 1989 and in Chapters 12 and 13.

[6] See Heyting 1931. Some of the Heyting-Becker correspondence is now published in Troelstra 1990. I thank Professor Troelstra for making the unpublished Heyting-Becker correspondence available to me.

'Euler's constant C is rational' expresses the expectation that one could find two integers a and b such that $C = a/b$. The assertion 'C is rational' would mean that one has in fact found the desired integers. Heyting suggests we distinguish an assertion from its corresponding proposition by the assertion sign '⊢' that Frege introduced and that Russell and Whitehead also used for this purpose. The affirmation of a proposition is not itself a proposition. It is the determination of a fact, that is, the fulfillment of the intention expressed by the proposition.

We should bear in mind that the notion of construction here is ambiguous between process and product. In one sense the 'construction' is the cognitive process carried out in time by the subject in fulfilling the intention; in the other sense it is the object obtained through this process. It is usually possible to determine which sense is at work by considering the context.

Heyting uses the distinction between intention and fulfillment to interpret the logical constants. He presents some clauses of the "BHK interpretation" of the logical constants mentioned in (ii). He comments on negation and disjunction in particular so that he can discuss the law of the excluded middle. Negation, he says, is something thoroughly positive: the intention of a contradiction contained in the original intention. The proposition 'C is not rational' signifies the expectation that one can derive a contradiction from the assumption 'C is rational'. In the BHK interpretation, there is a proof of ¬ C iff there is a proof of $C \rightarrow \bot$, where $\bot$ is an absurdity or contradiction. The negation of a proposition always refers to a proof procedure that leads to the contradiction, even if the original proposition mentions no proof procedure. 'A ∨ B' signifies that intention which is fulfilled iff at least one of the intentions A, B is fulfilled. As in (iii), but now using our new language, we say that 'A ∨ ¬ A' is a proposition that expresses an intention. It signifies the expectation of a mathematical construction (method of proof). 'A ∨ ¬' A can be asserted for a particular proposition A only if A has been either proved or reduced to a contradiction. A proof that it is a general law would consist in giving a method by which, when given an arbitrary proposition, one could always prove either that proposition or its negation. The question of its validity is a mathematical problem which, when the law is stated generally, is unsolvable by mathematical means. In this sense, logic is dependent on mathematics (as in (v)).

Troelstra has pointed out that in the Heyting-Freudenthal correspondence about the interpretation of intuitionistic logic, Heyting thinks it is not possible to give up the distinction between propositions and assertions (Troelstra 1982, p. 201). The distinction is required to make sense of hypothetical reasoning and of negation in particular. Freudenthal, however, is unwilling to consider propositions as such. What Freudenthal calls 'statements' in his correspondence are always asserted propositions, and Troelstra notes that Freudenthal does not acknowledge Frege's distinction. Freudenthal thinks the idea of hypothetical proof does not make sense in a constructive setting and says, quite surprisingly, that a proof of A → B should contain a proof of B. Heyting disagrees. Using his distinction, he gives the following example: take A to be the intention (proposition or 'problem') of finding a sequence 0123456789 in the sequence of digits of π, and B to be the intention of finding a sequence 012345678 among the digits of π. Then any fulfillment of (solution to) A can be trivially made into a fulfillment of (solution to) B. It is also possible to use the distinction to respond to Griss' criticisms of intuitionistic negation. It was in response to Griss' criticisms of intuitionistic practice, however, that Heyting began to distinguish different levels of intuitive evidence (see § 7). Indeed, Heyting (Heyting 1962) went on to say,

It seems to me that it is impossible to banish all unrealized suppositions from mathematics, for such suppositions are implicit in every general proposition.

It should be noted that in the 1931 paper and in the correspondence with Freudenthal, Heyting mentions the idea that a proposition can also be thought of as a problem (as in Kolmogorov's interpretation). A problem is posed by an intention whose fulfillment is sought. Heyting suggests that the intention-fulfillment distinction can be thought of in terms of the problem-solution distinction, or the expectation-realization distinction (see Chapter 13).

§ 6 Constructions and Manifestations of the Understanding of Meaning in Linguistic Behavior

Is the idea that constructions are fulfilled or fulfillable mathematical intentions really an alternative to Dummett's view? It is, and in more than one way. Dummett's analysis immediately takes a Wittgensteinian turn. The question can therefore be posed in the following way: are fulfilled mathematical intentions (i.e., intuitionistic constructions) adequately analyzed in terms of the manifestation of understanding of meaning in linguistic behavior? The remarks in § 4 already suggest that the answer is that they are not. I would now like to consider another dimension of this question that is addressed at length in recent work in the philosophy of mind.

First, consider the question of whether fulfilled mathematical intentions (i.e., intuitionistic constructions) are adequately analyzed in terms of Turing computations. It is probably not necessary to argue that there can be Turing computations that are not accompanied by a sequence of thoughts, or in which it is not the case that a person is fulfilling an intention. It does not follow from the fact that Turing computations exist that they are, so to speak, alive. In any such case, the computation cannot be a construction in the intuitionistic sense. It is not a form of consciousness. Note that it is not being claimed that Turing computations cannot manifest derived intentionality, for they are artifacts of our thinking that we put to work for our various purposes. Conversely, some processes in which intuitionistic mathematicians fulfill intentions are evidently not Turing computations. This may even be true in the case of the intuitionistic number-theoretic functions, as we noted in the earlier comment on (vii).

Now it also seems that a person could fulfill a mathematical intention on an occasion without there being a full manifestation of that fact in linguistic behavior. But might it never in principle be manifested at all? No. Insofar as constructions can play any role in scientific knowledge, such a claim is far too strong. The claim need only be that there is still a construction on that occasion. I am arguing that the intuitionistic notion of a construction depends on an experiential model in which experience or intuition is not to be identified with a manifested linguistic phenomenon. I mean this in the same sense in which, for example, a person would want to say that she *experiences* different things as she shifts from one perspective to another in the observation of a Necker cube, and yet would not describe this *experience* as a linguistic phenomenon, much less as manifested linguistic activity. I am not saying that it cannot be or will not be manifested in linguistic behavior in some way. Full manifestation in linguistic behavior, however, will typically be ruled out by the fact that our cognitive activity takes place against a vast background or context of beliefs, skills, cultural constructions, and so on.

Conversely, it seems that in principle a person could manifest an understanding of the meaning of some sentences in linguistic behavior without in fact fulfilling the intentions expressed by those sentences. In such a case there is no construction in the intuitionistic sense. There is no form of consciousness and no related work or learning of the appropriate type. In recent times, versions of this point have been made by Searle,

Jackson, and many other philosophers of mind, although none of these philosophers has noted the connections with intuitionism.[7] What these philosophers have done is to present arguments to show that a person could meet all of the conditions of an observability (or Turing) test for knowing or understanding the meaning of expressions without in fact knowing or understanding the meaning of the expressions, or at least without knowing something that would otherwise be known. The person (or possibly even a computer) can use the language but does not know what the expressions are about or does not have an inkling of their origins. It is easy enough, for example, to imagine a person or a computer using the natural deduction term calculus in this way.[8] What is lacking is a basic feature of human consciousness: intrinsic intentionality. What is absent in each case is the first-person experience with the meanings of the expressions and the objects the expressions are about. Intuitionism also helps us to speak directly to the issue of the determinateness of the objects involved in mathematical experience by claiming that logic is an applied subject that depends on mathematics. This has the effect of filling in the experience and giving specific objects. Logic is contextualized. The basic point at the moment, however, is that the link between meaning (or thought) and experience is absent in views that analyze meaning in terms of the behavioral manifestation of understanding of meaning. This link is missing or is sublimated in Dummett's view of intuitionism, for Dummett takes the kind of linguistic turn in which everything that is of interest in philosophy becomes a matter of the observable use of language.

The most plausible version of the view that an intuitionistic theory of meaning is one on which meaning is determined by use is the version suggested by Prawitz, that meaning is not determined by actual use but by total possible use (Prawitz 1977). The previous argument already applies to this version. Another way to respond to Prawitz's proposal is to note that meaning is only determined by total possible use, *other things being equal.* For example, a person who knows or understands the meaning of a statement P will manifest that knowledge only if he has *attended* to P, or *remembers* P. If he has not attended to P or does not remember P, he may not manifest that knowledge. Or he will manifest that knowledge only if he does not *believe* that doing so would conflict with other

---

[7] See, e.g., Searle 1980 and Jackson 1986; also, Tieszen 1997b, p. 254.

[8] For a formulation of the natural deduction term calculus see Troelstra and van Dalen 1988, pp. 556–563.

desires of more importance to him. If he believes otherwise, he may not manifest that knowledge. And so on. Assent to a linguistic expression in particular is prompted not by the occurrence of the expression but by its *believed* occurrence. But attending, remembering, believing, desiring, and expecting are all mental states of just the sort that exhibit intentionality. The qualifier "other things being equal" is illicit for the philosopher who holds that one is able only to refer to the *observable use* made of statements of a language. It is illicit because filling it in requires reference to mental states that exhibit intentionality. Thus, the set of possible uses that are supposed to determine the meaning under which we think of or see objects cannot be specified without invoking various mental states or processes that exhibit intentionality.

These arguments point in the direction of just the kind of meaning theory that Heyting invokes in his 1931 lecture: a theory of intentionality.

§ 7   Intentions and Constructions

Intuitionistic logic and mathematics is presented by Heyting as the logic and mathematics of fulfilled (or fulfillable) intentions, that is, of assertions. It is not presented by Heyting as the logic and mathematics of intentions per se. On my view, it need not be all there is to intensional logic and intensional mathematics. This is why I say in (ii) that the BHK interpretation gives us, in a general setting, the *constructive meaning* of the logical constants. (By itself, the BHK interpretation does not specify the nature of the constructions involved. One must look elsewhere in mathematics for that. In intuitionism, choice sequences are distinctive.) It does not, without qualification, give the meaning of the logical constants. Distinctions are needed here. One simple reason for distinguishing the constructive meaning of expressions from other possible types of meaning is that in mathematical practice expressions can have meaning apart from fulfillment. This is just part of the Fregean and Husserlian distinction between proposition and assertion, or thought and judgment.[9] In other words, it is possible to grasp a proposition or 'thought' without knowing whether it is true or false. An expression can have a sense for us at a given stage of experience even if we lack knowledge of its reference. Thus, a distinction should be drawn between what is known when a

---

[9] Husserl in particular distinguishes meaning as mere intention from the meaning associated with the fulfillment of the intention. See Husserl's *Logical Investigations*, Investigation I, §§ 14–15. This is also discussed in Chapter 12, § 5.

proposition (or intention) is grasped and what is known when the truth or falsity of the proposition is grasped.

The general claim that intentions determine extensions does not imply that intentions always determine extensions in a manner appropriate to constructivism. Dummett, however, seems to slip into such an understanding of intentions. He thinks that what one knows when one knows the meaning of an expression (or intention expressed by a sentence) is something that is like a method of verification or a construction procedure. Not only that, but in mathematics one knows just the kind of construction procedure recognized by intuitionism. Consider how he says the meaning theory underlying his argument is related to Frege's meaning theory (Dummett 1993, p. 128):

> Frege's argument is that the theory of reference does not fully display what it is a speaker knows when he understands an expression – what proposition is the object of his knowledge. I have here endorsed that argument, but have, in addition, gone one step beyond it, by maintaining that, since the speaker's knowledge is for the most part implicit knowledge, the theory of sense has not only to specify *what* the speaker knows, but also how his knowledge is manifested; this ingredient in the argument here used for the necessity of a theory of sense is not to be found in Frege.

This remark, when coupled with his claims about intuitionism and full manifestability, shows how Dummett slips into thinking that intentions must always be constructive means of or methods for arriving at extensions.

The matter, however, is more complicated and depends on other factors.[10] At a given stage of our knowledge, intentions in a particular domain may not be determinate enough to determine extensions constructively. This still allows for the possibility that these intentions condition our experience in the domain, even if they do so in a rather minimal way, as is the case when there may be thinking or theorizing about a domain but not much knowledge. If this is correct, then it should happen that one sometimes has to *find* the appropriate construction procedure. It is not always built in. Sometimes an appropriately determinate fulfillment procedure is present at the outset. It is known from the beginning. Then, in a sense, there need be no meaning that goes beyond the procedure. We often start with a problem (intention), however, and then find the fulfillment procedure later (if at all). It seems that for some intentions, taken at face value, it will not be possible to find a fulfillment procedure,

---

[10] See, e.g., Chapter 13.

that is, fulfillment for us humans. In other cases it is just not clear whether constructions will be found or not. Situations of these latter types arise in parts of modern mathematics, and especially in impredicative set theory. But this does not rule out the meaningfulness of these parts of mathematics, even if it does rule out constructive meaning. It only shows that the meaning of these parts of mathematics is less adequate, determinate, or clear.

Dummett argues that the problem with the extensional, *truth-conditional* view of meaning, insofar as it is used to support platonism or classical mathematics, is that it gives us a notion of meaning that is not recognizable by us, or that transcends our knowledge or understanding. I agree that there is a problem with purely extensional views of meaning. The fact that there is a problem does not, however, entail Dummett's version of the view that meaning is determined by use. The alternative I have presented defers to intensionalism and is thereby in sympathy with Dummett's comments about the extensional, truth-conditional view of meaning, but it does not identify all intentions with constructions. Dummett's argument against classical mathematics does not go through on the basis of his own meaning-theoretic preferences. If mathematical propositions far beyond the pale of constructivism are nonetheless meaningful, as I have argued, it is not necessarily because we know their truth conditions. The meaning theory that makes it possible to say this is not an extensional, truth-conditional meaning theory of the type that Dummett criticizes. It is a meaning theory based on intentionality. On this view, our grasp of the meaning of these propositions is less adequate, determinate, or clear, but we can have some grasp of their meaning.

If knowledge of meaning is separated from knowledge of extensions in the manner suggested, then saying that classical mathematics is incoherent, unintelligible, and illegitimate is too strong. Instead, a different perspective on classical mathematics emerges. Classical mathematics is not known to be contradictory, but it is also not constructive. One of the central features of intentionality is just that statements can be *about* objects without those objects' necessarily having to exist. Classical mathematics makes meaningful statements *about* certain objects, but without the same kind of direct evidence for the existence of the objects. There are, however, ongoing research programs in classical mathematics. I do not see how it can be claimed that these research programs are incoherent or meaningless. On the contrary, these programs must be determined by certain intentions. Thus, it can still be said that intentions in higher set theory determine how one can go on with the research (i.e., with unfolding the meaning that is already given) even if they do not determine this process through procedures that can be characterized exclusively as constructivist.

Let it be admitted that intuitionism is engaged in its own idealizations of human practice. It recognizes, for example, the existence of functions that are not feasibly computable, of objects that are 'abstract' relative to finitism, and of meaningful statements about such objects. The idealizations of classical mathematics, especially of set theory, extend even further. What we can say is that our grasp of the meaning of these propositions is even less adequate, less clear and distinct, relative to our own finiteness and to our other limitations in understanding the abstract, the transfinite, and the further reaches of what can be thought.

Suppose that constructive meaning is distinguished from other types of meaning on the grounds that mathematical expressions can have meaning apart from fulfillment. Then it is possible to count grammatically well-formed expressions of unfulfilled mathematical intentions as meaningful, along with all intentions that are more or less adequately fulfilled. It is possible to recognize degrees or types of fulfillment, or degrees or types of evidence. It is interesting that Heyting already recognized degrees of evidence in the context of intuitionism.[11] He suggested that we have the highest grade of evidence in the case of singular statements involving small natural numbers. In the case of such statements as $1002 + 2 = 1004$ we do not actually count, but we already use general reasoning that shows that $(n + 2) + 2 = n + 4$. Lower grades of evidence accompany the introduction of the notion of order type ω in connection with the constructible ordinals, the intuitionistic notion of negation, the use of quantification, the introduction of choice sequences, and the introduction of species. In a manner reminiscent of Bernays (Bernays 1935), I suggest that the evidence thins out even further in parts of mathematics that extend beyond traditional intuitionism. It is possible to allow for different types of evidence and, in particular, to allow for forms of indirect evidence. At the same time, we should distinguish the claim that intentions determine extensions from the claim that all intentions determine extensions constructively, recognizing that the general notion of construction is constrained by certain conditions but is not precisely defined. By way of contrast, it seems that Dummett does not want to say that what the BHK interpretation gives us is only the *constructive meaning* of the logical constants. He does not seem prepared to distinguish

---

[11] Heyting does this in the most detail in Heyting 1962, p. 195.

constructive meaning from other types of meaning. One of the consequences is that one always wants to know whether, on Dummett's view, we are supposed to understand the meaning of a statement (as opposed to its truth conditions) apart from effective decidability. What is laudable about Dummett's views are his critique of extensional, truth-conditional views of meaning in mathematics and his effort to make us think deeply about how well we really understand those parts of mathematics that are not more closely connected with our actual capacities and practices.

§ 8   Intuitionism with Intuition

It is worthwhile to make at least a brief comment about how the concept of intuition in intuitionism has disappeared as a result of a particular way of taking the linguistic turn. I think it is clear that its disappearance has resulted from the fact that the meaning-theoretic issues have been approached under the influence of Wittgenstein. Charles Parsons, who emphasizes the Kantian background of Brouwer's work, has argued that intuitionism needs a nonsolipsistic notion of intuition (Parsons 1986). His concern is directly related to the experiential aspect of constructions that I discussed earlier. To address this concern, several strands of the earlier argument can be brought together. First, simply substitute the term 'intuition' for 'construction' in (i)–(vii). Now define 'intuition' in terms of the fulfillment of intentions. Intuition can be understood in terms of the cognitive process carried out by the subject in fulfilling the intention; however, that in this process a particular object of intuition will be constructed is also true. This provides a good start on the basic statements of an updated intuitionistic theory of intuition. On this theory, intuition is a condition for knowledge of mathematical objects or states of affairs, but there can be mathematical thoughts or concepts that are not fulfilled in intuition. On the basis of (ii), it follows that the concern about intuition does not have to be understood as distinct from meaning theory. I conclude that intuition in intuitionism is not a useless shuffle. On the contrary, appeals to linguistic use are not intuitively fulfilling.

§ 9   Conclusion: An Alternative to Dummett's View

In conclusion, an alternative to Dummett's view that preserves (i)–(vii) may be formulated as follows: we start with the idea that mathematical cognition, like other forms of cognition, exhibits intentionality. The contents of our mathematical acts (i.e., mathematical intentions) are expressed in mathematical sentences. Mathematical intentions are not essentially private possessions locked in some Cartesian theater. They are or may be shared in the mathematical community. Scientific investigation is a matter of group intentionality. Many philosophers have held that human beings are so constituted as to have at least some isomorphic cognitive structure and that this characteristic is what makes learnability of language and communication possible. Learnability of language and communication should be possible in the case of both the intuitionistic mathematics and classical mathematics. Following Heyting and Becker, one can argue that intuitionistic constructions in particular may be thought of as fulfilled or fulfillable mathematical intentions. As such, they are in the first instance forms of consciousness or possible experience that are sharable but that are not completely exhausted in observable linguistic behavior. Not all mathematical intentions need to be fulfilled or to be fulfillable at a given stage in time. In the case in which mathematical intentions are fulfillable we have *constructive meaning*. Not all meaning or content, however, is to be identified with constructive meaning or content. In the case in which we do have constructive meaning or content we can recognize different degrees or types of fulfillment. That is, we can recognize different degrees or types of mathematical evidence. Finally, the concept of intuition in intuitionism can be defined in terms of the fulfillment of a mathematical intention. In this sense, mathematical constructions as cognitive processes carried out in time by human subjects are just mathematical intuitions.

On the alternative I have discussed in §§ 4–9, intuitionism continues to be important on epistemological and meaning-theoretic grounds, but the meaning theory does not preclude understanding the meaning of propositions of classical mathematics. Intuitionism offers a more detailed and refined view of the structures of mathematical cognition, of internal time, of mathematical intentionality and intensionality, and of mathematical evidence than can be found in classical mathematics. What stands out as one of the truly distinctive philosophical features of intuitionism, a feature central to epistemology but not present in other views of mathematics, is its conception of constructions as forms of consciousness or possible experience of a particular type, that is, fulfillable mathematical intentions. This philosophical assessment of intuitionism may not satisfy those who want their constructivism undiluted, but I believe it is more than enough to justify serious and careful consideration of intuitionism.

12

The Philosophical Background of Weyl's Mathematical Constructivism

My principal goal in this chapter is to describe the philosophical background of Weyl's mathematical constructivism. The most important phase of Weyl's work on the foundations of mathematics commenced with the development of his predicativism in *Das Kontinuum (DK)* in 1918 and proceeded through his alignment with Brouwerian intuitionism in the early twenties. The work of this period shows a deep commitment to constructivism. Around 1924 Weyl began to feel the need to account for parts of mathematics that could not be understood constructively. His views on constructivism would need to be supplemented in some way. I discuss this matter in § 7, but my primary focus will be on the central constructivist phase of Weyl's work. There is no indication in Weyl's comments on foundations that he was ever prepared to accept realism about mathematics, and, thus, the inclination toward constructivism is not completely absent even in Weyl's later comments. I will argue that Weyl's views on constructive foundations were shaped in a general way by a form of idealism, and in particular by a kind of transcendental idealism in the tradition of Kant.

I open the essay by reminding the reader of a few general themes from Kant's philosophy. Two other philosophers in this tradition are especially important for understanding Weyl's views: Fichte and Husserl. I will comment on a few themes in Fichte's work and then focus on Husserl's

This chapter is based on a talk I gave at the conference *Hermann Weyl: Mathematics, Physics and Philosophy*, held at UC-Berkeley in April 1999. I thank the conference organizer, Paolo Mancosu, for inviting me to speak. For helpful comments I am indebted to Paolo Mancosu, Solomon Feferman, Thomas Ryckman, John Corcoran, Martin Davis, and several other members of the audience. I also thank Mark van Atten, Dagfinn Føllesdal, and an anonymous referee for *Philosophia Mathematica* for their comments on a draft of this essay.

influence on Weyl. Weyl studied some of Husserl's writings and the two corresponded. Weyl's wife, Helene, was a student of Husserl's. Husserl's influence can be seen very clearly in a number of Weyl's works. One characteristic that makes Weyl so interesting is that he was able to take philosophical ideas from thinkers such as Fichte and Husserl and develop them in mathematical ways. I hope to show how the mathematical ideas of *Das Kontinuum* and Weyl's other works on the foundations of mathematics were shaped and guided by his philosophical preferences. It seems that Weyl's general philosophical outlook did not derive principally from Poincaré, Russell, Hilbert, or even Brouwer, although there are points of contact with each of these thinkers. I do not discuss the more technical side of Weyl's work in any detail in this essay. It would in any case be difficult to surpass Feferman's (Feferman 1988) excellent discussion of the technical work in *DK* and van Dalen's (van Dalen 1995) superb treatment of Weyl's technical variations on Brouwerian intuitionism.

I also hope to present Weyl's ideas in a way that will make apparent their relevance to recent issues in the philosophy and foundations of mathematics. The ideas on intuition, vicious circularity, the epistemic approach to the paradoxes, meaning, the continuum, and the intuitive-symbolic distinction are of special interest. Weyl's arguments on foundations are derived from an important tradition in philosophy, and, in some respects, they remain compelling.

§ 1 Weyl and Idealism

Idealism, in its simplest form, is the view that there are no mind-independent objects or truths. Realism, in its simplest form, is the view that there are mind-independent objects or truths. Transcendental idealism, in the Kantian tradition, is more subtle than either of these views. For our purposes here it will suffice to note that in transcendental idealism one considers the role of the mind, or the 'transcendental ego', in structuring, forming, or ordering our experience. (The transcendental or pure ego is to be thought of as the ego or self studied from the point of view of epistemology, not from the point of view of empirical psychology.) This ordering or forming is a function of the mind, but there is a sense in which not all of the material that is ordered or formed is itself constituted or constructed by the mind. At bottom, some material is given to us by the senses. Kant says that the senses are a faculty of receptivity, whereas the 'understanding' is a faculty of spontaneity or activity. Sensory 'objects' in transcendental idealism are, in a sense, both mind independent and mind dependent. They are mind independent in the sense that there are aspects of them that we do not constitute or construct, that is, their sensory component. They are mind dependent in the sense that their form is due to us. Kant says he is a transcendental idealist *and* an empirical realist. In sensory perception the mind is constrained in what it can form by the 'matter' of experience. Roughly speaking, Kant uses this form/matter distinction to find a way between the traditional, naive forms of idealism and realism that preceded him.

There are several additional features of Kant's philosophy (see, e.g., Kant 1973) to which I shall refer at later points in the essay. One of the most important features is Kant's distinction between intuitions and concepts. Kant says that concepts without intuitions are empty, and intuitions without concepts are blind. Knowledge is a product of concepts and intuitions. For Kant, *intuition* always refers to the faculty of receptivity: intuition is sensory intuition. There are two basic forms of sensory intuition, space and time. When Kant speaks of 'pure intuition' he has in mind these two basic *forms* of intuition. Space is the form of outer intuition and time is the form of inner intuition. Kant thought that the basic form of outer intuition – space – was captured by (Euclidean) geometry. He linked the form of inner intuition – time – to the natural number sequence and arithmetic; however, the exact nature of the link is not clear in Kant's writings. Concerning mathematics in particular, Kant says that mathematical knowledge requires the "construction of concepts in pure intuition." This too is not as precise in Kant's writings as one might like it to be, but clearly Kant's critique of pure reason, applied to mathematics, is meant to establish that where there can be no construction of concepts in pure intuition there can be no mathematical knowledge. Philosophers who work in the Kantian tradition generally try to make determinations about where intuition (and hence knowledge) ends and pure reason begins. These ideas are certainly part of the background of Weyl's views on foundations, although they take on a more specific form that can be traced to the influence of Fichte and especially Husserl.

Weyl mentions Fichte's work at several points in *DK*, in *Philosophy of Mathematics and Natural Science*, and in other essays (see Weyl 1925, 1955). Fichte was a post-Kantian transcendental idealist who, as did Kant, emphasized the role of the mind or of consciousness in the constitution of knowledge. Fichte retained many of Kant's basic ideas about epistemology and developed them in various ways (see especially the *Wissenschaftslehre* of 1794, contained in Fichte 1908–1912). For example, he agrees with

Kant that time and space are basic forms of intuition, but he is not as closely wedded to the idea that the form of outer intuition is captured in Euclidean geometry. This development would certainly have appealed to Weyl, for Weyl wrote an entire book, *Raum, Zeit, Materie* (*RZM*), on the "new amalgamation of time and space" required by relativity theory. Fichte's philosophy contains a detailed account of the structures and processes of the transcendental ego that are needed to explain how there could be knowledge of the 'nonego', that is, the 'objective world' with its various general and specific features. As do many post-Kantian transcendental idealists, Fichte rejects Kant's notion of the noumenon or 'thing-in-itself'. Fichte retains the objective world *as phenomenon*. In retaining the objective world Fichte seeks to avoid naive idealism, but he wishes to explain the genesis of experience from the side of the self. He recognizes that we have the belief that we are acted upon by things that exist independently of us, but he argues that the ego must have the capacity to produce this very belief. Fichte's explanation of how the ego could produce this belief is quite remarkable, and it contains far more detail than we can go into here. There are many remarks in Weyl's work that reflect the general claim of transcendental idealism that we must start with the transcendental ego and its acts and processes and determine in a critical fashion what can be built up from this basis. We must determine what kinds of structures and processes are involved in the production of knowledge. Weyl contrasts this view with what he calls "naive absolute realism." Naive absolute realism is not concerned with the ego and its acts or with what the ego can constitute or construct in its acts.

Similar themes in Husserl's writing clearly have great resonance for Weyl. Weyl studied with Husserl and throughout his career cited Husserl's work in his writings. In the Preface of *DK* he states flatly that he agrees with Husserl's views concerning the epistemological side of logic as they are expressed in the second edition of Husserl's *Logical Investigations* and in *Ideas I*. In the Introduction to *RZM* (published in the same year as *DK*), Weyl describes his general philosophical view in terms that are basically paraphrased from Husserl's discussions of the phenomenological reduction in *Ideas I*. He says, for example, that

we are only concerned in seeing clearly that the datum of consciousness is the starting point at which we must place ourselves if we are to understand the absolute meaning as well as the right to the supposition of reality. In the field of logic we have an analogous case. A judgment which I pronounce affirms a certain set of circumstances; it takes them to be true. Here again the philosophical question of the meaning of, and the justification for, this thesis of truth arises;

here again the idea of objective truth is not denied, but becomes a problem which has to be grasped from what is absolutely given. "Pure consciousness" is the seat of that which is philosophically *a priori* . . . what is immanent is absolute. (Weyl 1918b)

At several places in *DK* and *RZM* Weyl refers to a central notion in Husserl's transcendental idealism and epistemology: the notion of intentionality. In the context of physics, for example, he expresses the view that "the real world, and every one of its constituents with their accompanying characteristics, are, and can only be given as, intentional objects of acts of consciousness" (Weyl 1918b, Preface). In *DK* he says that "existence is only given and *can* only be given as the intentional content of the processes of consciousness of a pure meaning-bestowing ego" (Weyl 1918a, p. 94).[1] This concept of intentionality requires a brief explanation. When Husserl points out that consciousness exhibits intentionality he means that various kinds of conscious acts (e.g., thinking, believing, knowing, perceiving, desiring, remembering, willing) are directed toward objects or states of affairs. Thinking, for example, is always thinking of something or other. The object or state of affairs, however, need not always exist. A subject is directed toward or referred to an object by way of the content of an act. Husserl also calls the content the 'meaning' or 'intention' of the act. For an act of thinking we could express this in the form 'P thinks that S', where P is the transcendental ego, 'thinking' is the type of consciousness in this case, and S expresses the content or intention of the act. Thus, we can be referred to an object by way of the intention of an act even if the object does not exist. In philosophical logic one says that intension determines extension (if there is one), and in Husserl's philosophy 'intention' is just the intension by virtue of which cognitive acts are directed.

In order to *know* that there is an object or state of affairs corresponding to an intention one needs to have *evidence* for the object or state of affairs. The basic source of evidence for the truth of the judgments that express our intentions is *intuition*. We can merely think that something is the case, but in order to know it we must *experience* that it is the case. It is intuition that will either provide or fail to provide this experience. Husserl therefore distinguishes empty intentions from fulfilled or fulfillable intentions. An intention is fulfilled when we actually experience the object or state of affairs that the mere intention is about. Thus, in a manner

---

[1] Page numbers in citations of Weyl's works refer to the English translations of his works listed in the Bibliography.

reminiscent of Kantian epistemology, we could say that intentions without intuitions are empty and that intuitions without intentions are blind. The latter part of this formulation means that intuitions that are not informed by some intention or other are blind. Husserl also holds that mathematical knowledge is distinct from knowledge of sensory objects but is founded on sensory experience. It has its origins in *forms* that are already found in sensory experience. One additional feature that we should notice now is that our intuitions take place in time. In this sense, time will be a basic form of intuition. The pure ego is of course a finite being. We do not have unlimited powers of intuition.

§ 2 Weyl: Idealism and Epistemology in Mathematics

Transposing these ideas to mathematics gives us a general framework for understanding Weyl's view of the foundations of mathematics. Weyl's constructivism is motivated by transcendental idealism and the view that intuition is the central source of knowledge. Of course this does not mean that his views on foundations, especially in their technical details, are precisely those of Fichte or Husserl. Weyl, however, is clearly not motivated by some of the other positions that have been associated with constructivism: nominalism, conventionalism, empiricism, or Hilbertian formalism. This is why Weyl notes points of contact between his view and the views of Poincaré, Russell, and Hilbert but also directs critical comments toward each of these views. During his 'intuitionist' phase his agreement with Brouwer is also not complete. About formalism and empiricism, for example, Weyl has the following things to say in *DK*:

It is not the purpose of this work to cover the "firm rock" on which the house of analysis is founded with a fake wooden structure of formalism – a structure which can fool the reader and ultimately, the author into believing that it is the true foundation. Rather, I shall show that this house is to a large degree built on sand. I believe I can replace this shifting structure with pillars of enduring strength. They will not, however, support everything which today is generally considered to be securely grounded. I give up the rest, since I see no other possibility. (Weyl 1918a, Preface)

Although this is primarily a mathematical treatise, I did not avoid *philosophical* questions and did not attempt to dispose of them by means of that crude and superficial amalgamation of empiricism and formalism which still enjoys considerable prestige among mathematicians. . . . Concerning the epistemological side of logic, I agree with the conceptions which underlie Husserl. (Weyl 1918a, Preface)

Of the formalist conception of proof in particular Weyl says:

As if such an indirect concatenation of grounds, call it a proof though we may, can awaken any "belief" apart from our assuring ourselves, through immediate insight, that each individual step is correct! In all cases, this process of confirmation (and not the proof) remains the ultimate source from which knowledge derives its authority; it is the "experience of truth." (Weyl 1918a, footnote 19, Chp. 1)

As one would expect from a transcendental idealist who thinks knowledge is founded on intuition, the emphasis here is on the *experience* of truth. Weyl starts from the ego or self to see what can be built up from its cognitive structures and processes and, in particular, to determine what we can know.

In the case of mathematics we may use the schema mentioned previously – 'P thinks that S' – and simply substitute mathematical expressions for S. Depending on the expression that is substituted, the mind will be directed toward different objects or states of affairs, such as natural numbers, geometric objects, functions, real numbers, sets. In order to have mathematical knowledge the intention expressed by S will need to be fulfilled or to be fulfillable in principle. That is, intuition or intuitability will be required in order to make the object or state of affairs present to the mind. Intuition will involve some process carried out in time in which the object or state of affairs will come to be experienced. One might, for example, think that $254 + 303 = 557$, but one can also come to know it by carrying out a procedure to verify it. An intention might also be frustrated, as will be the case if one thinks that $254 + 303 = 554$. These are very simple cases; in Weyl's technical work in *DK* and elsewhere the basic ideas are developed quite precisely and extensively.

In addition to Weyl, Oskar Becker and Arend Heyting were developing some of Husserl's ideas in connection with constructive mathematics. Heyting and Becker straightforwardly identified mathematical constructions with fulfilled or fulfillable mathematical intentions (Becker 1927; Heyting 1931; and Chapter 13). Weyl also, in effect, thinks of construction or intuition in terms of fulfillable intentions, but he construes this somewhat differently from Heyting and Becker and his views change somewhat as he proceeds from the work in *DK* to his subsequent association with Brouwer and then on to his later writings. One of the central themes in Weyl's philosophy of mathematics is directly related to the distinction between concepts and intuitions, or empty intentions and intuitions: Weyl constantly distinguishes the theoretical, conceptual, or symbolic in mathematics from the intuitive, and he holds that genuine knowledge in mathematics requires intuition. He portrays his work in *DK* as directly related to the intuition/conception distinction:

Our examination of the continuum problem contributes to critical epistemology's investigation into the relations between what is immediately (intuitively) given and the formal (mathematical) concepts through which we seek to construct the given in geometry and physics. (Weyl 1918a, Preface)

We will look more deeply into this matter later.

§ 3   Intuition in Mathematics: Husserl and Weyl

When Weyl says in *DK* that he believes he can replace the shifting structure of real analysis with pillars of enduring strength, clearly he means to found analysis on intuition. Much of the secondary literature takes cognizance of the fact that Weyl has in mind a conception of intuition that derives from Husserl (da Silva 1997; Mancosu 1998; Pollard 1987; Tonietti 1988). Husserl's comments on a copy of *DK* sent to him by Weyl are worth noting: "At last a mathematician . . . finds himself again on the original ground of logico-mathematical intuition, the only ground on which a really authentic foundation of mathematics and a penetration into the sense of its achievement is possible" (letter of Husserl to Weyl, April 1918, in van Dalen 1984).

"Intuition" is a loaded term. There are many conceptions of intuition and people often seem to misunderstand comments about it by attaching their own particular conception to the expression. Weyl has a particular conception of intuition in mind, and many of its basic features can be identified and distinguished from features that are not part of this conception.

Some of the basic features of this conception of intuition are not difficult to discern. As we have been saying, there are forms of intuition. In particular, intuition takes place in a sequence in time. Since we are only concerned with the form of intuition, let us suppose we start with an intuition of something or other, which we denote schematically by '|'.

(*) Now we can have another act of intuition.

Since it occurs later in time, let us call it a *successor* act of intuition, which we denote schematically as (|)s. Iterate (*). After doing this a few times it is easy to have what Weyl calls the 'intuition of iteration' by simply disengaging from this process and reflecting on it. Weyl holds that the intuition of iteration is the foundation of the category or concept 'natural number': we have the form (((|)s)s)s. . . . In Weyl's words: "Regarding the *relation of number to space and time* we may say that time, as the form of pure consciousness, is an essential, not an accidental, presupposition for the mental operations on which the sense of a numerical statement is founded" (Weyl 1949, p. 36). Since Weyl starts with a philosophy according to which the ordering of experience is imposed by the mind, it is not surprising that he holds that the ordinal conception of number is basic (see Weyl 1949, pp. 34–35).

This form of intuition of course has many instantiations in ordinary sensory intuition. For example, I intuit a cup before me; then at a later stage in time I have a successor intuition of it, and then a successor intuition of that intuition. Another possibilitity is that I intuit a pencil; then at a later stage in time I have a successor intuition of it, and then a successor intuition of that intuition, and so on. Yet another possibility is that I intuit a particular physical object and then a different physical object in a later intuition, and then a different physical object in a still later intuition, and so on. The point is that I can have an intuition, then another intuition, then another, and so on. It does not matter what the sensory intuition is of.

From the perspective of Husserl's phenomenology we can make the following observations about this conception of intuition: at later (successor) stages of intuition the earlier stages sink back in time but are retained in an appropriately modified manner, indicating the importance of the role of memory in our constructions. Intuition is finite. We humans (as transcendental egos) carry out only finitely many acts of intuition. There are no completed infinite sequences of intuitions or, if you like, no completed infinite sets of acts of intuition. Infinite sequences of intuitions do not, as it were, have *being*. The notion of infinity for our intuitions can be thought of only as potential, in the sense of becoming. In the simple case under consideration it is clear that this will be a lawlike or rule-governed becoming. It is determinate. The future course of experience is fixed: it is simply the iteration of successor. In Husserl's language, we could say that the 'horizon' of possibilities of experience here is fixed. (Later we will contrast this with Weyl's reflections on choice sequences.) At any stage in this form we obtain a definite or determinate object; for example, the object expressed by $((((|)s)s)s$ is clearly distinct from the object expressed by $((|)s)s$.

It is because this sequence is lawlike and gives definite objects, I think, that Weyl says in *DK* that "the intuition of iteration assures us that *the concept 'natural number' is extensionally determinate*" (Weyl 1919; Weyl 1987, p. 110). The concept 'natural number' has a definite horizon. In fact, in

DK Weyl takes the natural numbers (but only the natural numbers) to be a closed infinite totality and he accepts the associated principles of proof and definition by induction for these objects. The 'definiteness' of the extension of the concept is reflected in the fact that the principle of the excluded middle is permitted in this case but is not permitted in the case of other mathematical categories. In light of his later (after 1919) views on lawlike and choice sequences, Weyl seems prepared to hold that even if we cannot carry out an intuition that gives all the natural numbers, we can still suppose that what is given by a lawlike sequence has 'being' whereas what is given by a choice sequence is only 'becoming'. He evidently thinks that we can shift from the notion of 'becoming' to the notion of 'being' in the case of lawlike sequences precisely because we can appeal to the fact that the future course of experience is determined by a law (see § 6).

The intention/fulfillment scheme is also at work here. There could be a mere intention in which I am directed toward a particular number, but I can also fulfill this intention by carrying out a process to obtain the number. If the number is quite large, I can bypass its actual construction by instead working with the law that generates it.

One of the most important features of this notion of intuition is not explicitly noted by either Weyl or Husserl, but I think it is clearly quite important for Weyl, especially in *DK*. It is that *there is no vicious circularity in intuition*. Carrying out a sequence of intuitions step by step to obtain an object (to obtain, for example, $(((|)s)s)s$, or an instantiation of this form in ordinary sense perception) does not require or presuppose that we construct some (infinite, completed) totality to which the object belongs. It is, as it were, a principle of perception or intuition that this should not be required. If it were necessary to refer to such a totality, then obtaining the object would require obtaining the totality, but obtaining the totality would require obtaining the object – a vicious circle.

In real analysis the least upper bound principle is one of the main casualties of adhering to this view of intuition. This principle states that every bounded set of real numbers has a least upper bound (l.u.b.). In this case the object – the l.u.b. – is defined in terms of the totality to which it belongs. One needs to know the set of upper bounds before one can determine the least among them. The l.u.b. therefore presupposes the set of all upper bounds of which it is itself an element. In *DK* Weyl of course gives up the l.u.b. principle for sets of reals, but he sees that one can often get by with the l.u.b. principle for sequences of reals (see also Feferman 1988). In the case of sequences bounded above by a real number the l.u.b. is definable by means of an existential quantifier that ranges only over the natural numbers. The l.u.b. principle for sequences thus holds at the first level of what Weyl calls the 'mathematical process' for forming sets and functions (see § 4) and Weyl is quite firm about not proceeding beyond this level.

I think that Weyl's critical view of impredicativity in mathematics, especially in *DK*, stems from this kind of view of knowledge and of what the mind can construct. In genuine cases of knowledge, founded as they are on intuition, we do not have vicious circles. There is no possibility of knowledge of an object, of the fulfillability of intentions, where there is a vicious circle. In contrast to the standard progression involved in obtaining knowledge in which an intention may terminate in an intuition (partial or complete), there is no possibility of the experience of truth where there are vicious circles. It is intuition that gives us a starting point or foundation of knowledge, understanding, and meaning. One can see why Weyl would be unhappy with the set-theoretic treatments of the natural numbers that were afoot. In such treatments one *starts* with a theory of infinite, completed totalities that is riddled with vicious circularity and then defines the natural numbers on the basis of these totalities. As does Poincaré, Weyl thinks this is exactly the wrong way to proceed (see also Chapter 14). The founding relation should be in the other direction.

In reading Weyl I have often wondered why he does not seem sensitive to the fact that impredicative specifications in mathematics do not always yield contradictions. On the grounds of their impredicativity, he sweeps away the set-theoretic version of the l.u.b. principle in the same stroke as the Russell paradox, the Grelling paradox, and others. The explanation, I think, is that for Weyl it is intuition that matters and on this basis we should avoid vicious circularity whether it is bound up with contradictions or not. Where there is vicious circularity the claim to knowledge is an illusion.

Weyl's view of vicious circularity shows his commitment to idealism and the importance of constructive intuition. He automatically embeds the vicious circle principle ("no totality can contain members definable only in terms of itself") in a theory of our cognitive acts and processes. 'Definability' is interpreted in terms of intuitive constructibility. The contrast with Gödel's realistic reading of the vicious circle principle, for example, could not be more stark. Gödel says that "if... it is a question of objects that exist independently of our constructions there is nothing in the least absurd in the existence of totalities containing members which can be described (i.e., uniquely characterized) only by reference to this totality" (Gödel 1944). On such a realist view, Gödel argues, one can deny the vicious circle principle.

§ 4 Intuition, Categories, Founding, and Existence

We do not have vicious circularity in our intuitive experience of objects, but we may have it in our *thinking* or *conceiving* that is not founded on intuition. As we survey our various cognitive productions we may find that in some cases they are riddled with vicious circularity. This is exactly what Weyl finds when he turns his attention to classical analysis. The category or concept 'natural number' has a foundation in intuition and seems to play a fundamental role in mathematics. In *DK* Weyl proposes to take it as the basic category. The category 'real number', as it is treated in classical analysis, is not founded on intuition. It is shot through with vicious circularity. The general categories 'set' and 'function' are in the same state. (Indeed, Weyl says in his 1921 paper that there can be no general theory of sets and functions.) This is why we should not start with either category, 'set' or 'real number,' and try to found the category 'natural number' on it. It should be precisely the other way around. Weyl says:

I became firmly convinced (in agreement with Poincaré, whose philosophical position I share in so few other respects) that the *idea of iteration, i.e., of the sequence of natural numbers, is an ultimate foundation of mathematical thought* – in spite of Dedekind's "theory of chains" which seeks to give a logical foundation for definition and inference without employing our intuition of the natural numbers. For if it is true that the basic concepts of set theory can be grasped only through this "pure" intuition, it is unnecessary and deceptive to turn around then and offer a set-theoretic foundation for the concept "natural number." Moreover, I must find the theory of chains guilty of a *circulus vitiosus*. (Weyl 1918a, p. 49)

Weyl's idea in *DK* is therefore to recast analysis without the viciously circular parts, that is, without the parts that cannot be founded on intuition.

Here we see the distinction between thought and intuition applied to mathematics, along with the idea that only the thinking that is founded on intuition can count as mathematical knowledge. The general idea of founding and founded structures in mathematical cognition is a fundamental part of Husserl's view. In the context of mathematics, Weyl takes the natural numbers to be the founding structure and then proceeds to determine what can be founded on it without forgoing intuition. Husserl's language of the independence and nonindependence of objects also seems to be in the background of Weyl's view (see also da Silva 1997). If we take the natural numbers to be independent objects that can stand on their own in our thinking, then the sets and functions that Weyl permits in *DK* are founded on the natural numbers and are nonindependent. Weyl construes this in a precise manner in *DK*: the sets and functions permitted are explicitly defined over the basic category, that is, the natural numbers.

Weyl is rather strict in *DK* in following what he calls the 'narrower procedure' for forming sets and functions. Using the 'mathematical process' that he describes in Chp. 1, § 4 of *DK*, one can form from the basic category a new object category consisting of the definable sets (sets of the first level). If one permitted iteration of the mathematical process, one could introduce sets of the second and higher levels. Weyl, however, will not allow higher types in the manner of Russell. He is focused on analysis and says that analysis with level distinctions is artificial and unworkable. On the other hand, we could in effect ignore the level distinctions by adopting Russell's axiom of reducibility, but, not surprisingly, Weyl finds the axiom of reducibility totally unacceptable. His conclusion is that the only acceptable procedure is to restrict attention to definable sets of the first level and simply accept the consequences (such as giving up the set-theoretic l.u.b. principle for the real number system). He says that the objects of the basic categories (natural numbers and rational numbers) remain "uninterruptedly the genuine objects of our investigation only when we comply with the narrower procedure" (Weyl 1918a, Chp. 1, footnote 24). Otherwise the profusion of derived properties and relations would be just as much an object of our thought as the primitive objects. In order to reach a decision about the judgments formed under the restriction of the narrower procedure we need only to survey these basic objects. We would otherwise be required to survey all derived properties and relations as well.

This is one point on which it seems to me that Weyl diverges from Husserl. Husserl's writings suggest that for mathematics in general we could allow sets of different levels provided they were formed in the right way. Husserl's notion of constitution or construction would presumably be wider than Weyl's. It is interesting to note that correspondence between Weyl and Becker shows that they already disagreed about the scope of the phenomenological conception of mathematical constitution and knowledge (see Mancosu and Ryckman 2002).

We have already seen that Weyl says that some categories or concepts are extensionally definite and some are not. The categories 'natural number' and 'rational number' are extensionally definite, for example, but the category 'property of a natural number' is not. The category 'real number', as it is understood in classical analysis, is not extensionally definite. In speaking of categories or concepts Weyl makes a distinction between intension or 'meaning' and extension. He says that

the failure to recognize that the *sense* of a concept is logically prior to its *extension* is widespread today; even the foundations of contemporary set theory are afflicted with this malady. It seems to spring from empiricism's peculiar theory of abstraction; for arguments against which, see the brief but striking remarks in Fichte (1908–1912) and the more careful exposition in Husserl (*Ideas I*). Of course, whoever wishes to formalize logic, but not to gain *insight* into it – and formalizing is indeed the disease to which a mathematician is most prey – will profit neither from Husserl nor, certainly, from Fichte. (Weyl 1919; pp. 110–111 in Weyl 1987)

The distinction between sense and extension would in fact be required by the notion of intentionality discussed earlier. Weyl says that the category 'real number' in classical analysis does have a sense or meaning and that this meaning leads mathematicians to think they understand modern analysis. Their understanding in this case is illusory, however, because this category is not extensionally definite. It is not founded on intuition.

It is important to note that for Weyl *existence* claims are reserved for the objects of extensionally definite categories. This principle is clearly at work in *DK* and it seems to persist in the later writings. The simpler view of *DK* is that what *exists* is what can be given in intuition, that is, the natural numbers. The use of the existential quantifier is restricted in this manner. In later writings in which Weyl appeals to Brouwerian choice sequences one has to distinguish, as it were, lawlike intuitions from choice intuitions. In these later writings Weyl (unlike Brouwer) restricts the existential quantifier to lawlike sequences. Thus, what exists is what is given in lawlike intuition, in the intuition of definite objects, whereas existential quantification over choice sequences is not permitted (see § 6).

§ 5 Meaning in Mathematics

Weyl makes a number of pronouncements about what has meaning and what does not have meaning in mathematics. These pronouncements can be understood against the background of the ideas about intentionality and intuition presented previously. We said that we are always directed toward an object or state of affairs by way of a meaning or intention. Let us call this meaning by virtue of which we are directed toward objects 'meaning$_i$'. This designates the meaning as mere intention. If the intention is fulfilled or fulfillable, we will say that we have 'meaning$_f$'.

This is the kind of meaning associated with knowledge and genuine understanding, the meaning that is founded on intuition. This distinction in made by Husserl in § 14, Investigation I, of the *Logical Investigations*. We should therefore also distinguish what is meaningless$_i$ from what is meaningless$_f$. Expressions containing mathematical terms may be meaningless$_f$ without being meaningless$_i$. Indeed, Weyl says that the category 'real number' in classical analysis has a sense or meaning, which gives modern mathematicians the illusion that they understand it. Weyl must be saying that it has meaning$_i$. This category, however, is not extensionally definite. It lacks the kind of meaning that is founded on intuition.

A string of signs might also be meaningless$_i$. In the opening pages of *DK* Weyl gives an example that seems to be of this type: the expression "An ethical value is green." The terms in this expression are incompatible with one another because they are from the wrong semantic categories. These kinds of issues about meaning and the (in)compatibility of categories are discussed at great length by Husserl in his work on pure grammar and logic in the *Logical Investigations*. Since this expression is meaningless$_i$ there is no possibility of its being fulfilled in intuition: that is, there is no possibility of its being meaningful$_f$. Husserl even distinguishes other grounds of meaninglessness that are prior to problems about mixing expressions from incompatible semantic categories. These latter forms of meaninglessness result from problems of pure formal syntax. In his analysis of conditions for the possibility of the meaningfulness and truth of expressions Husserl thus finds that there are conditions at different levels, starting with conditions on pure formal syntax and the compatibility of semantic categories, followed by conditions on consistency, followed in turn by conditions on fulfillability.

It can be concluded from what has been said that expressions involving vicious circularity must be meaningless$_f$. They express intentions that are in principle not fulfillable. I take it that this is what Weyl has in mind in *DK* when he declares questions such as "Is 'heterological' heterological or autological?" to be meaningless. This seems to be Weyl's general approach to the paradoxes. The paradoxes contain vicious circles, but we do not find vicious circles in knowledge that is founded on intuition. Hence, conceptions involving vicious circularity cannot have the kind of meaning associated with knowledge or understanding. There is no possibility of knowledge or of the experience of truth with such conceptions.

These matters about meaning, intuition, and paradox in Weyl and Husserl are also discussed by da Silva (da Silva 1997), but it seems to me that his account is not quite correct. He says, correctly, that Husserl distinguishes meaning with respect to form from meaning with respect to matter or content. He then says that "in Weyl, as in Husserl, if a judgment is meaningful with respect to matter then what it expresses can *in principle* be an object of intuition, which is only another way of saying that we can see *a priori* that its content does not bring together things which are *conceptually* incompatible." Here there is a problem. Consider, for example, the l.u.b. principle for the real number system. The l.u.b. principle is meaningful with respect to matter but what it expresses – the existence of the least upper bound – cannot in principle be an object of intuition. The judgment that "every bounded set of real numbers has a least upper bound" is surely not like "an ethical value is green." No one would be taken in by classical real analysis if it were. The contents of the l.u.b. principle are not semantically incompatible in the manner of "an ethical value is green." The latter is meaningless$_i$ because it combines terms from incompatible semantic categories. It follows that it is already meaningless$_f$. Expressions can, however, be meaningless$_f$ without being meaningless$_i$, as is the case with the l.u.b principle. What we should say is that if a judgment is meaningful$_f$ then what it expresses can in principle be an object of intuition. It does not follow that if a judgment is meaningful$_i$ then what it expresses can in principle be an object of intuition. Both meaning$_f$ and meaning$_i$ are types of meaning with respect to matter or content, as opposed to being types of meaning with respect to mere form.

One of the objections Weyl raises to Hilbert's formalism is that, in effect, it permits us to put together strings of symbols that do not have meaning$_f$. Meaning$_f$, however, is concerned precisely with knowledge of the truth of the judgments that possess it. As we have noted, Weyl says that what the purely formal conception of proof omits is the most important characteristic, the 'experience of truth'. If formalism is taken seriously, then one even abstracts from meaning$_i$. Given the role of meaning$_i$ in the theory of intentionality, this is to abstract from important aspects of the directedness of our thinking and from the notions of purpose and the goal of attaining truth that attend this cognitive directedness.

§ 6   Time Consciousness, the Continuum, and Choice Sequences

In all of his writings on the foundations of mathematics Weyl is concerned with the nature of the continuum and with the way it is represented in real analysis. Many of his central ideas on this topic grow out of an interesting section (§ 6) of Chapter II of *DK*, "The Intuitive and the Mathematical Continuum." Weyl's characterization of the experienced, intuitive continuum is related directly to his studies with Husserl. In particular, Husserl taught at Göttingen and lectured extensively on the consciousness of internal time when Weyl began his studies there. Weyl follows the idealist tradition discussed previously in holding that time is the basic form of the stream of consciousness. In *DK* he takes the consciousness of internal time as his model of the intuitive continuum:

> In order better to understand the relation between an intuitively given continuum and the concept of number..., let us stick to *time* as the most fundamental continuum. And in order to remain entirely within the domain of the immediately given, let us adhere to *phenomenal* time (rather than to objective time), i.e., to that constant form of my experiences of consciousness by virtue of which they appear to me to flow by successively. (Weyl 1918a, p. 88)

In his subsequent descriptions in *DK* of the consciousness of internal time Weyl cites Husserl's work on the subject and follows it closely.

The stream of consciousness flows in time and has a number of basic features that can be discerned readily. Consider as an example of the flow of consciousness in time the awareness of a long sentence as it is uttered. The utterance begins and as it continues the earlier parts of the utterance sink into the past and out of our immediate awareness even though they are *retained* and remain active in processing present parts of the utterance. Indeed, this must be the case, or it would not be possible to understand the utterance. While the utterance is in progress there will also be some more or less determinate *expectations* at any stage about the way it will unfold and about its completion. What a simple example such as this shows is that wherever we might slice into the flow of consciousness in time we do not obtain a durationless now point. We do not experience or intuit such a point (Figure 6). Rather, what we experience is a 'specious' or 'extended' present that we can picture as in Figure 7. The diagonal lines below the horizontal line indicate 'retentions' that are part of the awareness at the point chosen; the diagonal lines above the horizontal line indicate 'protentions' (expectations) about what is to come. The retentions are continuously modified as they sink back into the less immediate part of our present experience, and the protentions shade off into a less determinate awareness of what is to come. Retentions and protentions are associated with all acts of consciousness. Retention or

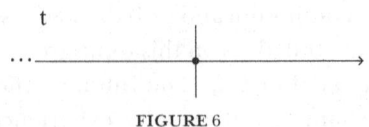

FIGURE 6

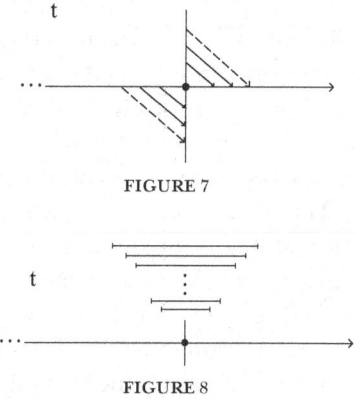

FIGURE 7

FIGURE 8

'primary memory' is not itself an *act* we undertake. It happens passively
or automatically and is to be distinguished from acts of remembering.
Acts of remembering are recollections or a type of *re*presentation. Acts
of remembering themselves have a retention-protention structure. Using metaphorical language, Husserl describes this specious present as a
'halo' or a 'comet's tail'. Its extension is indeterminate and shades off
continuously.

We could try to approach a durationless point in this intuitive continuum by an infinite sequence of nested rational intervals whose lengths
converge to 0 (Figure 8). Indeed, such a durationless point would just
be a real number according to the standard characterization of the real
numbers. Now Weyl already says in *DK* that a point or real number in
this sense is something that is only thought or conceived. It is an 'idealization'. It is not something that we intuit or experience. It is something
we can talk about 'in theory', but then we see that there is a significant
gap between theory and intuition. The same can be said about points in
space. We do not intuit extensionless points in space. Weyl sees clearly
that his mathematical apparatus in *DK* will not give us the intuitive continuum. Because it has at its disposal only arithmetically defined (and
lawlike) sequences it can give us only what he later calls an 'atomistic'
account of the continuum. Weyl therefore concludes that

> certainly, the intuitive and the mathematical continuum do not coincide; a deep
> chasm is fixed between them. . . . So one might say that our construction of analysis
> contains a *theory of the continuum.* (Weyl 1918a, p. 93)

Since the intuitive continuum is not captured by the mathematical
theory erected in *DK*, Weyl says that we need to look elsewhere for the

justification of his mathematical theory. His solution to this problem is to say that it must "establish its own reasonableness (beyond its mere logical consistency) in the same way as physical theory." He says that "a very essential part of such a rational justification is at hand. Evidence for this is that those numbers and functions [of DK analysis] allow us to give an exact account of what 'motion' means in the world of physical objectivity" (Weyl 1918a, p. 94). The idea is that the mathematics of *DK* suffices for applications in physical theory. Solomon Feferman and others have argued convincingly that Weyl's claim here has been largely vindicated by subsequent research (Feferman 1988).

A few years after the publication of *DK* Weyl returned to the problem of the continuum in his famous 1921 paper "On the New Foundational Crisis of Mathematics." Many of the ideas on the continuum in this paper are repeated in Weyl's other writings (see, e.g., Weyl 1925, 1926, 1949). In these later writings Weyl contrasts his own mathematical theory of the continuum in *DK* with the intuitive continuum. His arithmetical theory in *DK*, he says, is an atomistic theory of the continuum. It is not a theory of the 'fluid continuum'. Of his own theory he says, for example, that

existential questions concerning real numbers only become meaningful if we analyze the concept of real number in this extensionally determining and delimiting manner. Through this conceptual restriction, an ensemble of individual points is, so to speak, picked out from the fluid paste of the continuum. The continuum is broken up into isolated elements, and the flowing-into-each-other of its parts is replaced by certain conceptual relations between these elements, based on the "larger-smaller" relationship. This is why I speak of an *atomistic conception of the continuum*. (Weyl 1921, p. 91)

As is well known, around 1919 Weyl became enamored of Brouwer's views on the intuitive or fluid continuum. In the 1921 paper he in fact announced that he had given up his own earlier attempt at finding a foundation for analysis and had joined Brouwer. As it turns out, Brouwer and Weyl were not in complete agreement in their intuitionism, but there are some very interesting points of comparison.

As Weyl sees the matter, the intuitive continuum cannot be understood as a set of (durationless or extensionless) points. Classical analysis represents real numbers this way by identifying them with the points that are obtained in sequences of nested intervals. On this view, the real line would consist of the set of such points. This is also an atomistic, static view of the continuum. The problem is precisely that it makes the intuitive continuum disappear. We simply have a set of durationless points.

Weyl says that in order to capture the fluid continuum we should replace the element/set relation with the part/whole relation. There are in fact many kinds of part/whole relations, and it happens that Husserl distinguished many of these and discussed their logic in Investigation III of his *Logical Investigations*. The intuitive continuum would be what Weyl, following Husserl, calls an 'extensive whole' (Weyl 1949, p. 52). An extensive whole is a whole whose parts are of the same lowest genus as the undivided whole itself. If a particular temporal interval is an extensive whole, then its parts must themselves be temporal intervals. This idea conforms to our experience since we have no intuition of durationless points. (Similar observations can be made about extensionless points in space.) On the basis of this part/whole analysis we would not be able to obtain real numbers as idealized, extensionless points, as purely conceptual objects that are not given in intuition.

Weyl sees that his own theory in *DK* will not account for the intuitive continuum. He is also certainly sensitive to mathematical facts about the real continuum. The rational numbers, for example, are dense on the real line, and they may be all that is required for measurement in the physical sciences, but it is a mathematical fact that the rationals do not exhaust the reals. The arithmetically defined or lawlike sequences of *DK* will not suffice for the real continuum. One needs arbitrary sequences. Weyl believes that Brouwer has a constructively acceptable way to obtain this effect. Weyl thinks that through the introduction of choice sequences Brouwer has provided a mathematical account of the continuum that at least is closer to conforming to intuition. In *Philosophy of Mathematics and Natural Science* he states it as follows:

How then do assertions arise which concern, not all natural, but all real numbers, i.e., all values of a real variable? Brouwer shows that frequently statements of this form in traditional analysis, when correctly interpreted, simply concern the totality of natural numbers. In cases where they do not, the notion of sequence changes its meaning: it no longer signifies a sequence determined by some law or other, but rather one that is created *step by step by free acts of choice*, and thus remains in statu nascendi. This 'becoming' *selective sequence* represents the continuum, or the variable, while the sequence determined *ad infinitum* by a law represents the individual real number falling into the continuum. The continuum no longer appears, to use Leibniz's language, as an aggregate of fixed elements but as a medium of free 'becoming'. (Weyl 1949, p. 52)

Similar remarks are made in many places in Weyl's writings. For example:

A real number is not to be defined as a set, but as an infinite sequence of nested rational intervals whose lengths converge to 0. . . . The individual interval sequence, determined *in infinitum* by a law, then produces the *individual real number*, while the free choice sequence produces the *continuum*. (Weyl 1925, p. 134)

We can still think of a free choice sequence as a kind of intuition even though it is quite different in some respects from a lawlike intuition. It is still a sequence carried out in time by a subject (or transcendental ego), only part of which is actually completed. We would actually complete only a finite initial segment of it, it will have a horizon, and we should think of it as a 'medium of free becoming'. The contrast with lawlike sequences, however, is what Weyl highlights in these passages. A lawlike sequence is determinate and gives us a definite object at each stage. Its horizon is fixed. A choice sequence is more or less indeterminate or indefinite. Its horizon is not fixed in advance but is quite open (see also Becker's description in Becker 1923). Apart from some very general specifications, there is nothing in particular that should occur in it. There is a mere requirement to posit something further. This is quite formal and abstract compared with a lawlike sequence. With choice sequences we are considering intentions that can be fulfilled by any arbitrary postulation within the specifications, if any, that have been indicated.

Relative to a choice sequence, nothing essentially new occurs in carrying out a lawlike sequence. We can overview or survey lawlike sequences in a way that is not possible with choice sequences. With a lawlike sequence we are directed toward an endless lawlike succession. The infinite can in this sense be grasped by a finite mind (transcendental ego) by way of grasping the law that generates the sequence. The law can be finitely specified. It is evidently for this reason that Weyl thinks the lawlike sequence, and hence the real number produced or represented by the lawlike sequence, can be individuated. Since there can be no such overviewing in the case of choice sequences, we do not have individuation. But we should presumably not have it since by Weyl's sights the idea of choice sequences is to give us an analysis of the intuitive continuum in which there are no individuals. The experience of the temporal continuum, as we said, is experience of an indefinite or indeterminate temporal phase as an 'extensive whole'. We do not experience durationless points. Insofar as this is really experience of a continuum it is also an experience of *free* (as opposed to 'determinate') becoming.

With a lawlike sequence we need not continue carrying out the sequence. Since nothing essentially new occurs we can break off further acts and simply appeal to the law. With a choice sequence we have no grounds to break off further acts. In this sense, the idea of free choice sequences forces the idea of 'becoming in time', of infinity as potential, much more impressively than the idea of lawlike sequences. With lawlike sequences we can downplay the 'becoming' and incompletability by appealing to the law. In fact, in these writings Weyl seems to equate lawlike sequences with 'being' and choice sequences with 'becoming' (see the quotation from his 1921 paper that follows). Weyl seems to think, for reasons of this kind, that with choice sequences a real number (or 'point') *becomes* through its generation. The sequence generates a point only as long as it is a sequence in progress. The 'point' is not given as durationless. If a durationless point were to be obtained, the continuum would be disrupted. The condition that the point never *is* but always *becomes* preserves the continuum. Unlike a lawlike intuition, a free choice intuition conforms to the experience of the temporal continuum.

In his 1921 paper Weyl thus says that the continuum cannot be split into separate pieces. Anticipating Brouwer's work a few years later, he says that "there can be no other functions on the continuum than continuous functions." As van Dalen (van Dalen 1999, p. 325) has expressed it, Weyl was convinced of the unsplittability of the continuum and the continuity of the real functions on the basis of a phenomenological analysis. He did not prove a theorem on the continuity of the real functions. Rather, his statement is an immediate corollary of his definition of real function. By way of contrast, Brouwer proved such a theorem a few years later (see van Dalen 1995, pp. 160–162).

Weyl's view of choice sequences differs in some important respects from that of Brouwer. As is indicated by the preceding remarks, Weyl sharply separates lawlike sequences from the domain of choice sequences. Brouwer does not do this. Brouwer allows choice sequences in real analysis to be given by a law (see van Dalen 1995). In his 1921 paper Weyl says that "the notion of sequence oscillates, depending on the logical context in which it occurs, between 'law' and 'choice', between 'being' and 'becoming'" (Weyl 1921, p. 109). Brouwer's comment on this is instructive: "for me 'emerging sequence' is neither one; one considers the sequences from the standpoint of a helpless spectator, who does not know at all in how far the completion has been free" (Brouwer 1921; see Mancosu 1998, p. 120). As van Dalen (van Dalen 1995) has indicated, one has the impression that Weyl hesitates to attribute to choice sequences the sort of individuality that a natural number or lawlike sequence possesses. For Brouwer a choice sequence is an individual, and both existential and universal quantification over choice sequences are permitted. Brouwer and Weyl both reject the use of classical logic (in particular, the principle of the excluded middle) in reasoning about the intuitive or fluid continuum. In Weyl's treatment, however, the situation with quantifiers is more complicated. Weyl allows passage from universally quantified statements ($\forall x$)Ax to particular ones Aa only for lawlike a. Statements of the form ($\forall x$)Ax are allowed for choice sequences on the grounds of our 'insight into the essence' of choice sequences (see Weyl 1921, p. 100). The fact that Weyl withholds use of the existential quantifier with choice sequences seems to be a vestige of his earlier view that existential quantifiers can be used only with extensionally definite categories.

I will not enter into the issues about choice sequences that separate Weyl from Brouwer. There are a host of puzzling matters that deserve further investigation (see, e.g., van Atten, van Dalen, and Tieszen 2002). One central source of the difference between Weyl and Brouwer seems to be connected with Weyl's idea that although lawlike sequences are individuals, we should not have individuality with choice sequences since choice sequences can be used to analyze the intuitive continuum and there is no experience of individuals in the intuitive continuum. We can appeal only to the concept or essence of choice sequences, not to choice sequences as individuals.

Weyl seems to think that with the introduction of choice sequences we can obtain what eluded him in *DK*: a mathematical theory of the intuitive continuum. We have, as far as it is possible, a mathematics that conforms to Husserl's description of the consciousness of internal time. (In light of subsequent investigations of choice sequences, however, there are some issues about exactly what the technical features of this mathematics would be.) It is interesting to note that Weyl is not at all worried about the applications to physical theory of the mathematics of the intuitive continuum. Unlike the mathematics of *DK*, it wears its justification on its sleeve.

Since Weyl distinguishes between intuition and concepts, and since he believes that the intuitive basis of the concept 'real number' is to be found in choice sequences, it evidently follows that in Cantorian set theory we have an empty, purely conceptual (and possibly inconsistent) view of the continuum.

Weyl says that "Brouwer represents idealism thought through to the end." With Brouwer, Weyl says, mathematics gains its highest intuitive clarity. He continues, however, that "it cannot be denied ... that in advancing to higher and more general theories the inapplicability of the simple laws of classical logic eventually results in an almost unbearable awkwardness. And the mathematician watches with pain the larger part of his towering edifice which he believed to be built of concrete blocks dissolve into mist before his eyes" (Weyl 1949, p. 54).

§ 7   The Intuitive and the Symbolic in Mathematics

In this last quotation one might hear echoes of Hilbert's worries. Around 1924 Weyl begins to express ambivalence in his foundational writings about rejecting the parts of mathematics that are not founded on intuition. He feels that there is a fundamental gap between the intuitive and the theoretical or conceptual in mathematics. Parts of mathematics seem to involve concepts for which we have no constructions in intuition. They involve intentions for which we cannot find corresponding intuitions. It would then follow that we do not have *knowledge* in these parts of mathematics. In order to explain this gap Weyl often describes the parts of mathematics that cannot be founded on intuition as purely 'symbolic'. He never rejects the idea that with Brouwer mathematics gains its highest intuitive clarity, but he sees that something needs to be said about the parts of mathematics that would be put into question if we were to accept only what could be founded on intuition. One of his summaries of the situation is as follows:

> The stages through which research in the foundations of mathematics has passed in recent times correspond to the three basic possibilities of epistemological attitude. The set-theoretical approach is the stage of *naive realism* which is unaware of the transition from the given to the transcendent. Brouwer represents *idealism*, by demanding the reduction of all truth to the intuitively given. In axiomatic formalism, finally, consciousness makes the attempt to 'jump over its own shadow', to leave behind the stuff of the given, to represent the *transcendent* – but, how could it be otherwise?, only through the *symbol.* (Weyl 1949, pp. 65–66)

The part of mathematics that is not founded on intuition would be 'transcendent'. It would take us outside idealism. It would involve the attitude of realism, the view that there are mind-independent abstract objects or truths. There is no indication in Weyl's foundational work that he was ever prepared to accept mathematical realism. On the other hand, there is something here – classical mathematics – that needs to be explained. Rather than accept mathematical realism, Weyl suggests that through 'axiomatic formalism' consciousness attempts to "jump over its own shadow" and to represent the transcendent through symbols. It is at this point that Weyl appropriates some of the ideas of his other great teacher, Hilbert. (Although Weyl typically mentions Hilbert when he appeals to this role of axiomatic formalism it should be noted that in the *Philosophy of Arithmetic* and other early writings Husserl distinguishes between the intuitive and the merely symbolic and holds that most of our arithmetic knowledge, along with a lot of other mathematical knowledge, is merely symbolic.) Weyl says that "if mathematics is to remain a serious cultural concern, then some *sense* must be attached to Hilbert's game of formulae, and I see only one possibility of attributing it . . . an independent intellectual meaning." In theoretical physics we have before us an example of a kind of knowledge of a completely different character from the common or phenomenal knowledge that expresses purely what is given in intuition. In the case of phenomenal knowledge, every judgment has its own sense that is completely realizable within intuition, but this is by no means the case for the statements of theoretical physics. In theoretical physics it is rather *the system as a whole* that is in question if confronted with experience. Weyl says that

> theories permit consciousness to 'jump over its own shadow', to leave behind the matter of the given, to represent the transcendent, yet, as is self-evident, only in *symbols*. Theoretical creation is something different from intuitive insight; its aim is no less problematic than that of artistic creation. Over idealism, which is called to destroy the epistemologically absolute naive realism, rises a third realm, which we see Fichte, for example, enter in the final epoch of his philosophizing. Yet he still succumbs to the mystical error that, ultimately, we can nonetheless apprehend this transcendent within the luminous circle of insight. But here, all that remains for us is symbolic construction. (Weyl 1925, p. 140)

Weyl says that symbolic construction will never lead to a final result in the manner of phenomenal knowledge. Phenomenal knowledge is subject to human error but is nonetheless immutable by its very nature. Symbolic construction, Weyl says, is not a reproduction of the given, but it is also not the sort of arbitrary game in the void proposed by some of the more extreme branches of modern art. Consistency is one of the principles that govern symbolic construction in mathematics. Weyl says that

> if phenomenal insight is referred to as *knowledge,* then the theoretical one is based on *belief.* . . . If the organ of the former is 'seeing' in the widest sense, so the organ of theory is 'creativity'. If Hilbert is not just playing a game of formulae, then he aspires to a theoretical mathematics in contrast to Brouwer's intuitive one. But where is that transcendent world carried by belief, at which its symbols are directed? I do not find it, unless I completely fuse mathematics with physics and assume that the mathematical concepts of number, function, etc. (or Hilbert's symbols), generally partake in the theoretical construction of reality in the same way as the concepts of energy, gravitation, electron, etc. (Weyl 1925, p. 140)

Weyl concludes that

> beside Brouwer's way, one will also have to pursue that of Hilbert; for it is unde-
> niable that there is a theoretical need, simply incomprehensible from the purely
> phenomenal point of view, with a creative urge directed upon the symbolic repre-
> sentation of the transcendent, which demands to be satisfied. (Weyl 1925, p. 140)

Weyl thus appeals to some of Hilbert's ideas to try to accommodate 'transcendent' mathematics. In so doing he distinguishes *knowledge* in mathematics from *belief*. Note that in the passage just quoted he does not say that one will have to pursue Hilbert's way *instead* of Brouwer's way. Weyl never takes back his claim that with Brouwer mathematics gains its highest intuitive clarity. He never abandons the view that knowledge of truth in mathematics requires intuition, but he thinks that our 'theories' expressed in symbolic, axiomatic constructions are based on belief and we can at least inquire after the consistency of these symbolic constructions.

It should be noted that all of these reflections on the role that Hilbert's program might play are made prior to Gödel's incompleteness theorems. The incompleteness theorems suggest that it will not be possible to pre-serve classical mathematics in the manner hoped for by Hilbert. In par-ticular, questions about the consistency of the symbolic constructions expressing our beliefs pose special problems. Gödel himself evidently felt compelled to embrace realism in his reflections on these problems.

Of course Weyl is dealing with an issue that is just as difficult now as it has ever been: how to explain both constructive and classical mathe-matics. If, in particular, one starts from a constructive position, then how does one explain the parts of mathematics that cannot be brought into the fold? There are many possible responses to these questions. In the passages quoted earlier Weyl already says that Fichte "still succumbs to the mystical error that, ultimately, we can nonetheless apprehend this transcendent within the luminous circle of insight." But all that remains for us, according to Weyl, is symbolic construction. Some other sugges-tions about how to deal with the problem can be found in parts of Husserl's philosophy that Weyl does not discuss. In his later writings Husserl argues that we have intuitions of concepts or essences whether we intuit objects that fall under these concepts or not. The object of the intuition in this case is the concept or essence itself, not any particular object that falls under it (if there are such objects). This kind of intu-ition is not considered by Weyl, although there does seem to be some inkling of it in some of his comments. For example, earlier we cited

Weyl's remark that universally quantified sentences (∀x)Ax may be used with choice sequences because we grasp the concept or "essence" of these sequences. One cannot, however, proceed from essence to existence. Existential quantification over choice sequences is not permitted. Consider the following remarks by Weyl:

> It may well always be that the sense of a clearly and unambiguously determined object concept assigns to the objects of the nature expressed by the concept their sphere of existence. But this does not make the concept an *extensionally definite* one; that is, it does not ensure that it makes sense to consider the *existing* objects that fall under the concept as an ideally closed, in itself determined and delimited totality. This cannot be so if only because the wholly new idea of existence, of being-there [*Dasein*], is added here, while the concept itself is only about a nature, a being such-and-such [*So-sein*]. Seemingly the only reason why people have been tempted to adopt the contrary assumption is the example of the actual thing in the sense of the real external world, which is believed to exist in itself and to possess a composition determined in itself. If **E** is a property, clear in its sense and unambiguously given, of objects falling under a concept **B**, then "x has the property **E**" is a sentence that makes a claim about a certain state of affairs concerning an arbitrary object x of the sort. This state of affairs does or does not obtain; the judgment is in itself true or not true. . . . If the concept **B** is, in particular, extensionally definite, then not only will the question "Does x possess the property **E**?" have a clear and unambiguous sense for an arbitrary object x falling under **B**, but also the existential question "Is there an object falling under **B** that possesses the property **E**"? Based on the generating process of the natural numbers given to us in intuition, we hold on to the view that the concept of natural numbers is extensionally definite, and likewise for rational numbers. But concepts such as 'object', 'property of natural numbers', and the like are certainly not extensionally definite. (Weyl 1921, pp. 88–89)

Note that Weyl allows that "x has the property **E**" is a sentence that makes a claim about a certain state of affairs concerning an arbitrary object x of the sort under consideration and that this state of affairs does or does not obtain. The judgment is in itself true or not true. This suggests that even where concepts are not extensionally definite we can obtain decisions about them based on their 'sense' or meaning. Obtaining these decisions would not be an arbitrary affair; nor would it be a matter of establishing formal consistency; nor would it simply be a matter of artistic creation. Its organ seems to be "seeing" in a wide sense, not just "creativity." It is just this kind of decidability of judgments, based on an intuition of the concepts they contain, that the later Husserl has in mind in speaking of the intuition of concepts. Weyl's comments in this passage bear a resemblance to Husserl's ideas about meaning clarification and the process of intuiting essences through free variation in imagination. If Weyl is correct, we still cannot derive any existence claims from this process alone. Where does this leave us? I think it does leave us with a view that is worth exploring, but there is no space to do so here.

§ 8 Conclusion

In Weyl's own attempt to do justice to mathematics as a whole we seem to be left with an irreconcilable split between the intuitive and the purely conceptual, along with everything that is entailed by this split. Weyl thinks that Brouwer has done justice to the intuitive continuum with the introduction of choice sequences. His own arithmetic analysis in *DK* does not do justice to the intuitive continuum, but Weyl was prescient to suggest that it would suffice for applications in physics.

Weyl's work in foundations presents us with an uncommonly deep analysis of mathematical cognition and provides a vivid illustration of the kind of technical work that can result from taking idealism and constructive intuition seriously.

13

Proofs and Fulfillable Mathematical Intentions

What is a proof? One answer, common especially to some logicians, is that a proof is a finite configuration of signs in an inductively defined class of sign-configurations of an elementary formal system. This is the formalist's conception of proof, the conception on which one can encode proofs in the natural numbers. Then for any elementary formal system in which one can do the amount of arithmetic needed to arithmetize proof, and metamathematics generally, one can prove Gödel's incompleteness theorems. Gödel's theorems are often described as showing that proof in mathematics, in the formalist's sense, is not the same as truth, or that syntax is not the same as semantics. On this view "proof" is always *relative* to a given formal system. One might ask whether there is some common feature that these different purely formal 'proofs' share, a feature by virtue of which we are prepared to call them 'proofs'.

The formalist's conception of proof is quite alien to many working mathematicians. The working mathematician's conception of proof is not nearly so precise and well delineated. In fact, just what a proof is on the latter conception is not so clear, except that it is not or is not only, what a strict formalist says it is. In mathematical practice proofs may involve many informal components, a kind of rigor that is independent of complete formalization, and some kind of "meaning" or semantic content. Another philosophical answer to the question, that of the (ontological) platonist, is that a proof is a mind-independent abstract object, eternal, unchanging, not located in space-time, and evidently causally inert. On this view a proof is certainly not an inductively defined syntactic object, and one might think of a proof as something not created but rather discovered by a mathematician. For an intuitionist, a proof is neither a syntactic object nor a mind-independent abstract object. Rather a proof is a mental construction, a sequence of acts that is or could be carried out in time by the mathematician. For Wittgenstein and such philosophers as Dummett, a proof is a type of linguistic practice or usage. One could also distinguish empiricist, pragmatist, and other philosophical accounts of proof. On one kind of empiricist account, for example, an inference from premises to conclusion in mathematics would be part of the fabric of our empirical knowledge. It would not be different in kind from our knowledge in the physical sciences.

What I would like to do in this chapter is to discuss some aspects of an epistemological conception of proof. This epistemological or cognitivist conception of proof, as I shall understand it, concerns the role of proof in providing evidence or justification for a mathematical proposition, evidence that we would not possess without a proof. (I use the term *cognitivist* to indicate that the conception is concerned with the cognitive acts and processes involved in providing evidence. Although I shall emphasize the "cognitive" aspects of proof in what follows, I do not take this to imply that proof is completely independent of social or cultural determinations.) I mean to use the term *evidence* in such a way that there is no such thing as evidence outside the actual or possible experience of evidence. One can get some sense of the concept of "evidence" that I have in mind by reflecting on what is involved when one does not just mechanically step through a "proof" with little or no understanding, but when one "sees," given a (possibly empty) set of assumptions, that a certain proposition must be true. Anyone who has written or read proofs has, no doubt, at one time or another experienced the phenomenon of working through a proof in such a merely mechanical way and knows that the experience is distinct from the experience in which one sees or understands a proof. My contention is that one can only be said to have evidence in mathematics in the latter case, although there are clearly differences in degrees of evidence. What is the difference between the two experiences just mentioned? The difference can apparently only reside in our *awareness* in the two cases, for nothing about the string of symbols we write or read changes or has to change in order for the experience to occur. To give a rough description, one might say that some form of "insight" or "realization" is involved, as is, in some sense, the fact that the proof acquires "meaning" or semantic content for us upon being understood.

I would argue that a "proof" in this sense ought not to be viewed as just a syntactic object, or as a mind-independent object. It is also not something reducible to linguistic behavior along the lines of behaviorist psychology, for the sensory stimulus involved underdetermines the experience. Consider, for example, what is generally thought to be a difficult proof written in a book on number theory. No doubt people will experience this proof in quite different ways, even though the same stimulus irritates their nerve endings. (And if the stimulus is not the same, or does not occur at their nerve endings, then there are grave difficulties for the behaviorist about what it could be.) If this is possible, then there is more to their experience than meets their sense receptors. If follows that the "proof" could not be a function of only what they receive at their nerve endings. Their responses will instead be determined by what they receive at their nerve endings plus what they actually "see" or "experience." It is necessary to make allowances, that is, for differences in the mental states or processes that are involved between stimulus and response, and it is these mental states or processes that determine whether or not one in fact has a proof. One needs an account of cognition according to which mental states can themselves be the effects of stimuli and/or other mental states, in addition to being the causes of other mental states and/or responses.

My aim in this chapter then is to outline and argue for a theory of proof which does justice to what I am calling the epistemological or cognitive dimension of proof. I believe that neither formalism, empiricism, pragmatism, platonism, nor any of the other "isms" in the philosophy of mathematics, with the exception of intuitionism on some counts, does very well with the cognitive aspect of proof, even though this cognitive aspect is essential for understanding mathematical knowledge.

§ 1

A passage from some work of the logician Martin-Löf (1983–84, p. 231; cf. Martin-Löf 1987, p. 417) does a good job of capturing the cognitive or epistemological aspect of proof that I wish to discuss. Martin-Löf says that

the proof of a judgment is the evidence for it... thus proof is the same as evidence... the proof of a judgment is the very act of grasping, comprehending, understanding or seeing it. Thus a proof is, not an object, but an act. This is what Brouwer wanted to stress by saying that a proof is a mental construction, because what is mental, or psychic, is precisely our acts... and the act is primarily the act as it is being performed, only secondarily, and irrevocably, does it become the act that has been performed. (Martin-Löf 1983–84, p. 231)

In the work from which this passage is drawn Martin-Löf is emphasizing the intuitionistic view that a proof is a cognitive process carried out in stages in time, a process of engaging in some mental acts in which we can come to "see" or to "intuit" something. Proof, on this conception, is in the first instance an *act* or a process. Only secondarily does it become an *object.* The conception of proof that Martin-Löf is describing is found very strikingly, as he notes, in Heyting's identification of proofs (or constructions) with fulfillments of intentions in the sense of Husserl's philosophy (Heyting 1931). On this view, the possession of evidence amounts to the fulfillment of mathematical intentions. I agree with Martin-Löf's remark (Martin-Löf 1983–84, pp. 240–241) that Heyting did not just borrow these terms from Husserl but that, in the following sense, he also applied them appropriately: the distinction between the presence or absence to consciousness of particular mathematical objects is modeled appropriately by the constructivist view that we should be able to find objects if we wish to claim that they exist. Thus, as a first approximation, I propose to answer our opening question as follows:

(1) A *proof* is a fulfillment of a mathematical intention.

Likewise, provability is to be understood in terms of fulfillability. In one fell swoop this embeds the concept of proof in a rich phenomenological theory of mental acts, intentionality, evidence, and knowledge. To understand philosophically what a proof is one must understand what the fulfillment of a mathematical intention is. Let us therefore first consider the concept of mathematical intention.

The concept of an intention is to be understood in terms of a theory of intentionality. Many cognitive scientists and philosophers of mind believe that intentionality is a basic, irreducible feature of cognition, certainly of the more theoretical forms of cognition. Intentionality is the characteristic of "aboutness" or "directedness" possessed by various kinds of mental acts. It has been formulated by saying that consciousness is always consciousness *of* something. One sees this very clearly in mathematics, for mathematical beliefs, for example, are beliefs *about* numbers, sets, functions, groups, spaces, and so on. By virtue of their "directedness" or referential character, mental acts that are intentional are supposed to be responsible for bestowing meaning, or semantic content.

A standard way to analyze the concept of intentionality is to say that acts of cognition are directed toward, or refer to, objects by way of the "content" of each act, where the object of the act may or may not exist. We might picture the general structure of intentionality in the following way:

$$\text{Act (Content)} \longrightarrow \text{[object]}$$

where we "bracket" the object because we do not assume that the object of the act always exists. Phenomenologists are famous for suggesting that we "bracket" the object, and that we then focus our attention on the act (noesis) and act-content (or noema), where we think of an act as directed toward a particular object by way of its content (or noema). Whether the object exists or not depends on whether we have evidence for its existence, and such evidence would be given in further acts carried out through time.

We can capture what is essential (for our purposes) to the distinction between act and content by considering the following cases: a mathematician M might believe that ϕ, or know that ϕ, or remember that ϕ, where ϕ is some mathematical proposition. In these cases different types of cognitive acts are involved – believing, knowing, remembering – but they have the same content, expressed by ϕ. The act-character changes, but the content is the same. Of course the content may also vary while the act-character remains the same. The content itself can have a structure that is quite complex. Also, when we say that the content is "expressed" by ϕ we shall mean that the mathematical *proposition* ϕ is an expression of the content of a particular cognitive act. Thus, there is a direct parallel between intensionality, a feature of language or expression, and intentionality, a feature of cognition, insofar as we are restricting attention to those expressions which are expressions of cognitive acts. We should not necessarily expect, for example, substitutivity *salva veritae* and existential generalization to hold for inferences involving expressions of intentions.

We could start by regimenting our understanding of what forms ϕ can take by focusing on the syntax of first-order theories. Thus when we say that an act is directed toward an object we could express this with the usual devices of first-order theories: individual constants and bound variables. For example, a mathematician might believe that Sa for a particular S and a particular a, or that $(\exists x)Sx$ for a particular S and a particular domain of objects D. ϕ can also have the form $(\forall x)Sx$, $S \vee T$, and so on. The restriction to first-order theories is not necessary. In fact, there are good reasons for considering higher *types* of intentions and their fulfillments, but I shall not discuss this matter in any detail here.

The idea that an act is directed toward an object by way of its content has a direct analog in the thesis in intensional logic and mathematics that intension determines extension. We should comment on this thesis now in order to forestall some possible misunderstandings of the intentionality of cognitive acts. In particular, when we say that intension "determines" extension, or that an act is directed toward a particular object by way of its content, should we take this to mean that the act-content provides a "procedure," an algorithm, or a "process" for determining its extension? To answer yes, without qualification, would lead to a view of knowledge that is far too rigid. We shall say that the manner in which intension determines extension is one of degree and is a function of the intention itself, background beliefs, contextual factors, and the knowledge acquired up to the present time. In the growth of knowledge, intensions or concepts are themselves modified and adjusted at various stages as information and evidence is acquired. Thus, a "procedure" by which the referent of a given intention is fixed might be quite indeterminate or quite determinate, depending on how much knowledge one has acquired, or how much experience one has, with the object in question. A belief about an object may be quite indeterminate, but even so we usually have at least some conception of how to go about improving our knowledge of the object.

We said in (1) that proof is the *fulfillment* of a mathematical intention. One can think of mathematical intentions, or cognitive acts, as either empty (unfulfilled) or fulfilled. The difference between empty and fulfilled cognitive acts can be understood as the difference between acts in which we are merely entertaining conceptions of objects and acts in which we actually come to "see" or experience the objects of our conceptions. We shall elaborate further on the distinction in §§ 3 and 4. Philosophers familiar with Kant will recognize the similarity here with the Kantian distinction between conception and intuition (Kant 1973). On a Kantian view, knowledge is a product of conception and intuition. Of course this kind of distinction is not specific to Kant. It has a long history, in one form or another, in philosophy. We might modify somewhat a famous remark of Kant's, however, and say that in mathematics intentions (directed toward objects) without proofs are empty, but that proofs without intentions ("aboutness," and meaning or semantic content) are blind (as is perhaps the case in strict formalistic or proof-theoretic conceptions of proof). Knowledge is a product of (empty) intention and proof. In particular, the objects of acts of mathematical cognition need not exist and we would only be warranted in asserting that they do exist if we have *evidence* for their existence. Proof is the same as evidence. Thus, as we noted earlier, another way of expressing the distinction is to say that acts of cognition are "fulfilled" (or perhaps partially fulfilled) when we have evidence, and "unfulfilled" when we do not have evidence. Cognitive acts must be (at least partially) fulfilled if we are to have knowledge of objects. We have many beliefs and opinions in mathematics, but we only have proofs, that is, evidence, for some of these. Given our remarks about intentionality and fulfillment we should say that a necessary condition for a mathematician M to *know* that ϕ is that M believe that ϕ, and that M's belief that φ be produced by a cognitive process – proof – which gives evidence for it.

To say that proof is a "process" in which an intention comes to be fulfilled is to say that it is a process of carrying out a sequence of acts in time in which we come to see an object or in which other determinations relevant to the given intention are made. This obviously places some constraints on the notions of proof and evidence, and on what can count as justification for beliefs about objects in mathematics. The process must be one a human being can carry out, on the analogy of carrying out procedures to solve problems in the empirical sciences, or else we will have unhinged the concept of evidence for objects from anything that could count as evidence for human beings, that is, from anything that humans could experience. Thus, for example, human beings evidently cannot experience an infinite number of objects in a sequence of acts in a finite amount of time. One might think of mental acts, such as computations, as taking place in linear time of type ω. No doubt there is a sense in which we can perform classical deductions in linear time of type ω, but to leave the matter there would be to miss the point. For it might be asked whether we can construct the objects these deductions may be *about* in linear time of type ω, or whether we can execute operations (functions) involved in the propositions of the deduction in this time structure. The constraints on proof must be imposed all the way through the components of the propositions in a proof if we are to avoid a recognition-transcendent concept of evidence. Having said that much, however, I would argue that there are degrees and types of evidence provided by proof. We should pause over this point for a moment.

§ 2

We said that it is the function of a proof to provide evidence. But what kind of evidence? Intentions may be fulfilled in different ways or to different degrees. For knowledge in general we could consider the presence or absence of the following types of evidence: a priori evidence, evidence of "necessity," clear and distinct evidence, intersubjective evidence, and "adequate evidence." In mathematics a fine-grained approach to questions about the adequacy of evidence is already found in proof theory, where proofs are ranked in terms of computational complexity and other measures. The classifications of proofs that result are surely relevant to questions about the "processes" which produce M's belief that φ. As we are viewing the matter, if one insists on a "feasible" proof, a finitist proof, an intuitionistic proof, or a predicative proof, one is insisting on a certain kind of evidence.

There are many philosophical arguments about whether proofs provide a priori knowledge, and whether they provide knowledge in which a conclusion follows with "necessity" from its premises. One could also discuss degrees of clarity and distinctness of proofs, and other matters such as simplicity; length, as it relates to reliability; and elegance. I would like to set these questions aside here, however, since they do not pose problems specific to the conception of proof that I am discussing. Rather, I would like to comment on two issues that are especially important to the view of proof as the fulfillment of certain kinds of cognitive acts. First, one of the principal concerns about the kind of characterization of proof given in (1) has been that one then perhaps makes the concept of proof a private, subjective matter, that one flirts with solipsism. One often hears, for example, that Brouwer's conception of proof was solipsistic. Second, if there are constraints on what can count as evidence then what is the status of classical proof?

Concerning the first question, the identification of proof with the fulfillment of mathematical intentions is perfectly compatible with viewing mathematics as a social activity, and it need not entail solipsism. Arguments for the possibility of understanding cognition without falling into Cartesian difficulties go back at least as far as Kant. In spite of Brouwer's other references to Kant, it is a basic theme of Kant's philosophy that human beings are so constituted that their fundamental cognitive processes are isomorphic. This explains why there is intersubjective agreement in elementary parts of mathematics, and it shows that the investigation of cognition or intentionality need not entail commitment to personal, introspective reports, in the style of introspectionist psychology. Proof, on this view, is not a species of introspection. Recent work in cognitive science depends on a similar approach to cognition.

Thus, it could be argued that there is no such thing as a "proof" that could in principle be understood by only one person, for that would contradict the hypothesis of a universal, species-specific mental structure which makes scientific knowledge (in particular, mathematics) possible in the first place. It is not as if a proof is some kind of subjective mental content to which only one person has direct access. There have been views according to which a proof is just such a subjective mental content, but philosophers such as Frege and Husserl went to great pains to refute them, correctly I believe, in their critiques of psychologism. The view of proof I would like to defend is definitely not psychologistic. Thus, as I construe (1), supposing that an intention is fulfillable for a particular mathematician without the possibility of fulfillment of the same intention for other mathematicians is incoherent. Note that this does not, however, entail that there *in fact* has to be intersubjective agreement at all times and all places on all mathematical propositions. It entails only that intersubjective agreement is possible. Where de facto intersubjective agreement about a proof exists, knowledge is thought to have a firmer evidential foundation. This is one source of the "objectivity" of proof. The import of these remarks is perhaps best appreciated in connection with phenomena such as Dedekind's famous "proof" of the existence of infinite systems, Hilbert's original solution to Gordon's problem, and the history of "proofs" of Fermat's last theorem.

Concerning the second question, one might ask, for example, whether the view described here entails that classical proofs do not deserve to be called proofs at all. It is a delicate matter to state a position on this question that will not immediately lead to objections. Let us consider a traditional (weak) counterexample of the type that constructivists raise for the classical notion of proof. These counterexamples proceed by showing how the assumption that we have a (classical) proof of a certain kind would imply that some unsolved mathematical problem has been solved. And of course even if the particular problem chosen were to be solved, the significance of the counterexamples arises from the fact that one can produce reductions to an endless number of other unsolved mathematical problems. Thus, for example, consider the decimal expansion of π, and let An be the statement that "the nth decimal of π is 7 and is preceded by 6 7s." Then what could it possibly mean to say that $(\exists x)Ax \vee \neg (\exists x)Ax$ is provable? We would need to have a proof that either provides us with a natural number n such that An or shows us that no such n exists. Since no such evidence is available it appears that in a logic that is supposed to conform to the idea of intuiting particular mathematical objects we ought not to accept the principle of the excluded third. That is, we do not have evidence of its necessity in these cases. Or one could say that to accept it is to suppose that we have evidence or knowledge that we in fact do not possess. In some of the constructivist literature the principle is referred to as the "principle of omniscience." It is a principle that can lead us outside the domain of intuitive knowledge. In my book *Mathematical Intuition* (Tieszen 1989), where I wanted to focus on a logic that would be compatible with constructive intuition, I likened the principle to an "ideal of reason" in a Kantian sense. Thus, one could view it as an epistemically illegitimate (when applied beyond certain bounds) but unavoidable postulation of human reason, one which might nevertheless sometimes serve a purpose in human affairs. It would be illegitimate when it leads us to believe that we can intuit particular objects that are in fact not intuitable.

If $(\exists x)Ax \vee \neg(\exists x)Ax$ does not hold in all contexts, we can also see that $\neg\neg\ \phi \rightarrow \phi$ ought not to be acceptable. Thus, in a system of proof such as the Gentzen-Prawitz intuitionistic system of natural deduction (e.g., Prawitz 1965), one does not have indirect proofs of the form

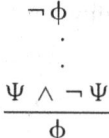

$$\neg\phi$$
$$\vdots$$
$$\frac{\Psi \wedge \neg\Psi}{\phi}$$

If proof is to be defined as in (1), then, following our description of the difference between empty and fulfilled intentions, one must be directly presented with an object, or at least possess the means for becoming so presented. In an indirect proof in the form of an existentially quantified proposition one has nothing more than the contradiction obtained by assuming that an object satisfying some condition does not exist. This is not the same as seeing the object itself, and so cannot count as fulfillment of the intention directed to the object. Might one think of such "existence proofs" as at least providing "indirect evidence" for the existence of an object? It seems that this would be reasonable in some contexts in mathematics because one might later find a constructive proof. Thus, in such contexts one might regard propositions proved by such means (as opposed to provably objectless propositions) as meaningful assertions about objects for which intuitive evidence might be found at some later point. There are some indirect proofs, however, for which it appears that constructive proofs could not in principle be found. Consider, for example, the prospects for a constructive version of Cantor's indirect proof of the existence of nondenumerable totalities. Some philosophers might be tempted to adopt a form of fictionalism in the case of such "objects." However, we should not rush headlong into mathematical fictionalism in the presence of such purported proofs. Perhaps the best thing to say, in answer to our question, is not that such proofs are not part of some meaningful deductive structure, or of some systems of rules of reason, but rather that the intuitive evidence for objects simply thins out so significantly in these kinds of "proofs," as in some proofs that employ the excluded third, that one cannot hope for them to provide assurances of reliability, or consistency in our reasoning. They represent reason unbound or unconditioned and hence are subject to the possibility of antinomies or paradoxes. Here the idea of looking for relative consistency proofs for the parts of mathematics in question again suggests itself. In the more remote parts of impredicative set theory, however, such proofs will not be easy to come by. Interestingly, there have been serious concerns about evidence even in the highest reaches of transfinite set theory. Thus, Gödel (Gödel 1964), for example, distinguishes the "iterative" (or "quasi-combinatorial") from the "logical" conception of set and suggests that proofs in set theory about sets as objects in the iterative hierarchy do possess some degree of reliability.

§ 3

The concept of proof described in (1) admits of substantial investigation and elaboration. Among other things, the distinction between empty and fulfilled intentions can be looked at in a number of ways. Fulfillments of intentions, in Husserl's philosophy, are also understood as realizations of expectations (Tieszen 1989). One can already see the foundations for this in ordinary perception. In ordinary perception I may have a conception of some object without actually seeing it, so that the intention directed to the object is empty. Even so, I could not help but have some expectations about the object, given the set of background beliefs I would have acquired up to that stage in time. The empty intention, that is, can be viewed as a set of anticipations or expectations about the object which are determined by background beliefs, memories, and so on, at that stage in time. Having such expectation is a fundamental feature of human cognition, especially in contexts where one is attempting to acquire knowledge. Then the expectation(s) may either be realized or not be. Attempting to gain knowledge about an object, that is, is like realizing certain expectations about the object. Of course one's expectations may have to be corrected or refined in the growth of knowledge.

The same structure is present in mathematical experience. Thus, we can think of mathematical propositions (which we are viewing as expressions of intentions) as expressions of expectations. The fulfillment of the intention is the realization of the expectations. Hence, another possible way to think about an epistemological or cognitivist conception of proof is to hold that

(2) A *proof* is a realization of a mathematical expectation.

Speaking of a proof as a realization of an expectation, that is, should not be separated from the meaning that these terms have in a theory of intentionality. Note that although provability here is the same as "realizability," we do not necessarily mean realizability in Kleene's sense. Kleene's realizability interpretations are distinct from the intended interpretation of intuitionistic logic and arithmetic, and possibly from an appropriate cognitivist conception of proof, even though they do give some insight into what it means for a proof to be a realization of an expectation.

Note, incidentally, that on the view we are discussing there is a similarity in cognitive structure which cuts across perceptual and mathematical knowledge, so that there are some interesting analogies between evidence and justification in these two domains, even if some different types of evidence might be involved.

Kolmogorov's interpretation (Kolmogorov 1932) of propositions as problems or tasks, and proofs as solutions, provides another way to look at the difference between unrealized and realized expectations, or unfulfilled and fulfilled intentions. Heyting and Kolmogorov later took their interpretations to be equivalent, although it is possible to question whether they really are in fact equivalent. Attempting to acquire knowledge about an object is, in general, like solving a certain problem about the object. One solves the problem, or fails to solve it, by carrying out certain acts in time that will improve one's knowledge. This is done in a way that is evidently rule governed at some level, for we do not just go about it randomly. The way one solves a problem has parameters which are based on the beliefs one has acquired up to the present time. One might then, by reflection, develop more insight into the structure of the process. Thus, we might also say

(3) A *proof* is the solution of a mathematical problem.

I think it would not be possible to do justice to the epistemological or cognitivist dimension of proof, however, if one were to separate the concepts of problem and solution in (3) from a theory of intentionality.

The point about intentionality is especially important to keep in mind in the final remark I would like to make about the proof/proposition relation. It has been suggested by Martin-Löf (Martin-Löf 1982) that the characterizations of proof in (1)–(3) are equivalent to the statement that

(4) A *proof* is a program that satisfies a particular specification.

That is, the procedure or method by which I fulfill an intention can be viewed as a program that satisfies a given specification. Viewing (1) and (2) in terms of (4) puts us directly in the middle of many interesting issues at the intersection of constructive mathematics, cognitive science, and artificial intelligence. Note especially that if (4) iff (1), then one would be begging some important philosophical questions in cognitive science and artificial intelligence in identifying the notion of a program in (4) with that of a *machine*-computable program, for example, in terms of Turing machine computability. If we did so, for example, it could be argued that our conception of proof is reducible to the formalist's conception that we briefly described at the beginning of the chapter. One might also have other reservations about the conception in (4). For example, it might rule out choice sequences. If this is the case, then it would not allow us to recognize a notion of proof suitable to intuitionistic real analysis. In any case, it seems that if we want to think of a program as a method for fulfilling an intention, then using Errett Bishop's idea of "person programs" (Bishop 1967) would be better. Of course intuitionists have never been willing to identify human computation with machine computation, and this unwillingness is reflected by the status of Church's Thesis in intuitionistic mathematics. Human programs (that is, methods for fulfilling at least arithmetical intentions) may involve some kind of intrinsic intentionality, irreducible semantic content, consciousness, representational character, Gestalt qualities, indeterminateness, implicit content, qualitative content, and so forth. With the proper kinds of qualifications, however, (4) might also be taken as an answer to the question put in this chapter.

One might ask what the implications of this view are for the kinds of "proofs" provided by automated or mechanical theorem proving. One answer would be to point out how counterintuitive it would be to suppose that machines could possess "evidence" in the sense in which we have been using this term. For that would imply that machines must be able to "experience" evidence. Machines, however, do not experience anything or have intentionality, or at least they do not have intrinsic intentionality. One might therefore argue that automated proofs provide evidence only derivatively, only insofar as the proofs are interpreted and understood by beings that have intentionality. Another answer would be to say that the problem whether machines might possess intentionality or not, or be capable of experience or not, is undecided at this stage in time: that is, that the principle of the excluded third is not provable for these propositions.

§ 4

It is possible to develop some of the foregoing ideas in detail in formal theories of proofs or constructions. I do not mean to suggest, however, that any particular formal theory will adequately capture the concept of proof in (1). In Martin-Löf's work the ideas are developed by way of the conception of "propositions-as-types" (e.g., Martin-Löf 1984). Martin-Löf's work is in the tradition of work by Howard, Curry, Girard, Troelstra, Lambek, Läuchli, and others. There has been a great deal of research along these lines, especially in the computer science community, by de Bruijn, Reynolds, Coquand and Huet, and others. Kreisel and Goodman also did some early work on a theory of constructions, using a different approach. What these approaches have in common is the effort to interpret proofs directly as algorithmic processes (acts) or objects. In this respect they are similar to nonstandard interpretations of intuitionistic logic and arithmetic such as Kleene's realizability interpretations of Gödel's *Dialectica* interpretation. The algorithmic theories can be contrasted with the semantic or model-theoretic interpretation of intuitionistic logic and mathematics, starting with the earliest topological models up through Beth and Kripke models (see, e.g., Troelstra and van Dalen 1988). Although the model-theoretic interpretations are thought to be more or less artificial by intuitionists, they nonetheless capture some of the aspects of proof that we have been discussing. Let us briefly consider an example of how some of the ideas we have been discussing can be (partially) formalized in each type of approach.

In his work on intuitionistic type theory Per Martin-Löf has developed what can be looked at as a theory of proofs which is motivated by the kind of philosophical account of proof discussed previously. The concept of proof is formalized in a typed λ-calculus, which can be viewed as an abstract programming language. The λ-calculus is especially natural in this setting since it is about functions as rules, rather than as graphs, and thus represents the idea of a process of going from argument to value as coded by a definition, an idea which also preserves something of the "intensional" flavor of the notion of proof. Martin-Löf's system uses four basic forms of judgment, among which are the two that "S is a proposition" and "a is a proof (construction) of the proposition S." One can read these equivalently as, respectively, "S is an intention (expectation)," and "a is a method of fulfilling (realizing) the intention (expectation) S." Let us abbreviate the latter as "a:S." On this reading the meaning of the logical constants, for example, can be explained as follows: we use $\perp$ for propostions where there is no a such that a: $\perp$. That is, $\perp$ is a false or absurd intention such as $1 = 2$. A method of fulfilling the intention

    (a) S $\wedge$ T consists of (a, b) where a:S and b:T;
    (b) S $\vee$ T consists of i(a) where a:S or j(b) where b:T;
    (c) S $\rightarrow$ T consists of $(\lambda x)b(x)$ where b(a):T provided a:S;

(d) $(\forall x)Sx$ consists of $(\lambda x)b(x)$ where $b(a){:}Sa$ provided a is an individual;

(e) $(\exists x)Sx$ consists of (a, b) where a is an individual and $b{:}Sa$;

(f) $\neg S$ is an abbreviation of "$S \to \bot$."

Martin-Löf, like Dummett (Dummett 1977), Prawitz (e.g., Prawitz 1978), and others, distinguishes canonical from noncanonical proofs and notes that this list shows methods of canonical form only. That is, it gives an explanation of what constitutes a "direct" proof of a proposition formed by means of one of the constants. The meaning of a proposition is thought to be given by what counts as a canonical proof of it (see, for example, Sundholm 1986; Prawitz 1978). In the case of conjunction, for example, a canonical proof of $S \wedge T$ consists of a proof of S and a proof of T. A noncanonical ("indirect") proof in intuitionistic mathematics is a method or program for obtaining a canonical proof. Every introduction rule, in Gentzen's sense, gives a canonical proof, whereas elimination rules give noncanonical proofs of their conclusions. Thus, think of the right-hand side as the conclusion of an introduction rule. A canonical proof is one which has a form by which one can directly see that it follows from one of the rules. Martin-Löf's type theory as a whole is developed as a set of formation, introduction, elimination, and equality rules. Thus, for example, the introduction rule for natural numbers is

$$0 \in N \qquad \frac{a \in N}{a' \in N}$$

10^{10} is not obtainable by the rule even though it is an element of N. But we know we can bring it to the form a' for some $a \in N$. Or consider a proposition such as $10^{10} + 5^2 = 5^2 + 10^{10}$. We know how to obtain a canonical proof. A noncanonical, and shorter, proof would be to show that $(\forall x)(\forall y)(x + y = y + x)$ by mathematical induction (which is an elimination rule in Martin-Löf's system) and then instantiate (which is also an elimination rule).

Consider how one could justify the following conditional by using the clauses (a)–(f): $((S \wedge T) \to U) \to (S \to (T \to U))$. It might be helpful to think of how the proof of the conditional would look in a proof tree or in a Fitch-style system of natural deduction. Let $a{:}((S \wedge T) \to U)$. That is, a is a proof that coverts any proof (b, c) of $S \wedge T$ into a proof $a((b, c))$ of U. We would like a method of fulfilling (realizing) the intention (expectation) $(S \to (T \to U))$, so let $d{:}S$ and $e{:}T$. Define a construction k such that $k(d){:}T \to U$. That is, $(k(d))(e){:}U$. We should set $(k(d))(e) = a((d, e))$, so that, using the functional abstraction operator, $k(d) = \lambda e.a((d, e))$ and

$k = \lambda d.\lambda e.a((d, e))$. The proof needed for the conditional is a construction which carries a into k, that is, $\lambda ade.a((d, e))$.

As another example consider $(\exists x) \neg Sx \to \neg (\forall x)Sx$. Let $(a, b):(\exists x) \neg Sx$. That is, b is a proof of $Sa \to \bot$. Suppose $c:(\forall x)Sx$. Then $c(a):Sa$, and hence $b(c(a)): \bot$. Thus $\lambda c.b(c(a)): \neg (\forall x)Sx$. So the proof needed for the conditional is $\lambda (a,b)\lambda c.b(c(a))$.

In order to do some mathematics one can carry this approach further by developing, as Martin-Löf does, formation, introduction, and elimination rules for finite sets, natural numbers, lists, and various predicates defined by transfinite induction and recursion. There are a number of variations on this kind of theory in the literature, as well as a type-theoretic interpretation of constructive Zermelo-Frankel set theory, and connections with other theories.

As an example of the semantic or model-theoretic approach, consider the notion of proof in (1) in the context of Kripke models. Kripke models are especially nice for providing the kinds of weak counterexamples that are the staple of intuitionism. A Kripke model is a triple $K = <A, D, I>$, where A is an inhabited partially ordered set (species), D is a mapping from A into a collection of inhabited sets, and I is a mapping defined on pairs $<\alpha, P>$ of elements of A and predicate symbols, or pairs $<\alpha, c>$ of elements of A and constants such that (for $\alpha, \beta \in A$)

(a) $\beta \le \alpha \Rightarrow D(\alpha) \subseteq D(\beta)$,

(b) $I(\alpha, P) \subseteq (D(\alpha))^k$ for k-ary P

   $I(\alpha, c) \in (D(\alpha)$,

(c) $\beta \le \alpha \Rightarrow I(\alpha, P) \subseteq I(\beta, P)$

   $\beta \le \alpha \Rightarrow I(\alpha, c) = I(\beta, c)$.

For 0-ary predicate symbols $I(\alpha, P) \in \{1, 0\}$ where

$$\beta \le \alpha, \qquad (I(\alpha, P) = 1 \Rightarrow I(\beta, P) = 1).$$

D(I) is the "domain function." The interpretation of a first-order language by a Kripke model is defined inductively. We suppose that the language contains constants for all elements of $\cup_{a \in A} D(\alpha)$, and suppose a is denoted by $\bar{a}$. Where only closed formulas are considered, we have the following definition:

(a) $\alpha \Vdash P(\bar{a}_1 \ldots \bar{a}_k)$ iff $< a_1, \ldots, a_k > \in I(\alpha, P)$

   $\alpha \Vdash P$ (0-ary) iff $I(\alpha, P) = 1$

(b) $\alpha \Vdash S \vee T$ iff $\alpha \Vdash S$ or $\alpha \Vdash T$

(c) $\alpha \Vdash S \wedge T$ iff $\alpha \Vdash S$ and $\alpha \Vdash T$

(d) $\alpha \Vdash S \rightarrow T$ iff for all $\beta \leq \alpha$, $\beta \Vdash S \Rightarrow \beta \Vdash T$

(e) for no α, $\alpha \Vdash \bot$

(f) $\alpha \Vdash (\forall x)\, Sx$ iff for all $\beta \leq \alpha$ and for all $b \in D(\beta)$, $\beta \Vdash S\bar{b}$

(g) $\alpha \Vdash (\exists x)\, Sx$ iff there is an $a \in D(\alpha)$ such that $\alpha \Vdash S\bar{a}$

(h) $\alpha \Vdash \neg S$ iff for all $\beta \leq \alpha$, $\beta \not\Vdash S$.

$\alpha \Vdash S$ is usually read as "α forces S," or "S is true at α." On the basis of (1), we can read it as "S is fulfilled at α." It is worthwhile to work through each of the clauses (a)–(h) to see what they mean on this interpretation. Intuitively, we think of mathematical research as progressing in stages in time. We view S as the expression of an intention, and the elements of A as *stages* in time at which we may or may not have *evidence* about such intentions. We should have a partial ordering of the stages and not, for example, a linear ordering, because at a given stage there will typically be for a mathematician M various possibilities about how his/her knowledge might progress. M might even stop attempting to gain knowledge altogether. We might think of M as not only having evidence for *truths* at given times, but also as possibly acquiring evidence for *objects* as time progresses. The models also postulate a monotonicity condition on fulfillment: intuitively, M does not forget at later stages when S is fulfilled at some earlier stage, and once a conclusion is reached, no additional information will cause it to be rejected. This represents a certain idealization of human knowledge. If the intention expressed by S at stage α is not fulfilled, $\alpha \not\Vdash S$, then we may think of it as *empty*. To say the intention is empty does not mean that S is not directed for M, or not meaningful for M, but only that at α it is not fulfilled. Hence we can speak of empty and fulfilled intentions or cognitive acts at a given stage in time. Intentions that are empty at one point in time may later come to be fulfilled, or frustrated. By viewing the models as tree models in the standard way one gets a graphic representation of fulfilled and empty intentions in alternative courses of possible experience. As in our discussion of (2), intentions S that are empty can be understood as *expectations* or anticipations. To say that $\alpha \Vdash S$ is thus to say that the expectation expressed by S is realized at α.

Now define $K \Vdash S$ iff for all $\alpha \in A$, $\alpha \Vdash S$; and $\Vdash S$ iff, for all K, $K \Vdash S$. Also, $\Gamma \Vdash S$ iff in each K such that if, for all $T \in \Gamma$, $K \Vdash T$, then $K \Vdash S$. It is known that $\Gamma \vdash S$ in intuitionistic first-order logic iff $\Gamma \Vdash S$. For Kripke models, proof of a proposition amounts to fulfillment in all models. One might relate this to the sense of "necessity" that a proof is supposed to provide, that is, not just fulfillment in some models but fulfillment in every model, or in every model in which assumptions are fulfilled. The countermodels provided by Kripke semantics to some propositions of classical logic show that we do not have evidence for the necessity of those propositions. However, the proof that $\Gamma \Vdash S \Rightarrow \Gamma \vdash S$ is classical. A constructive completeness proof has not been forthcoming. One could view Beth models and related models as variations on capturing some of the ideas we have been discussing.

14

Logicism, Impredicativity, Formalism

Some Remarks on Poincaré and Husserl

Poincaré is well known for claiming that logicist and formalist attempts to provide a foundation for mathematics are misguided. He is also known for holding that any mathematical results that depend on viciously circular impredicative definitions are illusory. In Poincaré's work these views are linked to a form of constructivism about the natural numbers, to the idea that it is appropriate to reason about only those objects that can be defined in a finite number of words, and to the related idea that in order to be meaningful mathematical theorems must be finitely 'verifiable'. Poincaré's arguments on these matters have been influential, and I believe we can still profit from thinking about them today. My own viewpoint on the foundations of mathematics has been influenced by Husserlian transcendental phenomenology, but I find that Poincaré and Husserl have similar positions on a number of general epistemological issues, and that some of their views reinforce one another.

In this chapter I shall discuss Poincaré's views on logicism, impredicativity, and formalism from the perspective of some central themes of Husserl's transcendental phenomenology. Among the most important themes in phenomenology are that mathematical cognition exhibits intentionality, and that the meaning and reference of mathematical propositions must be understood accordingly. This is accompanied by the view that there are various 'meaning categories' in our thinking, and

A shorter version of this chapter was presented as a plenary lecture at the International Congress Henri Poincaré, Nancy, France, May 1992. I benefited from the helpful comments of a number of the audience members, especially Moritz Epple, Gerhard Heinzmann, Jairo da Silva, and Jan Wolenski.

corresponding 'regional ontologies'. Other important themes are that a form of intuition is required for mathematical knowledge, that there are founding and founded acts of mathematical cognition, and that an analysis of the origins of mathematical concepts is needed. Finally, I claim that in transcendental phenomenology we see a more subtle relationship between subjectivity and objectivity in mathematics than has heretofore been recognized in standard forms of constructivism and platonism. I shall draw some comparisons and contrasts with Poincaré on these themes as I proceed.

In the phenomenological view of mathematics there is, as in Poincaré's work, a resistance to logicism and some types of formalism. This is one theme on which some of the views of Poincaré and Husserl appear to reinforce one another, insofar as both hold that the natural number concept has a basis in intuition, and that it is in some sense primitive and cannot be adequately defined in the style of logicists or set theorists. Both also hold that there are some serious problems with strict formalism. Husserl has a place for a kind of formalism but he certainly does not take an exclusively formalistic view of logic and mathematics. I shall argue that some of the views of Poincaré and phenomenology are correct on certain points about logicism and formalism, and in the ensuing paragraphs I say how I think they are correct. The matter of impredicativity is somewhat more complicated. In retrospect, it appears that some of Poincaré's views on impredicativity were not correct. I shall argue that a phenomenological view does better on the matter of impredicativity in at least one way, even though a number of Poincaré's insights about impredicativity are still valuable.

We live at a point in history in which basic philosophical conceptions of mathematics such as those of logicism and Hilbertian formalism have been found wanting, and in which new conceptions are needed. Since Poincaré and Husserl were among the earliest critics of logicism and strict formalism, it is natural to look to their views in the current reassessment of the foundations of mathematics. By continuing to study their work we can, I hope, deepen our understanding of the nature of mathematics and logic and arrive at a better positive characterization of these disciplines.

§ 1 Logicism

Poincaré and Husserl were both critical of logicism, although they were involved in the debate at different stages. Husserl's target was Frege's *Grundlagen der Arithmetik*, which he subjected to criticism in the *Philosophie der Arithmetik* (1891). Poincaré, writing slightly later, targeted Couturat and Russell. I shall focus mostly on Frege's version of logicism in this chapter, but I think that a number of the points I make also apply to other forms of logicism.

Frege (Frege 1884, 1893) held that the natural number concept was definable in and, in effect, reducible to a system of (impredicative) higher-order logic in which one would appeal only to extensions of concepts. The notion of analyticity would be understood in terms of this system of logic, so that one would be able to show that the propositions of arithmetic were analytic a priori, and not synthetic. One would thereby be able to see that the propositions of arithmetic depend upon reason alone, instead of on pure intuition or sense experience, and that they do not have some irreducible mathematical content.

Poincaré and Husserl saw arithmetic quite differently than Frege. Let us first briefly review Poincaré's ideas. Poincaré denies nearly everything in this brief characterization of Frege's view. The only thing he does not explicitly question is Frege's extensionalism about mathematics and logic. In this Poincaré is unlike Husserl, as we will see later. Poincaré's standard complaint against those who would derive arithmetic from logic or set theory was that their derivations already presupposed the concept of the natural numbers. The *petitio principii* of which they were guilty could involve a very simple use of the natural number concept, as in Poincaré's objection to Couturat (Poincaré 1908, p. 158),[1] or a more sophisticated use in the form of mathematical induction. What is valuable in Poincaré's objection, as I read him, is the idea that our knowledge of the concept of number has its origin in intuition and is primitive. Directly contradicting Frege's logicism, Poincaré says that mathematical induction is an expression par excellence of the primitive basis of our synthetic a priori arithmetical knowledge in intuition. In a formulation that anticipates Brouwer, Poincaré describes mathematical induction as "the affirmation of the power of the mind which knows it can conceive of the indefinite repetition of the same act, once the act is possible" (Poincaré 1902, p. 13). Mathematical induction expresses an a priori property of the mind itself. It is distinct from empirical induction, it is not a matter of convention, and it is not subject to 'analytic proof'.[2] Poincaré seems to think of the

---

[1] Page numbers in citations of Poincaré's works refer to the English translations of his works indicated in the Bibliography.

[2] Given these remarks on the a priori nature of mathematical induction, and on how it is distinct from empirical induction, it does not seem charitable to claim, as Goldfarb (Goldfarb 1988) does, that Poincaré's views are psychologistic.

criterion of analyticity in more strictly Kantian terms than Frege, that is, in terms of identity or the principle of contradiction. Thus, he argues (Poincaré 1908) that induction ('the rule of reasoning by recurrence') is irreducible to the principle of contradiction. Analytic proof, which is what Poincaré takes to be found in pure logic, is characterized as 'sterile' and 'empty', when contrasted with mathematics. Logic remains barren, Poincaré says, unless it is fertilized by intuition.

Poincaré also criticizes logicist and Cantorian definitions of number on the grounds that they are impredicative and subject to vicious circularity. Although it is true that Frege's system contained a contradiction and that various paradoxes had appeared in set theory, I think that we must now see the role of impredicativity in mathematics differently from the way Poincaré saw it. The presence of impredicativity in mathematics, or the failure to apply the vicious circle principle, does not necessitate paradox. We can, however, preserve something of Poincaré's point even if we agree that impredicativity itself does not necessitate paradox. It has been argued that impredicative specification in the context of very abstract and comprehensive systems of the sort put forward by logicists and set theorists is incompatible with the idea that mathematical propositions must be based on constructive intuition. If these arguments are correct and if the natural numbers could be adequately defined in such comprehensive systems, then one might argue, as Frege did, that propositions of arithmetic do not depend on intuition. We can therefore see, from another perspective, why Poincaré might criticize logicist and set-theoretic definitions of the natural numbers on account of their impredicativity, even if he overstated the case by declaring impredicative specifications in these theories to be meaningless. I will return to the topic of impredicativity later.

The phenomenological view also entails the denial of nearly everything in our brief characterization of Frege's position. Husserl explicitly criticized Frege in the *Philosophie der Arithmetik*. I will discuss one part of this early criticism by way of some of Husserl's more mature views in the *Logical Investigations* (second edition), *Ideas I*, *Formal and Transcendental Logic*, and *The Crisis of the European Sciences and Transcendental Phenomenology*.

A central feature of the phenomenological view is the claim that mathematical cognition exhibits intentionality. This means that mathematical cognition is directed toward (or 'referred' to) mathematical objects or states of affairs by way of the intentional contents (or 'senses') of its acts, where the objects or states of affairs need not exist. This is, in some respects, a cognitive counterpart to Frege's idea that an expression can have a sense (or 'meaning') but lack a reference.

There can be little doubt that we in fact think of objects in mathematics under different meanings or under different types: natural numbers, real numbers, complex numbers, and so on. These are meanings under which our experience in different areas of mathematics is organized. Although both Husserl and Frege have a meaning theory in which they distinguish sense from reference, Husserl (Husserl 1913) holds, unlike Frege, that in our thinking and experience we find various irreducible 'meaning categories' and corresponding 'regional ontologies'. This view applies quite nicely within mathematics itself. For example, it applies to the domains of objects just mentioned. In particular, it seems that mathematical induction is central to the category of the natural numbers. It is closely related to the meaning of 'natural number', but it does not play the same role in the case of other kinds of mathematical objects. It is also held in phenomenology that the meanings under which we originally think objects are informal or quasi-formal. Formalization has its origin in these meanings, as Frege might have agreed, and we must therefore recognize the use of informal rigor in mathematics. The syntactic features picked out in the formalizations then emerge relative to the meanings, even if they do not exhaust them. I relate this last point to Poincaré's critique of formalism later.

Husserl also discusses the idea of a purely formal ontology and promotes the idea of formalized mathematics. It must be by virtue of the meanings of our acts, however, that we have genuine insight into and understanding of these different areas of mathematics. The horizons of mathematical acts are determined in a rule-governed way by these meanings. The meanings of our acts fix our expectations and our understanding of how to work toward the ideal of completing our knowledge through posing and solving open problems. They determine a set of possible ways of filling in our knowledge. Only certain possibilities are compatible with the meaning under which we think about a mathematical object. Hence, to abandon meaning, or meaning as we actually have it in mathematics, is to lose our aims and to undermine reason itself. From this perspective, Frege's particular effort to found arithmetic on reason alone actually appears as a subversion of reason. Logic, for Frege, is extensional and is supposed to be topic neutral. Fregean logicism levels all meaning categories.

Given his emphasis on intentionality and the meaning/reference distinction, Husserl holds that we must always start with objects precisely as they are intended. This makes his view more faithful to the actual practice of mathematicians than the views of those who wish to define or derive the principles of number from logic or set theory. If we take objects such as natural numbers as they are intended in mathematical practice, then we can see why Husserl claimed early on in *Philosophie der Arithmetik* that it is a problem that the analyzed sense in Frege's definition of number is not the same as the original sense (see Tieszen 1990 and Chapter 15). Number terms need not be taken as referring to Frege's infinite equivalence classes, nor to any other specific kinds of sets that can be used to model the principles of arithmetic.

The phenomenological view generally implies that we should be cautious about eliminative reductionist schemes in the foundations of mathematics. Reductionism is the effort to collapse the differences between the meanings under which we think objects by introducing translation schemes and the like. Problems can arise when reductionism is coupled with eliminativist motivations. The logicist and set-theoretic efforts to provide explicit, extensional definitions of numbers constitute one such form of reductionism. In Husserl's work there is a resistance to such a one-sided reductionism and extensionalism about mathematics and logic.[3] The discovery of set-theoretic definitions of the natural numbers is important, but there are infinitely many extensional definitions of the natural numbers that satisfy the principles of arithmetic and do not preserve extensional equivalence. A comment of Poincaré (1908, p. 154) comes to mind here, even if it is somewhat overstated: "you give a subtle definition of number, and then, once the definition has been given, you think no more about it, because in reality it is not your definition that has taught you what a number is, you knew it long before, and when you come to write the word number farther on, you give it the same meaning as anybody else." Poincaré thinks the problem is compounded by the fact that logicism and Cantorism start with notions that are unclear and unfamiliar instead of starting with clear and familiar notions.

As we consider these ideas of phenomenology in relation to Poincaré, we need to keep in mind several important points about meaning theory. Both Husserl and Frege distinguish the meaning of an expression from its reference, but there does not seem to be such a distinction in Poincaré's writing. Poincaré does not write extensively on meaning theory, but he

---

[3] Husserl in fact favored the development of intensional logic, and in this he is perhaps unlike Poincaré. Frege also came up against the idea of developing an intensional logic, but he of course did not pursue the matter.

does say (1913, p. 62) that if a mathematical theorem is not finitely verifiable, then it is meaningless. 'Verifiable' evidently means that it must be possible in principle to provide an instance for which the proposition holds. General propositions for which we do not possess proofs, for instance, Goldbach's conjecture, would then be meaningful insofar as we can at least verify the statement for an instance. Without attempting to pursue Poincaré's views on meaning any further, we can already draw a contrast with the phenomenological view. On the phenomenological view, propositions can have meaning quite independently of whether or not we can find an instance for which the statement is true. Even if we happen to be working under an illusion at a particular stage in our mathematical experience, our experience at that stage is still meaningful (as Frege discovered). It is clear from his *Grundgesetze* and from his responses to Husserl and others that Frege wanted to prevent intensionality from obtruding in the science of mathematics itself. In matters of meaning theory, however, Frege and Husserl both held that we should not confuse lack of reference with lack of meaning, nor even logical inconsistency with lack of meaning. In the same vein, I think we must say that meaningfulness could not be a function of the distinction between the finite and the infinite, nor of the distinction between the predicative and the impredicative. Poincaré's view can probably not accommodate these points. There appears to be no explicit mention of intentionality or of a sense/reference distinction in Poincaré's writing. This simple point about Poincaré's meaning theory has some important consequences which are noted later.

For Poincaré, Husserl, and the post-logicist Frege, the view that arithmetical propositions have an irreducible mathematical content or meaning accompanies the claim that arithmetical knowledge depends upon an a priori form of intuition. As the objections of Poincaré and Husserl suggest, arithmetic truths are not known to be true by virtue of purely logical (or analytic) relations between propositions alone. As Husserl would put it, there must be intuitions of objects or states of affairs corresponding to the meanings under which we think objects if we are to have arithmetic knowledge. There must be some way of providing a referent corresponding to the meaning. In Husserlian language, our arithmetic intentions must be fulfillable. Husserl defines intuition in terms of the fulfillment of our intentions. In order to support a knowledge claim about a particular object, there must be an intuition of that object. This bears a certain similarity to Kant's claim that all existence statements are synthetic. Fulfillable means fulfillable *for us*, so that the phenomenological view of the

knowledge of particular numbers, as does Poincaré's, takes on a constructivist slant. In phenomenology, the question of the intuitive basis of the
natural numbers is this: are intentions directed toward particular natural numbers fulfillable? This is just a way of asking whether the natural
numbers are constructible, and the answer is yes.

In some other work (Tieszen 1989, and Chapters 13 and 15 of this
book) I characterized a nonlogicist conception of proof in terms of these
ideas. The conception is also nonformalistic in Hilbert's sense. The idea
is that a mathematical proof is just the fulfillment of a mathematical intention. Mathematical intentions are organized in different, irreducible
categories. A proof in mathematics must therefore depend upon a mathematical intention that remains invariant through different cognitive acts
in time in which this intention is (partially) fulfilled. We cannot expect
either the mathematical intention or its corresponding proof to be reducible to pure logic alone; nor can we expect them to be reducible to the
meanings under which we think of such objects as sets. I have described
mathematical intentions, alternatively, as expectations, problems, or specifications. Even if they are presently unfulfilled, they nonetheless have a
regulative function in our thinking. In fact, I have argued (Tieszen 1994a)
that one can regard the principle of the excluded middle in mathematics as having such a regulative function in the belief that mathematical
problems are solvable, even if we do not presently possess solutions. If
all of this is correct, then Frege's idea of capturing all of mathematics in
a system of purely formal, gapless 'logical' proofs is doomed. The syntax involved in the formalizations will generally not be adequate to the
meanings. In logic, as it is understood by Frege, we also do not have the
specificity and contextual background to capture our intentions in different parts of mathematics. Hence, we cannot expect logical gaplessness
in mathematical reasoning.

This view appears to be closely related to an explication of Poincaré's
antilogicism which has been put forth by Detlefsen (Detlefsen 1990,
1992). Detlefsen says that a cornerstone of antilogicism in the style of
Poincaré is that rational thought is divided into irreducibly heterogeneous local domains, with each such domain governed by its own 'architecture' or 'universal'. This is similar to my claim that there are various
irreducible meaning categories and regional ontologies in our thinking.
One important difference, however, is that I have described rationality
in terms of the intentions or meaning contents of cognitive acts, and it
is not clear that Detlefsen sees it this way. If he does not see the matter
this way, then there might be some problems about how to understand

these universals or architectures in the framework of Poincaré's thought, for the appeal to 'universals' suggests a form of realism that may be difficult to reconcile with Poincaré's other views. Perhaps Poincaré would not reject the idea of adopting a transcendental view of these 'universals' in a theory of intentionality. The problem, however, is that it is not clear that this would be consistent with the elements of his meaning theory described earlier. It is basic to the concept of intentionality, for example, that there can be beliefs (intentions) about things that are not known to exist or that do not exist.

Much of what Detlefsen says about these 'architectures' or 'universals', however, does parallel what I have said about intentions. In fact, it is useful simply to exchange these terms in Detlefsen's account in order to see the similarities. Detlefsen says that logicism seeks to do away with heterogeneous domains in reasoning. Logicism holds that what is remarkable about mathematics is its homogeneity with the rest of rational thought. For Frege, the denial of an arithmetical law is taken to result in a global failure of rational thought. As Detlefsen explicates Poincaré, however, the grasp of different parts of mathematics depends on the grasp of such universals (or, as I say, intentions). They are essential to a truly scientific understanding of a given domain of inquiry. Indeed, Detlefsen argues that a distinctively mathematical proof requires the presence of a comprehending universal (intention) which persists through the differences involved in states of the development of this universal (intention).

Detlefsen suggests that distinctively mathematical proofs are different from logical inferences because they are based on different kinds of universals (intentions). These universals serve to organize or unify mathematical thinking. The difference between a distinctively mathematical proof and a logical inference centers on the choice of universals under which premisses and conclusion are united. A mathematical proof unites these as a development under a particular mathematical architecture or theme. There is also a parallel to my view in Detlefsen's remark that these universals (intentions) serve not only as standards with which our epistemic choices are to accord, but as guides which direct those choices. They have a regulative function in our mathematical thinking.

Detlefsen goes on to say that the elimination of gaps between steps in a proof does not call for the exclusion of topic-specific information in an inference, which is what logicism demands. Instead, it calls for inclusion of a mathematical universal to fill what would otherwise be a mathematical gap between premisses and conclusion. The key feature of a Poincaréan proof, according to Detlefsen, is the grasping or intuiting of a mathematical architecture between premises and conclusion.

Although much of this parallels the phenomenological view I favor, I question how well it accords with the thinking of Poincaré. For example, I do not find the notion of mathematical architectures or universals in Poincaré's work; nor do I find the idea that universals could themselves be objects of intuition. It is also not clear to me that Poincaré has the idea that there are states of development of a universal. Perhaps an object falling under a universal could be intuited, but even here we seem to be importing into Poincaré's view language that makes it difficult to know what he would say. Moreover, how does Detlefsen's account give us Poincaré's constructivism or predicativism? It perhaps sets into clearer relief Poincaré's antilogicism, but we do not necessarily get constructivism or predicativism from the idea that a mathematical proof requires a universal which persists through the differences involved in states of development of the universal. Such a development presumably also takes place in highly nonconstructive proofs in set theory. My approach to this is to claim that in very abstract parts of set theory there may still be a development based on reflection on the meanings or intentions expressed by our mathematical propositions, even if these intentions are not themselves fulfillable. This position seems to be quite different from what we find in Poincaré, as noted in our earlier comments on meaning theory. It is not clear whether Detlefsen recognizes universals or architectures for which there are no corresponding instances, or for which we cannot find such objects. Are there meaningful mathematical propositions which refer to objects that we cannot construct? If so, how would this fit in with Poincaré's other views? A number of other questions could also be raised about Detlefsen's construal.

On my view, some of the major objections to logicism are based on the meaning theory that accompanies the theory of intentionality. It is not clear that Detlefsen sees the matter this way. I claim that if we keep in mind the intentionality of human cognition, then we are quickly led to many other important insights, some of which are discussed later.

Another suggestion about these matters has been made by Heinzmann (Heinzmann 1997). Heinzmann argues that ideas of Peirce can be used to give a semiotic reconstruction of the mathematical architectures involved in mathematical inferences. This is an interesting approach, which, in my opinion, would lead to some fruitful comparisons with what I have said in this essay.

I would like to mention one other feature of the phenomenological theory of intuition, which, in my view, bears an interesting similarity to Poincaré's antilogicism and anti-Cantorism. Husserl thinks that arithmetical knowledge is originally built up in founding acts from basic, everyday intuitions in a way that reflects our a priori cognitive involvement. In order to understand the sense of the natural number concept one must analyze its origins and the manner in which it is built up in acts of counting, collecting, correlating, abstraction, reflection, and formalization.

This gives us another way to elaborate on what motivates Poincaré's *petitio* argument and his claim (Poincaré 1908, pp. 144–145; 1913, p. 64) that logicists and Cantorians are involved in reducing clear or familiar notions to notions that are less clear or familiar. Poincaré says that the procedure of the logicists and Cantorians is 'contrary to all healthy psychology'. His claim here need not be viewed as falling into the trap of psychologism. We can take the point to be epistemological, and to be about the nature of mathematical evidence. It is a basic epistemological fact, for example, that we are finite beings. Elementary finitary processes such as counting objects in everyday experience, collecting them, or correlating them one to one, are clear and familiar to nearly everyone. It is not necessary to know any set theory in order to be able to do these things. And recall Poincaré's characterization of mathematical induction: it is "only the affirmation of the power of the mind which knows it can conceive of the indefinite repetition of the same act, once that act is possible." On Husserl's view, number theory is founded on processes of this type. We then develop higher parts of mathematics, such as Zermelo-Fraenkel (ZF) set theory, by further abstraction, idealization, reflection, and formalization. We do not develop elementary number theory in this way by starting with ZF, and we certainly do not carry out transfinite processes. To suppose that we do is to get it backward, from the viewpoint of epistemology. The processes on which number theory is founded are more basic, familiar, and clear in the structure of founding and founded forms of meaning, understanding, and evidence.

'Founding', in this sense, is an epistemological relation. It figures into an analysis of the conditions for the possibility of knowledge of a phenomenon. Given the epistemological asymmetry of this relation, one then quite naturally holds that any attempt to *found* number theory on set theory or logic will beg the question. We have already made note of arguments about the irreducibility of various meaning categories and, in particular, about how arithmetical propositions have an irreducible content or meaning. We now add that these other theories will at least presuppose these clear and familiar processes in which number theory has its origins. Elementary finitary activities such as iterating an act, collecting, correlating, and counting are conditions for the possibility of much that we do in mathematics and logic.

It appears, independently of anything that Poincaré says, that there just are parts of mathematics that are more or less intuitive or idealized, or that are better understood or not understood very well. The infinite is, in some sense, less clear, and not as well understood as the finite. But then the Fregean and Cantorian reductionists consistently start on the wrong side of these epistemological divisions. They start with the idea of the actual infinite and admit impredicative specifications of transfinite objects in their systems, and *then* proceed to the finite. Poincaré says that these would-be reductionists start in the field of the infinite and eventually show us a small, remote region in which the natural numbers are hidden. Logicists in particular wish to derive the principles of arithmetic from logic. Whereas first-order logic may be relatively clear and familiar, however, there can be no question of deriving arithmetic from it. On the other hand, is Poincaré not correct to hold that higher-order logic is less clear and familiar than first-order logic?

Distinctions of the sort we are depending on here run deep in philosophy. There is something similar, for example, in Kant's view of the movement of reason from the conditions of sensibility to the unconditioned or absolute. The unconditioned or absolute is not understood very well and unconditioned reason may be involved in antinomies. In Husserl's philosophy we have the similar view that mathematics is built up in founded acts of abstraction, collection, reflection, idealization, and formalization from our basic experience in the lifeworld. That mathematics has its origins in the lifeworld also helps to explain why parts of this subject have applications.

Poincaré and Husserl seem to me, for these reasons, to be correct in holding that the natural number concept has a basis in intuition, that it is primitive, and that it cannot be defined once and for all in the style of logicists or set theorists.

§ 2   Impredicativity

Poincaré associated the paradoxes with the maladies he saw in logicism, Cantorism, realism about sets, and an overreliance on a formal treatment of sets. There could be nothing worse in science than to adopt principles that led to contradictions, and, as did Russell and Weyl, Poincaré thought the problem resided in the use of impredicative definitions that were viciously circular. The way to avoid the paradoxes was to avoid such impredicative definitions, to observe the 'vicious circle principle' (VCP), which, in one formulation, says

no totality can contain members definable only in terms of itself.

We do not now see the set-theoretic paradoxes as a necessary consequence of impredicativity, or believe that we must impose the VCP to avoid paradox. The study of predicativity that emerged from this period is generally taken to have a different significance. It is significant in constructivism, proof theory, computability theory, and the study of the epistemology of mathematics. In this guise I think it still preserves the general sensibilities discussed about starting with what is more familiar and clear – the intuitive and the finite – and building from this basis. It is not clear, however, that impredicativity can or should be excised from mathematics. Indeed, it has been argued by Charles Parsons (Parsons 1992) that some forms of impredicativity are unavoidable in the foundations of mathematics, and that a form of impredicativity is involved in explaining the concept of natural number itself.

Unlike Poincaré, Husserl did not write in any direct way about the paradoxes, impredicativity, or role of the VCP in mathematics. It is curious that Husserl was not exercised by any of the paradoxes, and there is virtually no mention of them in his published work. However, both Poincaré and Husserl emphasize the role of intuition in arithmetical knowledge. As noted, there are arguments purporting to show that the predicative/impredicative distinction marks a limit on the (idealized) constructive intuitability of mathematical objects when the distinction is viewed in the context of very comprehensive systems of the sort put forward by logicists and set theorists. This is not to say that impredicativity must always be inadmissable for Poincaré. Folina (Folina 1992) has argued that Poincaré accepted impredicative specifications involving real numbers because he thought we have a kind of geometric intuition of the reals. In systems that are as abstract and comprehensive as those of the logicists and set theorists, however, we have no independent way of gaining epistemological access to objects that are impredicatively specified.

Gödel (Gödel 1944) is widely quoted as pointing out the bind between constructivity and impredicativity in his discussion of Russell's formulation of the VCP. He does this in the context of discussing reasons for holding the VCP, rather than classical mathematics, to be false. Gödel says that in the form in which it has been stated earlier the VCP is true only if we take the entities involved to be constructed by us. In that case there must clearly exist a definition (i.e., the description of a construction) which does not refer to a totality to which the object defined belongs, because the *construction* of an object cannot be based on a totality of things to which the object to be constructed itself belongs. If the construction of an object were to be based on such a totality, we would simply never get the object. To construct the object we would have to construct the collection of which the object is a member, but constructing this collection requires constructing the object. The circularity here is vicious.

Michael Hallett (Hallett 1984) has described this situation by saying that there is no way to 'complete' an impredicative process, no matter what powers of surveillance or of 'running through in a finite time' are granted to a constructing agent. Folina (Folina 1992), explicating remarks of Poincaré, characterizes the matter by stating that in the case of impredicative definitions that are viciously circular there is no way to obtain a 'determinate' object. What it means for something to be an object, however, is that it be determinate, that it have definite boundaries. Poincaré's objection to impredicative definition is that it does not give us definite objects, objects with determinate boundaries. Predicativity gives us determinate objects, and that is what constructivism requires. Folina argues that impredicative definitions are therefore unfaithful to our understanding of what an object is. I would argue that the situation is more complicated than this, however, and that some additional distinctions need to be taken into account. For example, one might distinguish between what is determinate for us and what is determinate in itself.

In any case, since the VCP holds in the case of constructed entities, it appears to provide one natural cutoff point for what should or should not be admitted in constructive mathematics. Gödel argues, however, that if we are realists about mathematical objects and meanings, then impredicative definitions by themselves are not a problem. Zermelo (Zermelo 1908) had similarly appealed to a form of realism to defend his axiomatization of set theory against Poincaré's charge that it allowed impredicative specification of sets. As Gödel (Gödel 1944) puts it, "If . . . it is a question of objects that exist independently of our constructions there is nothing in the least absurd in the existence of totalities containing members which can be described (i.e., uniquely characterized) only by reference to this totality." Gödel wants to recognize a difference between the definition of an object, in the sense of a unique description of an object *already presumed to exist*, and constructive definition, that is, the description of a process whereby we are to form a new object. Relative to such a realist view, VCP can be denied. But there is reason, prima facie, to think that we have here a limit on the idealized constructive intuition of objects.

As I said, I shall not try to settle the question of whether the predicative/impredicative distinction marks a limit on (idealized) constructive intuitability in set theory. Nor shall I address the matter of whether Husserl is some kind of a predicativist concerning knowledge of the particular objects of our mathematical intentions. Instead, I would simply like to note that there are impredicative specifications in set theory which are so far not known to lead to contradictions. A rich practice in which there are many methods and results has developed in impredicative set theory. It seems to me that we are at least owed an explanation of the conditions for the possibility of this practice.

If Poincaré's thinking about impredicativity and intuition can be faulted, it is not for failing to embrace a form of realism about the objects of impredicative set theory in the style of Zermelo and Gödel, so much as for ruling out the possibility of consistent, meaningful mathematical thinking apart from constructive intuitability. Poincaré suggests, at times, that mathematical propositions must either be founded on intuition or be contradictory. In his criticism of Zermelo's separation axiom in the "La logique de l'infini," however, he appears to back away from this strong position. He notes (Poincaré 1913, p. 59) that Zermelo indeed does not allow himself to consider the set of all the objects which satisfy a certain condition because such a set is never 'closed'. It will always be possible to introduce new objects. Zermelo has no scruples, however, about accepting a set of objects which satisfy a certain condition and which are part of a certain set s, for he believes that we cannot possess a set without possessing at the same time all of its elements. Among these elements, we choose those that satisfy the condition, without fear of being disturbed by new and unforseen elements, since we are supposed already to have these elements at our disposal. Poincaré says that Zermelo has erected an enclosing wall with his separation axiom by positing the set s beforehand to keep out intruders who would enter from without, but he objects that Zermelo does not worry about whether there could be intruders from within his wall (Poincaré 1913, p. 59–60) If s is infinite, then its elements cannot be conceived of as existing beforehand, all at once. It is possible for new elements to arise constantly, and they will simply arise inside the wall instead of outside. Thus, although Zermelo has "closed his sheepfold carefully, it is not clear that he has not set the wolf to mind the sheep." What I would like to note is that Poincaré is not arguing here that a proposition such as the separation axiom is contradictory because it is not founded on intuition. Rather, his position seems to be that we are still in the unenviable position of not knowing that it will not lead to contradictions.

There is, however, a problem for Poincaré at this point. How could he account for what appear in practice to be *meaningful* impredicative specifications of objects in set theory for which we might agree that there can be no constructive intuition, but which are so far not known to be contradictory? Poincaré says that if a mathematical theorem is not verifiable, then it is meaningless, and that verifiability requires the possibility, in principle, of providing an instance for which the statement holds. Poincaré could presumably not mean that a mathematical statement is 'meaningless' because it is ill formed according to the syntax of a particular *formal* predicativist theory. This would appear to be incompatible with his views on the relation of formalism to informal meaning.

It is on this meaning-theoretic point concerning impredicativity that the phenomenological view has the advantage over Poincaré's. Even if it is true that it makes no sense to speak of idealized constructive intuition of impredicatively specified objects in systems such as ZF, it does not follow that we cannot *meaningfully think* or reason about such purported objects without contradiction, even if these purported objects can only be indeterminate for us. Here I am simply appealing to the distinction between thought and intuition, or between meaning and reference, that is part of the phenomenological theory of intentionality. In the case where there is no way to intuit a referent corresponding to a meaning we have only pure conception, but this is not to say that our experience in such cases is meaningless. We allow that there may be meanings (or perhaps 'architectures') that do not have, or for which we cannot find, corresponding referents. This view may be contrasted with constructivist positions such as strict finitism, traditional intuitionism, and even Poincaréan predicativism, which tend to hold that all expressions lying beyond some particular evidential cutoff point are 'meaningless'. I think that such a theory of meaning cannot be correct. For some intentions we may expect to provide a referent at some point, or to find a procedure that will do this, even if we do not have one now. In the case of impredicative definitions in set theory that are viciously circular, it appears that this cannot be expected. We can only hope to eliminate the impredicativity in the given context in favor of a predicative (or perhaps a generalized inductive) characterization if we are to preserve the claim to knowledge of the individual object.

This situation is to be contrasted with cases in which there is a contradiction and impredicativity is involved, as in the definition of Russell's set of all non-self-membered sets. In the case of Russell's set, a meaningful proposition purporting to refer to such an object can be formed, but the assertion that such an object exists leads us to a contradiction. We know that there can be no object corresponding to this thought.

As an alternative to Poincaré, one could claim that in impredicative set theory, as in other domains of mathematics, we also think of objects under certain meanings (or architectures), and that it is by virtue of our insight into these meanings, such as it is, that we have an understanding of impredicative set theory of the sort that could not be explained if we were strict formalists about set theory. It need not be argued that there can be no further clarification of these meanings, or that there will be no further corrections in our understanding.

The realism of classical mathematics or of present-day impredicative set theory can be regarded as the result of idealizing, abstracting, and reflecting even further from the basic processes in which mathematics has its origins. On the present construal, classical mathematics abstracts away from idealized constructive intuition of individual objects when it employs impredicative specifications, but it does not abstract away from meaning. On this view, we do not ignore important epistemological points. Frege and some of the early Cantorians made the mistake of boldly proceeding from the actual infinite, from the less clear and familiar in all of the respects we have mentioned, irrespective of epistemological concerns. The mistake is to suppose that concerns for intuition, origins, founding, acts of abstraction and reflection, and so on, can be set aside. This supposition reflects an unconditioned or uncritical realism (and rationalism) about mathematics, and we see this manifested in the naive comprehension principle. We can therefore preserve some of Poincaré's insights. On the other hand, the paradoxes need not be seen as an inevitable consequence of realism or rationalism, although they might be more likely to result from an unconditioned or uncritical realism or rationalism of the sort found in Frege. The problems with naive realism lie in forgetting origins and in not respecting intuition. They stem from having an ungrounded epistemology for realism.

§ 3   Formalism

I cannot treat everything that Poincaré and Husserl say about formalism, and so I will focus on one general theme that emerges from their writings.

Instead of entering into issues about Poincaré's effort to wield his *petitio* argument against Hilbert's formalism (see, e.g., Parsons 1965; Steiner 1975), I will focus on some of the problems alluded to earlier about the role of meaning in mathematics. These problems are common to the versions of formalism found in Thomae, Hilbert, Curry, and others.

In his discussion of Hilbert, Poincaré (Poincaré 1908, p. 147) notes that for the formalist it is not necessary or even useful to know the meaning of a theorem in order to demonstrate it. We might as well put axioms into one end of a machine and take out theorems at the other end, a procedure Poincaré disdainfully likens to a legendary sausage machine in Chicago. He complains that, on this view, it is no more necessary for a mathematician to know what he or she is doing than it is for a machine. Poincaré (Poincaré 1908, p. 148; also p. 32, p. 51ff.) contrasts this kind of formalism and mechanism with the case of "arguments in which our mind remains active, in those in which intuition still plays a part, in the living arguments, so to speak." A theme that runs through many of Poincaré's papers is that on the basis of logic, formalism, or mechanism alone, it is possible to combine mathematical ideas or entities in many different ways. It is possible to create indefinitely many formal systems. The problem is that most of these formal systems are useless and totally devoid of interest. There must be a process of selection or discernment in order to obtain the relatively small number of combinations that are useful. The nature of mathematical discovery lies in this process of selection, and formalism, mechanism, and logic alone cannot account for it.

We have already seen a suggestion about what the phenomenological explanation of all of this would be. Husserl holds that where there is reason there is thinking of objects under certain intentions or meanings. These meanings govern our expectations about objects and, hence, our investigations. Intentionality just is directedness. The horizons of possibilities associated with mathematical acts are determined in a rulelike way by these meanings. The meanings of our acts fix our expectations and our understanding of how to work toward the ideal of completing our knowledge, and it is on this basis that we set and (nonmechanically) solve problems. Poincaré's "living arguments in which the mind remains active" are precisely the arguments in which this kind of meaning plays a role. If this kind of meaning or intentionality were not involved, there could be no selection or discovery. There would be no directedness, no goal-oriented or purposive activity of the sort that we associate with reason. There would be no particular set of possibilities associated with a meaning, as distinct from some other set of possibilities. The meaning-giving acts of human subjects make this possible. This is at least a necessary condition for selection of the sort that is missing in mechanism or strict formalism. Machines, or formal systems by themselves, have no intentionality or understanding. To abandon meaning, or meaning as we actually have it in mathematics, is to lose our aims and in some sense to subvert reason itself. An exclusionary formalism, taken seriously, involves just the kind of forgetfulness of meaning or of origins that Husserl decries.

I also see no reason why Poincaré's (Poincaré 1908, Chp. 3) emphasis on the role of unconscious cognitive processes in mathematical discovery could not be accommodated by the phenomenological claim that the meanings under which we think objects are a function of a sedimented background and a context of beliefs and practices that are largely unconscious.

A response to the claim that formalism cannot adequately deal with the problem of selection has been offered by Curry (Curry 1951). Curry claims that the systems one should study and the theorems one should be concerned to prove are determined by the needs of other disciplines. Mathematics is saved from sterility and vacuity by its interplay with other sciences. It receives direction and meaning through its application, not through its own distinctive informal content. It seems to me, however, that Curry has not responded to the objection that Frege raised years earlier to Hilbert: formalism makes the applicability of mathematics inexplicable. For example, at what point is sense assigned to the uninterpreted signs of mathematical systems? The scientist presupposes an interpretation of these signs, but Curry holds there is no assignment of meaning to the signs within mathematics. So, as Frege says, the task of assigning sense to mathematical signs "falls into a void between the sciences." Moreover, the same mathematical assumptions are employed in different sciences. An explanation of how mathematics is useful to science will have to account for this general applicability. Mathematics must in large part receive direction through its own informal content, that is, meanings. Applications of mathematics are inseparable from these meanings, and appealing to applications in the style of Curry will therefore not help.

These comments on formalism need not be taken to imply that formalization has no value. We must distinguish strict formalism as a philosophical viewpoint about mathematics from the practice of formalization. Husserl thinks formalization has great value as long as it does not preclude the notion of meaning and degenerate into 'mindless' or 'meaningless' technical work. Indeed, the 'crisis' of the sciences described in works such as the *Krisis* (Husserl 1936) is supposed to involve just such a degeneration. On a strict formalist view, theoretical science amounts to nothing more than the construction and mastery of formal systems. Science becomes nothing but technique. The virtues required for its pursuit are narrow in character. They are the virtues of the technician who acquires familiarity with formal systems. The internal goals of science for such a technician shift from understanding and discovery to formal elegance and pragmatic success. The professional scientist is reduced to a theoretical technician. Formalism, taken strictly, distances the sciences from the informal problems and resources that are central to scientific growth. Speaking of both formalism and logicism in *La valeur de la science* (Poincaré 1905, p. 21), Poincaré says that "in becoming rigorous, mathematical science takes on a character so artificial as to strike everyone; it forgets its historical origins; we see how the questions can be answered, we no longer see how and why they are put."

15

The Philosophy of Arithmetic

Frege and Husserl

The work of Frege and Husserl on logic and mathematics might, from a modern perspective, be compared under three main headings: mathematical logic, philosophical logic, and the foundations of mathematics. Under the first heading there is little room for comparison. Frege surpassed nearly everyone in the history of logic in his technical achievements and discoveries. Husserl contributed virtually no technical work to the development of mathematical logic. Under the second heading, however, there is a great deal of room for comparison. Many of the issues raised by Frege and Husserl involving language, meaning, reference, judgment, platonism about logic, and other matters are still actively debated in research in philosophical logic.

Although there is some overlap among the three areas, the grounds for comparison are different again in the foundations of mathematics. Frege had a technical program for the foundations of mathematics in his logicism and, as we know from the *Grundlagen der Arithmetik*, it was a program shaped by certain philosophical ideas. The program was formulated so precisely in the *Grundgesetze der Arithmetik* that it could be seen to fail. That is, it was possible to derive a contradiction from the basic laws of arithmetic as these had been formulated by Frege, and subsequent attempts to repair the damage led to developments that were further and further removed from Frege's effort to derive the principles of number from logic clearly and decisively. Very late in his career, but still several years before Gödel established the incompleteness theorems

I would like to thank Bill Tait for some helpful critical comments on my views about Frege and Husserl.

for *Principia Mathematica*, Frege came to abandon his logicism completely and to develop some work based on his new 'geometrical' ideas about arithmetic.

Husserl was far less interested than Frege in technical work in logic and foundations. Although many of his ideas lend themselves to mathematical development, Husserl himself did not pursue the details. As a philosopher, he cautioned against the 'blind' or uncritical development of formal work. In the *Logische Untersuchungen (LU)*, for example, he says that mathematicians are primarily technicians and as such they tend to lose sight of the meaning or essence of their theories, and of the concepts and laws that are the conditions for their theories (Husserl *LU*, "Prolegomena to Pure Logic," § 71). On the other hand, he says that philosophers overstep their bounds when they fail to recognize that the only scientifically legitimate development of mathematics requires technical work. In *LU* and other writings Husserl argues that although the philosopher's critique of knowledge and the mathematician's technical work require a fundamental division of labor, they are nonetheless mutually complementary scientific activities.

Not surprisingly, Husserl's strongest contributions are to be found in the philosophy of mathematics. I shall in fact argue that, in its general outline, Husserl's postpsychologistic, transcendental view of arithmetic is still a live option in the philosophy of mathematics, unlike Frege's logicism. It is also superior to Frege's late views on arithmetic in several important respects. I hope to show, in the process, that we still have something to learn by comparing and analyzing the ideas of Frege and Husserl on arithmetic, all the more so because Husserl's ideas are still largely unknown to many people in the analytic tradition of philosophy. In spite of the fact that his logicist program failed, Frege contributed many important arguments on foundations and raised many interesting objections to views such as Husserl's. The tension from some of these objections has not yet dissipated and I shall remark on them at different points in the chapter.

§ 1   Frege on the Foundations of Arithmetic: Logicism

Frege and Husserl approached the foundations of arithmetic in very different ways, although they agreed on several general points about mathematics. Neither was content with formalism in the style of Hilbert, although Husserl's early work in Part II of *Philosophie der Arithmetik* was somewhat closer to that of Hilbert. Both Frege and the later Husserl argued against psychologism; both objected to conventionalism, pragmatism, and naturalism about mathematics; and both appeared to uphold some form of mathematical realism. In order to deepen our understanding of their differences, let us start by recalling, in outline, some of the distinctive features of Frege's views on the foundations of arithemtic.

Already in the *Begriffsschrift* Frege distinguished between two kinds of truths that require justification: those for which the justification must be supported by facts of experience, and those for which a proof can be carried out purely by means of logic alone (Frege 1879, pp. 5–6).[1] Frege tells us that in order to sort arithmetic judgments into one of these two classes he had to make every effort to keep the chains of inferences involved in proofs free of gaps. This led him to develop the formal logic of the *Begriffsschrift*. Only by developing such a rigorous, formal 'concept notation' would it be possible to eliminate appeals to intuition from proofs, and hence to decide finally whether arithmetic judgments were purely logical or were supported by facts of experience.

The ideal motivating Frege's concern for gapless formal proofs is expounded more fully in the *Grundgesetze*, where Frege cites Euclid's axiomatic method as a forerunner of his own idea of a strictly scientific method in mathematics (Frege 1893, pp. 2–3). Frege's idea is to state explicitly in his formal language all of the propositions that are to be used without proof, the basic laws, and to start with the smallest possible number of these propositions. Frege goes beyond Euclid, however, in demanding that all of the methods of inference to be employed in the formal system be specified in advance. It must also be possible to see, once their meanings are explained, that the basic laws of arithmetic express truths of pure logic, and that the rules of inference really are rules of logic and that they are sound. Thus, if the steps of proof are split up into logically simple steps, if we have started from purely logical propositions, and if we can actually derive the standard propositions that are already given in arithmetic from the basic laws by using only the rules of inference specified, we will finally see that the foundations of arithmetic lie in logic alone, and that we need not rely upon intuition or facts of experience at any point. Frege also tells us that by eliminating all gaps in reasoning we will achieve more than an empirical or inductive justification for the truths of arithmetic, and that doing so will be an important advance in mathematics, for as long as we fail to eliminate the gaps there is a

---

[1] Page numbers in citations of Frege's works refer to the English translations of his works indicated in the Bibliography.

possibility that we will have overlooked something that might cast doubt upon the proof of a proposition. Thus, we will finally be able to eliminate the possibility of error in our reasoning, and to establish arithmetic on a secure foundation.

The formal system from which Frege intended to derive the principles of arithmetic is a classical higher-order logic that includes a theory of extensions of concepts (or classes). Within this structure Frege wanted to give an explicit definition of number. The intended interpretation of the system of the *Grundgesetze* was by all indications to be extensional. Frege had of course made the sense/reference distinction (Frege 1892) and had used it to give an account of identity, but he was thereby also enabled in the *Grundgesetze* to set aside the notion of sense, and to prevent intensional aspects of meaning from obtruding. Numbers were to be defined in terms of the extensions of concepts, but concepts in Frege's philosophy were themselves taken to be the references of concept-words. The difference between a concept and the extension of a concept for Frege does not coincide with the distinction between sense and reference. Rather, it amounts to a difference in saturation: concepts are not saturated and so are not objects, but extensions of concepts are saturated and are objects. Frege knew that dealing with oblique contexts would require an intensional logic, but oblique contexts were not to occur in the *Grundgesetze*, and Frege did not go on to develop an intensional logic. The argument of the *Grundlagen* and his exchanges with Husserl also indicate that, on the whole, he sided with extensionalist approaches to logic.

In any case, the definition of number would be an extensional definition and because the definiens would involve extensions of concepts it would be reductive in nature, 'reducing' numbers to infinite equivalence classes. Of course Frege thought that this reduction involved no notions that were not purely logical; that he was appealing only to the acknowledged 'logical' notion of extensions of concepts in his definition. Numbers would thus be logical objects. It is interesting, however, that in the Introduction to the first volume of the *Grundgesetze* Frege had expressed some concern that his basic law involving the notion of the course-of-values of a function (the ill-fated Basic Law V), which includes the notion of the extension of a concept as a special case, might not be viewed as a law of logic. Basic Law V is of course the source of the contradiction in Frege's system, given his other assumptions. In the Appendix on the Russell paradox in the second volume of the *Grundgesetze* (Frege 1903), Frege says that he never concealed from himself the fact that Basic Law V lacked the self-evidence possessed by the other basic laws.

As we know from the *Grundlagen*, Frege thought that many old and fundamental philosophical questions about arithmetic were intimately linked to his logicist program. The principal dispute the *Grundlagen* is meant to settle, he says, is whether the concept of number is definable or not (Frege 1884, p. 5). He argues that if there are independent grounds for believing that the fundamental principles of arithmetic are analytic (as opposed to synthetic), then these would also tell in favor of the principles' being provable and of number's being definable. If his program could be carried out, we would also see that true arithmetical propositions do not have some irreducible mathematical content, that they are not synthetic a priori and do not depend on (pure) intuition as Kant had held, that they are not empirical in nature as Mill had held, that various theories of numbers as aggregates are mistaken, that efforts to identify numbers with perceivable signs are mistaken, and so on. In addition, Frege's vehement attacks on psychologism are meant to show that numbers are not mental entities, that various efforts to understand numbers by appealing to psychological processes such as abstraction are mistaken, and that true arithmetical propositions cannot be construed as psychological laws. Frege thought it crucial to the development of the science of logic to establish the objectivity of logic, and hence of number, against the subjectivity of our ideas about number.

We should note that although in many of his writings Husserl uses the term *logic* in a very wide sense, even going so far as to subsume mathematics under logic in his general theory of deductive systems, we do not find in his work anything like the picture that Frege presents. Husserl often tends to think of logic in the tradition of Bolzano and others as something like a general theory of science. Although Husserl speaks of 'logic' in this very wide sense, he could not of course be a Fregean logicist because he tells us in later works such as *Formale und transzendentale Logik (FTL)* that a phenomenological-constitutional foundation of formal logic is needed, and that the philosophical basis of logic is to be found in transcendental phenomenology. Moreover, in his own early work on number in *Philosophie der Arithmetik (PA)* he had argued that the concept of number cannot be reduced to logical notions in Frege's sense.

§ 2   The Frege-Husserl Dispute over Arithmetic

In the *Philosophie der Arithmetik*, published in 1891, Husserl had criticized the views on number that Frege expressed in the *Grundlagen*. Frege and Husserl had begun a correspondence in 1891, several years before Frege's well-known 1894 review of *PA* (Frege 1894). A central point in their disagreement, from Husserl's perspective, concerned the central question of the *Grundlagen*: is the concept of number definable or not? Frege thought that it was and Husserl thought that it was not. I shall argue, however, that there is a sense in which each philosopher missed the other's point. From Frege's perspective, judging from his review of *PA*, the locus of the disagreement concerned the intrusion of psychology into virtually every aspect of logic. Indeed, Frege claims that Husserl is even confused about the nature of definitions because he fails to keep psychology distinct from logic. In recalling this claim, however, we need to keep in mind the vicissitudes of Frege's own views on definition and the problem of how definition should be understood in his own late philosophy of arithmetic. I shall claim, in any case, that there is a point to Husserl's argument on definitions that is quite independent of psychologism. I shall not recount specific elements of Frege's criticisms here, for I agree that there is much that is misleading, unclear, or even wrong in *PA*. In *PA* Husserl does blur the distinction between the subjective and the objective to detrimental effect at times, his remarks on one-to-one correspondences are confused, and a host of difficulties of interpretation surround his view that numbers are aggregates of featureless units. But as Husserl himself reminds us in *FTL*, long after he had repudiated psychologism, there is a core of ideas in *PA* that he never abandoned (Husserl *FTL*, § 27). In the section "The First Constitutional Investigations of Categorial Objectivities in the *Philosophie der Arithmetik*" he states it this way:

I had already acquired the definite direction of regard to the formal and a first understanding of its sense by my *Philosophie der Arithmetik*, which, in spite of its immaturity as a first book, presented an initial attempt to go back to the spontaneous activities of collecting and counting, in which collections ("sums", "sets") and cardinal numbers are given in the manner characteristic of something that is being generated originally, and thereby to gain clarity respecting the proper, the authentic, sense of the concepts fundamental to the theory of sets and the theory of cardinal numbers. It was therefore, in my later terminology, a phenomenological-constitutional investigation; and at the same time it was the first investigation that sought to make "categorial objectivities" of the first level and of higher levels (sets and cardinals of a higher ordinal level) understandable on the basis of the "constituting" intentional activities.

Husserl makes it clear in his later writings that his phenomenological-constitutional investigation is an epistemological, not a psychological, undertaking. And it is just this kind of investigation, I believe, that contains important insights that are not found in Frege. I shall return to the ideas in this passage in § 4.

In *PA* Husserl argued that the concept of number was not definable, that it was a primitive concept, and he criticized Frege's construal of natural numbers as (infinite) equivalence classes. Husserl generally thought of concepts as intensional entities, so what he means when he says that the concept of number is not definable, in Frege's terms, is that one cannot define the 'content', or the sense (intension) of the concept of number (Tieszen 1990). The sense of the concept of number is logically simple or primitive, it cannot be reduced to any other sense, and so we must investigate it by some other means. In *PA* Husserl raises, in effect, an early version of the paradox of analysis for Frege's view (Resnik 1980; Tieszen 1990; Dummett 1991b). The problem, simply put, is this: logic, for Frege, is supposed to consist of analytic propositions, is not supposed to depend on intuition, and so on. If we are trying to define the *sense* of an expression of number, however, then the sense of the definiens will either be the same as or different from the sense of the definiendum. If the sense is the same, the definiens simply repeats the definiendum so that the definition is pointless. But if the sense of the definiens is different from that of the definiendum, the definiens adds something to the definiendum, in which case the definition is false. Thus, we cannot hope to define the concept of number.

Husserl argued, on the other hand, that our understanding of the extension of the concept of number is not problematic since we apply the concept with no difficulty. Since Frege characterizes only the extension of the concept of number, his work falls short of the goal of a philosophical analysis of the concept. Frege, however, argues that the concept of number is definable, that an explicit, extensional definition of the concept can be given. Further, he argues that Husserl's criticisms really apply to all of the concepts of mathematics, but that these criticisms miss the point since extensional definitions of concepts suffice for mathematical purposes.

The disagreement here reflects a more general division between Frege and Husserl on issues involving intensionality and extensionality in logic and mathematics, and to some extent on the different methods and purposes of mathematics and philosophy. Husserl, from the beginning of his career, sided with intensional logicians in taking logic to be concerned with senses or meanings as such. Frege's view, on the other hand, is succinctly expressed in the following comment (Frege 1892–95, p. 122):

They [the Umfangs-logicians] are right when, because of their preference for the extension of a concept to its intension, they admit that they regard the reference of words, and not their meaning, to be essential for logic. The Inhalts-logicians only remain too happily with the meaning, for what they call "Inhalt" (content), if it is not quite the same as Vorstellung, is certainly the meaning (Sinn). They do not consider the fact that in logic it is not a question of how thoughts come from thoughts without regard to truth-value, that, more generally speaking, the progress from meanings (Sinne) to reference (Bedeutung) must be made; that the logical laws are first laws in the realm of references and only then mediately relate to meaning (Sinn).

Logic for Frege is, in the first instance, concerned with truth. It is not concerned with (Fregean) thoughts or senses as such, but with thoughts insofar as they are true. As Mohanty has pointed out, Frege agrees with Husserl that the extension of a concept presupposes the intension of a concept, but he also takes the concept itself to be the reference of a concept-word (Mohanty 1982). Concepts and truth-values are both references for Frege. Husserl, on the other hand, favored the development of intensional logic, and he took the intensionalist outlook to be especially important for philosophy. Indeed, the concept of intentionality, which is central to his philosophy, calls for an intensional logic, for the intentionality of acts of believing, knowing, and so on, creates oblique contexts. In particular, an analysis of belief or knowledge involving number is ipso facto an analysis of intensionality. The marks of intensionality in the failure of substitution *salva veritate* and existential generalization to preserve validity of inference are seen throughout Husserl's later philosophy. Frege, on the other hand, devoted little attention to the analysis of beliefs or knowledge about number. He seemed unable, at least in his logicist period, to grasp the possibility of a nonpsychologistic analysis of this type.

Thus, one senses that there were some deeper issues behind the Frege-Husserl dispute and that, to some extent, each philosopher missed the other's point. Frege claims that he is not after the intension of the concept of number anyway, so that Husserl misses the significance of his project for mathematics. But Husserl is interested, in effect, in a logic of meanings, or a logic of oblique contexts generated by knowledge and belief about numbers, so that Frege misses the point about how important the development of such an intensional approach might be for deeper questions in the philosophy of arithmetic. Since Frege characterizes only the extension of the concept of number, his work falls short of the goal of a philosophical analysis of arithmetic.

It might be argued on Frege's behalf that we do have definitions of numbers in various standard set theories, so that something of Frege's original claim about the possibility of providing an extensional, reductive definition of number remains, even if the reduction is not to 'pure logic' in Frege's sense. I think this cannot be denied, and that this set-theoretic reduction constitutes a very important development in the foundations of mathematics. Even so, something of Husserl's point that the sense of the concept of number is logically simple or primitive, that it cannot be reduced to any other sense, also remains. Consider, for example, the fact that there are infinitely many nonequivalent extensional definitions of natural numbers in set theory. A view such as Frege's is saddled with the problem that there are many reductions. The existence of nonequivalent extensional definitions in mathematics is of course not confined to elementary number theory. To take a different example, real numbers may be defined as Dedekind cuts, as the upper or lower members of such cuts, as equivalence classes of convergent sequences or rational numbers, and so on. A real number defined as a Dedekind cut, for example, is not identical with a real number defined as the lower member of a cut since the former is an ordered pair whereas the latter is a member of such a pair. Thus, no particular definition of this type suffices to capture the sense of the concept of a real number. Each such definition is 'reductive' and fails to include the other definitions that are supported by the meaning of the concept. So we cannot suppose that we know what real numbers are or know that we have gotten to the essence of real numbers on the basis of any one of these definitions, even if some of the definitions do seem more 'natural' than others.

Looking back on the Frege-Husserl dispute I think we can say that the upshot of Husserl's objection is that we must seek to understand the *meaning* or intension of the concept of number in some other way, and that it is philosophically important that we do so, even if various kinds of explicit, reductive definitions can be given. I think that subsequent work in the foundations of mathematics bears this out. The same point can be made about other basic concepts of mathematics. This is of special interest in recent times in the case of set theory since mathematics can be 'reduced' to set theory, but no one understands what having an explicit definition of the concept of set would mean.

Frege was of course not persuaded by Husserl that we must try to understand arithmetic and the meaning of arithmetic concepts in a different way. *PA* was just too clouded with psychologistic confusions to be persuasive. What eventually did convince Frege was Russell's paradox.

§ 3   Frege's Later Philosophy of Arithmetic

After attempting for a while to repair the damage done to his logicism by Russell's discovery of the paradox in 1902, Frege turned his attention to other matters. When he returned to the foundations of arithmetic near the end of his life he reached the conclusion that he had started from incorrect philosophical assumptions in his earlier work on arithmetic. Frege had already held in the *Grundlagen* (§ 89) that traditional geometry was synthetic a priori and he now began to assimilate arithmetic to geometry. In several unpublished papers in 1924 and 1925 Frege now distinguished three sources of knowledge: sense perception, the geometrical/temporal sources of knowledge, and the logical source of knowledge (Frege 1924–25a and 1924–25c). In these papers he continued to hold that sense perception cannot be a source of knowledge in arithmetic or geometry, citing the fact that "sense perception can yield nothing infinite" (Frege 1924–25c, p. 274). The idea of the infinite is derived rather from the geometrical/temporal sources of knowledge. Frege argues that it is absurd that the series of natural numbers might come to an end, but not that there might be only finitely many physical objects, and so the mode of cognition involved in the geometical/temporal sources of knowledge must also be a priori (Frege 1924–25b, p. 277). He now also argues that although logic must be involved in knowledge whenever inferences are drawn, apparently logic cannot on its own yield us any objects. He even gives a brief analysis of how illusions about objects can arise in logic, just as illusions about objects arise in sense perception (Frege 1924–25c). In logic we rely heavily on the use of language and there is a tendency in language, even in strictly scientific language, to form proper names to which no objects correspond. Frege says that he himself fell under this illusion in trying to construe numbers as sets. But if an a priori mode of cognition must be involved in the awareness of number, and logic cannot be the source of our arithmetical knowledge, then our arithmetical knowledge must have a geometrical source. It is of some interest to note that Frege does not pursue the temporal source of knowledge in relation to arithmetic, perhaps because geometry already exists as a mathematical discipline but there is nothing similar to it for the concept of time. Thus, he says that "the more I have thought the matter over, the more convinced I have become that arithmetic and geometry have developed on the same basis – a geometrical one in fact – so that mathematics is in its entirety really geometry" (Frege 1924–25c, p. 277). In this remark we see that Frege's impulse to reduce arithmetic to something else has not abated but has only found a different target.

Frege continues to make it clear in these late papers that he is antipsychologistic and is an antiformalist (in Hilbert's sense) about mathematics. He still holds that a statement of number contains an assertion about a concept. He says that whereas he asserted in the *Grundgesetze* that arithmetic does not need to appeal to experience in proofs, he would now hold that it does not need to appeal to sense perception in its proofs. He now says, "I have had to abandon the view that arithmetic does not need to appeal to intuition either in its proofs, understanding by intuition the geometrical source of knowledge, that is, the source from which flow the axioms of geometry" (Frege 1924–25a, p. 278). Note that since Frege rejects sense perception but now accepts the role of intuition as a source of knowledge in arithmetic, he must be committed to the view that some form of 'pure' or nonsensory intuition is required.

Frege's new position raises a host of questions that he never answered. First, what kind of account could be given of this nonsensory form of intuition? Now that Frege thinks arithmetic (via geometry) does need to appeal to nonsensory intuition in its proofs, does it follow that he must abandon his idea of providing gapless formal proofs? Or does he think it is still possible to provide such proofs, except that now they will involve an element that is not purely 'logical'? Does Frege believe that this intuition is somehow fully formalizable or axiomatizable or, as Gödel later came to believe, that our arithmetical intuition is somehow inexhaustible and not fully formalizable? In the latter case, Frege would indeed be making a radical departure from his earlier scientific ideal of finding gapless formal proofs and it would again appear to follow on his premises, for example, that we could have no more than empirical justification in mathematical proofs. Since this seems to contradict the idea that the geometrical source of knowledge yields a priori knowledge, we might expect Frege to reject the idea that our geometrical intuition is not fully formalizable. It is not clear, however, that this would be Frege's view. There are many other questions about how the concept of number should now be understood with respect to the various philosophical issues Frege had discussed in the *Grundlagen*. For example, arithmetic is now presumably synthetic, not analytic, but does this mean that the concept of number is not definable? Frege presumably now requires an account of definition that is different from any of his earlier accounts. And if intuition is necessary, how do we avoid psychologism? How will Frege explain the role of intuition in our knowledge of large numbers, and of the infinity of numbers? Do we again need an account of arithmetic or of geometry which requires the idea of abstraction from sense experience? And so on. Many of these questions are addressed in Husserl's philosophy, and the fact that they are, I shall argue later, represents an advance over Frege's thinking on the subject.

When we consider what Frege discarded and retained from his logicism we see that in some respects he has come to a position that is much closer to Kant's view that the source of arithmetical knowledge lies in a pure a priori form of intuition. There are of course some notable differences: Frege separates the geometrical from the temporal source of knowledge only to set aside the temporal source, whereas for Kant the temporal source is linked to arithmetic, as distinct from geometry. Also, there might be disagreement between Frege and Kant about the claim that a statement of number contains an assertion about a concept, and about whether the concept of number is definable and statements of number are provable. There is also no trace of realism about mathematical objects in Kant's philosophy. As we shall see, Frege's postlogicist position on the foundations of arithmetic is also closer to Husserl's views in some respects, although here too there will be some notable differences.

§ 4   Husserl and the Philosophy of Arithmetic: What Is the Origin of Our Knowledge of Ideal Objects?

Husserl's transcendental phenomenology suggests that we should try to understand arithmetic and the meaning of arithmetic concepts by starting with the question of how arithmetic knowledge is possible. In particular, Husserl suggests that a genetic account of the conditions for the possibility of arithmetic knowledge is needed. There are far too many details of such an account to be considered here, and so I will simply give an overview of some points that I believe are germane to what has happened in the foundations of mathematics since Husserl's time.

In *PA* Husserl had already argued that since the concept of number is a primitive concept all we could hope to do is to indicate the concrete phenomena from which the concept is 'abstracted' and clearly describe the abstraction process used. Husserl's analyses in *PA* represent an early and rather primitive attempt to provide what he later referred to as a 'genetic' analysis of the concept. Husserl has a lot to say about genetic analysis in various later writings (e.g., Husserl 1929, 1936, 1939). One of his best-known later essays is entitled "The Origin of Geometry," and the subtitle of *Erfahrung und Urteil* is "Investigations in a Genealogy of Logic." What Husserl points out in these later works is that a genetic account of the conditions for the possibility of arithmetical knowledge will have to explain how it is possible to know about 'ideal' objects, that is, objects that are not located in space-time, but that are immutable and acausal.

In his analysis of the source of knowledge about number starting around 1900 Husserl agrees with Frege that the intended objects of acts directed toward numbers are neither objects of sense perception nor mental objects. But they are also not 'logical objects' in Frege's sense. They are, however, 'ideal' or abstract objects, even though the objects of the underlying, founding acts in which they have their origin may be objects of sense perception. Husserl has a more subtle analysis than Frege of what abstract objects are, and of how we could be aware of them. In "The Thought" (Frege 1918) Frege struggles to make sense of how we could come to know about a particular kind of abstract, eternal, immutable object – a thought – and his comments leave the matter shrouded in mystery. Husserl, on the other hand, gives a phenomenological (and also a transcendental) solution to this problem: abstract objects, and numbers in particular, are to be understood as invariants in our mathematical experience, or in mathematical phenomena. Even if we have not clarified the meaning of the concept of number completely, we can still say that numbers are identities through the many different kinds of acts and processes carried out at different times and places and by different mathematicians, and this is analogous to the fact that physical objects are identities in our experience even though we do not see everything about them. Now some phenomena simply do not sustain invariance over different times, places, and persons. In the case of logical or sensory illusions, for example, what we take to be present at one point in our experience is not sustained in subsequent experience. Parts of our arithmetic experience, however, are not at all like this, and they have in fact become quite stable across times, places, and persons, just as other parts of our experience have stabilized. So even if we have not brought the meaning of the concept of number to full clarity, we can still say that numbers are 'objects' in the sense that they are identities through the multiplicities of our own cognitive acts and processes. If this were not so, the science of mathematics as we know it would not be possible.

The sense of the 'abstractness' of numbers is derived from several facts. Numbers could not be objects (identities) of sensation because objects of sensation occur and change in space and time. They could not be mental in nature because what is mental occurs and changes in time. One could expand on these claims equally well by using either Fregean or Husserlian arguments. Numbers are also identities that 'transcend' consciousness in the sense that there are indefinitely many things we do not know about them at a given time, on the analogy with our knowledge of perceptual objects, but at the same time we can extend our knowledge of them by solving open problems, devising new methods, and so on. They transcend conciousness in the same way that physical objects do. And, similarly, we cannot will them to be anything we like; nor can we will anything to be true of them. They are mind independent. On the *object* side of his analysis Husserl can therefore claim to be a kind of realist about numbers. Numbers are not our own ideas. At the same time he is also a kind of idealist on the *subjective* side of his analysis because he has a constitutional account of our awareness and knowledge of numbers and a critical perspective on classical metaphysical realism.

Let us now focus on what the constitutional account of the awareness and knowledge of numbers looks like. How is knowledge of these ideal objects possible? To understand how the awareness and knowledge of any kind of object is possible we must realize, Husserl argues, that various forms of consciousness, such as believing and knowing, are intentional. So numbers, as ideal objects, must be understood as the objects of acts that are intentional. Intentional acts are directed to objects by way of their contents or 'noemata'. We can think of the contents associated with acts as the meanings or senses under which we think of the objects. So in the parlance of recent work on cognition and meaning, Husserl wants to provide a theory of content (specific to arithmetic) in which the origins of arithmetic content are taken to lie in more primitive, perceptual 'founding' acts and contents, where the idea is to determine the a priori cognitive structures and processes that make arithmetic content possible. So it is argued, for example, that 'founded' acts of *abstraction from* and *reflection on* such underlying, founding acts and contents are an a priori condition for the possibility of arithmetic content, and hence for the awareness of number. To understand or to clarify the sense of the concept of number, therefore, is not (or not only) to find an explicit, reductive definition of number in some other mathematical theory. It is rather to provide, among other things, a genetic account of the a priori conditions for the possibility of arithmetic knowledge.

Husserl argues that the sense of the concept of number has its origin in acts of collecting, counting, and comparing (i.e., placing objects into one-to-one correspondence). Note that these are acts that are appropriate to number and not, prima facie, to geometry. Insofar as we are aware of numbers in these acts the acts must involve a kind of abstraction from and reflection on our most primitive perceptual experiences with everyday objects. They are 'founded' acts in the sense that they presuppose the existence of more straightforward perceptual acts. The latter kinds of acts could exist if there were no arithmetic, but the genesis of arithmetic presupposes such straightforward acts (see Tieszen 1989). This is what Husserl has in mind when, in the passage quoted previously, he speaks of the "attempt to go back to the spontaneous activites of collecting and counting, in which collections ("sums", "sets") and cardinal numbers are given in the manner characteristic of something that is being generated originally, and thereby to gain clarity respecting the proper, the authentic, sense of the concepts fundamental to the theory of sets and the theory of cardinal numbers," and when he adds that he seeks to make "'categorial objectivities' of the first level and of higher levels (sets and cardinals of a higher ordinal level) understandable on the basis of the 'constituting' intentional activities."

In fact, a condition for the possibility of all of our higher, theoretical, or scientific modes of cognition is that there be a hierarchy of acts, contents, and intended objects. This means that at various levels in the hierarchy we have acts directed toward objects by way of their contents. In the growth of knowledge over time these contents either may be corrected through further experience or may not be. The constitution of content in founded and founding acts is a function of the interplay over time between the existing contents of acts and the experiences provided by further intuition, and this will be governed to some extent by the a priori rules or structures of cognition. It is not just arbitrary. The processes by which layers of content are built up, refined, and extended in relation to intuition or experience are described by Husserl as 'synthetic'. The analysis of the origin of arithmetic is therefore not the same as empirical historical investigation along either an individual or a social dimension, but it is also not simply a matter of analytic inference. Husserl would argue against historicist accounts of arithmetic knowledge, but also against any kind of ahistorical rationalism about arithmetic knowledge.

In his early work Frege of course always objected to accounts of number that required a process of abstraction, and in his review of *PA* he takes Husserl to task for this too. If we are careful, however, always to distinguish mental acts and processes from the objects toward which acts are directed, then we can skirt Frege's objection. We can distinguish numbers as abstract objects in their own right from the cognitive acts and processes that make it possible to know about numbers, and argue that we do not create the numbers themselves by abstraction, but only our knowledge or awareness of numbers. How else would arithmetical *knowledge*, as opposed to more primitive forms of perceptual knowledge, be possible? It is true that Husserl does not give us a detailed account of abstraction, but he does at least try to establish the claim that acts of abstraction must be involved in arithmetic knowledge. I think this point by itself will enable us to set aside Frege's early objection about confusing the subjective and the objective. And Frege may actually need a similar epistemological view of abstraction to support his later account of number.

I also suggest, in response to views such as those expressed by Dummett (Dummett 1991c), that Husserl's remarks about numbers as aggregates of featureless units in *PA* must be understood in the context of the effort to provide a genetic analysis of the concept of number. That is, in speaking of numbers as aggregates of featureless units Husserl is describing a stage, perhaps even a fairly early stage, and one that is closer to sense perception, in the genesis of our consciousness of numbers. The general project here bears comparison to Quine's description of the genesis of set theory in *The Roots of Reference*. We need not assume that Husserl's description constitutes the final or highest stage in our consciousness of number, especially in light of the clarification and development that has taken place in number theory, mereology, and set theory since the time of Husserl and Frege. I agree, however, that there are difficulties about precisely how numbers are to be understood in *PA*, including those due to obscurities surrounding the relationship between Husserlian 'aggregates' and Fregean 'concepts'.

§ 5   The Role of Intuition in Arithmetic Knowledge

From this brief sketch we already obtain a very different picture of the foundations of arithmetic from the one presented in Frege's logicism. To fill out the picture in relation to Frege we need to keep in mind the role that intuition would have to play in Husserl's later conception of arithmetical knowledge. Intuition is understood in terms of the 'fulfillment' of empty act-contents. Our act-contents are fulfilled when we are not merely directed toward objects in our thinking but actually experience the objects in sequences of acts in time, for it is experience that gives us evidence for the objects. Some of our act-contents can be fulfilled or verified in intuition and some cannot. We might say that act-contents without intuitions are empty, whereas intuitions without act-contents are blind.

Husserl's distinction between empty and fulfilled act-contents (or intentions) is closely related to Frege's distinction between thoughts (contents) and judgments. As early as the *Begriffsschrift* Frege had drawn a distinction between 'content' and 'judgment' strokes in his formal notation. A proposition set out with the content stroke (as in —A) is supposed to lack the assertoric force of the same proposition set out with a judgment stroke (⊢ A). Drawing a parallel with Husserl, David Bell has argued that the shift from a content to a judgment stroke also marks a different kind of subject matter (Bell 1990). The content stroke is used simply to indicate what Frege later calls a thought, devoid of any assertoric force, parallel with Husserl's idea of the content or sense of an act. The shift from judgment to content can be eludicated by way of Frege's idea that in intensional contexts a sentence no longer expresses a judgment or possesses assertoric force, and no longer refers to any object or property in the natural world; rather it refers to a sense. Its reference is now the sense that it expressed before it was embedded in an intensional context.

What little Frege has to say about the concept of knowledge is explicated by way of the distinction between thoughts and judgments. As late as 1924 (Frege 1924–25c) the view is expressed succinctly:

When someone comes to know something it is by his recognizing a thought to be true. For that he has to grasp the thought. Yet I do not count the grasping of the thought as knowledge, but only the recognition of its truth, the judgment proper. What I regard as a source of knowledge is what justifies the recognition of truth, the judgment.

Of course Frege's account of grasping a thought, and the role that such grasping is supposed to play in securing reference or knowledge, differs from Husserl's view (see, e.g., Dummett 1991d). On Husserl's view it is not necessary that the thought itself be an object of an act of consciousness in order for us to refer to either ordinary perceptual objects or to objects such as numbers, although it is possible to reflect on the thought. The path to an object does not require a detour through a different object of consciousness.

Now a critical problem for Frege is this: how, and under what conditions, is it possible to proceed, epistemically, from thought (or content) to judgment in the case where the thought is about numbers, or other mathematical objects? Frege says that a source of knowledge is what justifies the recognition of a truth. On what grounds, however, can the recognition of a truth about numbers be justified? Husserl's views on founding and founded acts and contents, and on the role of intuition in knowledge, are meant to answer precisely these questions. Husserl defines the concept of intuition in terms of fulfillment of (empty) act-contents and argues that a condition for the possibility of knowledge is that there be intuitions of objects at different levels in the hierarchy of acts, contents, and intended objects. Husserl thus argues that there is intuition of abstract or 'ideal' objects such as natural numbers, although this intuition will of course be a form of founded intuition (Tieszen 1989). It is built up from our straightforward perceptual forms of intuition in acts of abstraction and reflection that involve counting, collecting, and comparing.

In the ideas described earlier, Husserl is therefore beginning to give us an analysis of the source of knowledge about number, and what justifies the recognition of a truth about numbers is a certain form of founded intuition, a form in which intentions directed to numbers are fulfilled or are fulfillable. This clearly suggests a view of arithmetic that is more like Frege's later view: namely, that arithmetic is synthetic a priori and that we should not expect to be able to derive the principles of arithmetic from logic. To say that a number is *intuitable* on Husserl's view means that it is possible to carry out a sequence of acts in time in which the intention to the particular number would be fulfilled, that is, in which the number itself would be presented. I have argued elsewhere that this corresponds to and in fact deepens our understanding of the constructivist requirement that we must be able to find the number (Tieszen 1989). This is a founded form of intuition which is nonsensory insofar as sensory qualities play no essential role in making number determinations in acts of counting, collecting, and comparing. So we could think of '—A' as designating the proposition, content, or intention A, and of '⊢ A' as meaning that we have a proof, judgment, or fulfillment regarding A, except that now the proof, judgment, or fulfillment is not based on a purely logical source of knowledge, nor directly on sense experience, but rather on a founded form of intuition of the sort embodied in a mathematical construction (see Tieszen 1989 and Chapter 13).

In his late writings Frege also thinks that a nonsensory or founded form of intuition is the source of mathematical knowledge, but he has no account of such a form of intuition. Husserl, on the other hand, has an elaborate theory to explain how it is possible to have knowledge in acts directed toward natural numbers, where numbers are nonetheless understood as 'abstract' objects. Unlike the later Frege, however, he does not try to reduce this to geometric intuition. Although there may be connections between arithmetic and geometrical intuition, Husserl is not reductionistic from the outset. This is supposed to be an analysis of the sense of the concept of number, and so acts, contents, and intuitions of the appropriate type must be involved.

§ 6   A New View of the Foundations of Arithmetic

Let us now fill in a little more of the detail about the foundations of arithmetic on Husserl's later views. Husserl says that what justifies holding a proposition about numbers to be true is the evidence provided by (founded) intuition. As we said, what is given in intuition may be corrected or refined in subsequent intuition. The evidence provided by intuition has different degrees and types: adequate, apodictic, a priori, clear, and distinct. Husserl would agree with Frege that our arithmetical knowledge is a priori, and that at least a core of arithmetical statements must be understood as necessary (or apodictic) truths. Husserl's position implies that our evidence for large numbers and for general statements about numbers is inadequate, in the sense that we cannot actually complete the processes of counting, collecting, or comparing in these cases. Husserl's position also implies that we have not yet brought to full clarity and distinctness our understanding of the sense of the concept of number. In fact, perfect clarity and distinctness, and perfect adequation, are really ideals that we can only approach in our knowledge. So although arithmetic has a 'foundation' in intuition for Husserl, one can argue that this does not commit us to being absolutists or 'foundationalists' about arithmetic in any objectionable sense.

We have been saying with Frege that a source of knowledge is what justifies recognition of a truth, and we add that a founded form of intuition is a necessary condition for arithmetical knowledge. We have many arithmetic intentions but not all of these are fulfilled. Husserl's idea of "recognition of a truth" involves *evidence of truth*, or what he calls "truth within its horizons," whereas for Frege there is just an absolutized or idealized conception of truth, as it were, shorn of any relationship to a knowing subject. Consider, for example, the judgment that A v ¬A (i.e., ⊢ A v ¬A). How do we recognize the thought to be true in this case? We do not find a satisfactory answer to this question in Frege's work. In connection with his notion of "truth logic" in *FTL*, however, Husserl points out that this judgment involves a rather substantial idealization (Husserl *FTL*, § 77):

> The law of the excluded middle, in its subjective aspect, . . . decrees not only that if a judgment can be brought to an adequation . . . then it can be brought to either a positive or a negative adequation, but . . . that *every judgment necessarily admits of being brought to an adequation* – "necessarily" being understood with an *ideality* for which, indeed, no responsible evidence has ever been sought.

These remarks have obvious implications for the way we should understand our intuitive *knowledge* of the 'necessity' of certain statements of logic, and they bear a striking resemblance to some constructivist views of logic. In *FTL* and other works Husserl develops a 'critique', in a Kantian sense, of such idealizations in mathematics and logic. This is, correlatively, a critique of formalization, in the sense not only of Hilbert but also of Frege. For Husserl points out in various writings that formal systems idealize and abstract from our experience, so that some of the data of experience will always be unaccounted for in a formal system.

Thus, we do not find Husserl, in his *phenomenological-constitutional investigations*, insisting on Frege's point about eliminating all informal or intuitive gaps in our mathematical reasoning in order to have a foundation for arithmetic. Nor do we find him insisting that such a foundation can be had *only* on the basis of formalization. For Husserl then, the foundations of arithmetic cannot be based on anything like Frege's earlier view, but in fact must depend on the evidence provided by founded intuition. On this view it is not surprising that Frege could not find his Basic Law V to have the same kind of self-evidence as the other basic laws, for it involves a degree of abstraction or idealization from experience that does not allow it to be understood as expressing knowledge of a necessary truth based on founded intuition. It allows us to form proper names to which no objects correspond. Husserl, on the other hand, has at least a start on a theory of mathematical evidence, a theory based on the founding/founded structure of our more theoretical forms of cognition. Husserl's appeal to such a structure amounts to conditioning rationalism with empiricism in a way that we do not find in Frege's logicism, and it gives his view of the intuition of particular mathematical objects a constructivist slant. Frege's logicist view, by contrast, appears to embody just the kind of unbounded rationalism one might expect to issue in antinomy.

Frege's early ideal of gapless formal proofs in which all inference rules are specified in advance, and in which intuition is completely eliminable, appears to be attainable, at best, in predicate logic, or in only those formal systems for which we have soundness and completeness theorems. Of course Frege's conception of logic did not include anything like the idea of metamathematical completeness proofs. But, in any case, we know from Gödel's incompleteness theorems that it will not be possible to specify finitely all of the basic laws and rules of inference of arithmetic in any one axiomatic, formal system. So we cannot exhaust the sense or content of arithmetical statements in such a system. Contrary to Frege's view, we also cannot make all of our arithmetical intuitions explicit in a formal system of arithmetic. In the early twenty-first century we have come a long way from Frege's ideal of the strictly scientific method in mathematics, but this is a development that is quite compatible with

Husserl's later views on intuition. In fact, we know that Gödel himself (Gödel *1961/?) has positioned Husserl's philosophy at the very center of recent foundational research for similar reasons.

What does all this mean for the ideal in Frege's logicism of eliminating all error in arithmetical reasoning and securely establishing arithmetic once and for all? And how are we to justify arithmetic? On the Husserlian view we have been describing, beliefs about natural numbers are justified by a form of intuition appropriate to the fulfillment of our arithmetical intentions, and we have been arguing that this intuition as a process must be understood constructively. We nonetheless have gradations of evidence within arithmetic itself, proof itself may have different degrees of adequation, and so we cannot rule out the possibility of corrections of our arithmetical knowledge in the future. At the same time we appeal to the objects and facts that constitute the ideal invariants in arithmetic and to the stability over time of parts of our arithmetical experience. Arithmetic is a priori and a core of its statements must be understood as necessary truths, in the sense that they cannot be assimilated to truths established by empirical induction, even if we do not have adequate evidence everywhere in arithmetic. On top of this intuitive basis of arithmetic we can seek relative consistency proofs for arithmetic and engage in conceptual analysis of various arithmetical notions.

Husserl argues that by analyzing the genesis of our mathematical concepts in a phenomenological framework we will inevitably deepen our understanding of their meanings. Such an analysis will bring into a more explicit and direct awareness the implicit presuppositions underlying our concepts. The senses assumed by the concept, along with the 'horizons' of possible experience associated with the concept, will be uncovered so that genetic analysis will lead to the clarification of various hidden implications present in the concept. In addition, Husserl thinks we can clarify the contents of our acts through the procedure of free variation in imagination (Husserl 1911; 1913, §§ 66–70; 1939, §§ 87–93). This procedure, along with genetic analysis, amounts to a form of what Kreisel has referred to as 'informal rigor', that is, informal but rigorous concept analysis (Kreisel 1967). Through the application of the procedure of free variation to the concept of number we are supposed to be able to clarify our understanding of the essential features of number. Using the procedure we would see, for example, that the Dedekind-Peano axioms pick out essential (not accidental) features of the concept of number, even if the axioms capture only part of the meaning of the concept of number. More importantly, we can expect to extend our knowledge through further clarification of the meaning of mathematical concepts in such a way as to solve open problems, develop new methods, and so on.

§ 7 Conclusion

In his logicist period Frege portrayed genetic analysis as being fruitless for mathematics and logic and as actually leading us away from mathematical work. He repeated this charge in many contexts, applying it to Mill for example in the *Grundlagen* and to Husserl in his review of *PA*. From the preceding discussion, however, we see that what genetic analysis amounts to in Husserl's later work is quite different from what Mill had in mind, and also from what Frege took it to be. It is, first of all, not a psychological investigation. Rather, it is an investigation into the a priori conditions for the possibility of the consciousness and knowledge of number. So Frege would have to argue that genetic analysis, as we described it, has no place in the philosophy of arithmetic. But on what grounds could he make such an argument? It is true that genetic analysis might lead us away from *formal* work, but this is because we cannot forget about the intuitive and philosophical foundations of the concept of number, and because we cannot exhaust the sense of the concept of number in a particular formal system. There is a sense in which Frege's ideal of a strictly scientific method in arithmetic must be given up. We cannot have a complete system of gapless formal proofs for numbers. Husserl's later view places additional emphasis on the informal, rigorous concept analysis which is the source of formal work. Informal rigor and critical analysis on the one hand, and the mathematician's technical work on the other, are complementary scientific activities. Thus, technical and formal scientific work is not at all incompatible with the analysis of the origins of concepts. It is just that such work is itself now seen to have its origins in informal rigor, so that we cannot expect to supplant such informal rigor with some formal system. On this basis technical work can be encouraged, extensionalism need not be challenged, and so on.

A case can be made for the claim that the elements of a Husserlian transcendental philosophy of arithmetic discussed are compatible with the post-Fregean, post-Hilbertian, and post-Gödelian situation in the foundations of mathematics. And they are compatible in a way that still makes a kind of rationalism (or antiempiricism) about mathematics possible, thus preserving something of Frege's own antiempiricism about number. Moreover, they also preserve something of the realism or objectivism about mathematics that Frege championed, but in a way that does not make the very possibility of arithmetical knowledge a mystery. The view allows and even encourages formalization but does not demand an exclusively formalistic attitude, for there is also a role for informal rigor. It includes an account of the way to understand the primitive terms and rules of mathematical theories, and an account of mathematical evidence into which is built a more critical perspective on appeals to self-evidence. It gives a more balanced picture of arithmetical knowledge, and of the role of formalization and axiomatization in such knowledge. At the same time, however, it is in many ways only a schema for a philosophy of arithmetic which itself needs to be filled in and improved.

Bibliography

Barwise, J. (ed.), 1977. *Handbook of Mathematical Logic*. Amsterdam: North Holland.

Becker, O., 1923. "Beiträge zur phänomenologischen Begründung der Geometrie und ihrer physikalischen Anwendungen," *Jahrbuch für Philosophie und phänomenologische Forschung* 6, 385–560.

1927. "Mathematische Existenz: Untersuchungen zur Logik und Ontologie mathematischer Phänomene," *Jahrbuch für Philosophie und phänomenologische Forschung* 8, 439–809.

Bell, D., 1990. *Husserl*. London: Routledge.

Benacerraf, P., 1973. "Mathematical Truth," *Journal of Philosophy* 70, 661–679.

Benacerraf, P., and Putnam, H. (eds.), 1983. *Philosophy of Mathematics: Selected Readings*, 2nd ed. Cambridge: Cambridge University Press.

Bernays, P., 1935. "Sur le platonisme dans les mathématiques," *L'Enseignement Mathématique* 34, 52–69. Translated by C. Parsons in Benacerraf and Putnam (eds.) 1983, 258–271.

Bishop, E., 1967. *Foundations of Constructive Analysis*. New York: McGraw-Hill.

Boolos, G., 1971. "The Iterative Conception of Set," *Journal of Philosophy* 68, 215–232.

1987. "A Curious Inference," *Journal of Philosophical Logic* 16, 1–12.

Brouwer, L. E. J., 1921. "Comments on Weyl 1921." Translated by W. van Stigt in P. Mancosu (ed.) 1998, 119–122.

1975. *Collected Works*, Vol. I. Edited by A. Heyting. Amsterdam: North Holland.

Carnap, R., 1922. *Der Raum. Ein Betrag zur Wissenschaftslehre. Kant-Studien Ergänzungsheft* 56. Berlin: Reuther & Reichard.

Curry, H., 1951. *Outline of a Formalist Philosophy of Mathematics*. Amsterdam: North Holland.

da Silva, J., 1997. "Husserl's Phenomenology and Weyl's Predicativism," *Synthese* 110, 277–296.

Dennett, D., 1971. "Intentional Systems," *Journal of Philosophy* 68, 87–106.

Deppert, W. (ed.), 1988. *Exact Sciences and Their Philosophical Foundations*. Frankfurt am Main: Lang.

Detlefsen, M., 1990. "Brouwerian Intuitionism," *Mind* 99, 501–534.

1992. "Poincaré Against the Logicians," *Synthese* 90, 349–378.

Drummond, J., 1978–79. "On Seeing a Material Thing in Space: The Role of Kinaesthesis in Visual Percpeption," *Philosophy and Phenomenological Research* 40, 19–32.

1984. "The Perceptual Roots of Geometric Idealization," *Review of Metaphysics* 37, 785–810.

Dummett, M., 1973. *Frege: Philosophy of Language*. New York: Harper & Row.

1977. *Elements of Intuitionism*. Oxford: Oxford University Press.

1978. "The Philosophical Basis of Intuitionistic Logic," in M. Dummett, *Truth and Other Enigmas*. Cambridge, Mass.: Harvard University Press, 186–201.

1981. *The Interpretation of Frege's Philosophy*. Cambridge, Mass.: Harvard University Press.

1991. *The Logical Basis of Metaphysics*. Cambridge, Mass.: Harvard University Press.

1991a. *Frege and Other Philosophers*. Oxford: Oxford University Press.

1991b. "Frege and the Paradox of Analysis," in Dummett 1991a, 16–52.

1991c. *Frege: Philosophy of Mathematics*. Cambridge, Mass.: Harvard University Press.

1991d. "Thought and Perception: The Views of Two Philosophical Innovators," in Dummett 1991a, 263–288.

1993. "What Is a Theory of Meaning? (II)," in M. Dummett, *The Seas of Language*. Oxford: Oxford University Press, 34–93.

1994. *Origins of Analytic Philosophy*. Cambridge, Mass.: Harvard University Press.

Feferman, S., 1988. "Weyl Vindicated: *Das Kontinuum* 70 Years Later," *Atti del Congresso Temi e prospettive della logica e della filosofia della scienza contemporanee*, Vol. I. Bologna, CLUEB, 59–93. Reprinted in Feferman 1998, 249–283.

1996. "Penrose's Gödelian Argument," *Psyche* 2, 21–32.

1998. *In the Light of Logic*. Oxford: Oxford University Press.

Feferman, S., et al. (eds.), 1986. *Kurt Gödel: Collected Works*, Vol. I. Oxford: Oxford University Press.

1990. *Kurt Gödel: Collected Works*, Vol. II. Oxford: Oxford University Press.

1995. *Kurt Gödel: Collected Works*, Vol. III. Oxford: Oxford University Press.

Fichte, J. G., 1908–1912. *Werke*, 6 vols. Edited by F. Medicus. Leipzig: Meiner.

Field, H., 1989. *Realism, Mathematics and Modality*. Oxford: Blackwell.

Folina, J., 1992. *Poincaré and the Philosophy of Mathematics*. New York: St. Martin's Press.

Føllesdal, D., 1988. "Husserl on Evidence and Justification," in R. Sokolowski (ed.), *Edmund Husserl and the Phenomenological Tradition*. Washington, D.C.: Catholic University of America Press, 107–129.

1995. "Gödel and Husserl," in J. Hintikka (ed.), *Essays on the Development of the Foundations of Mathematics*. Dordrecht: Kluwer, 427–446.

Frege, G., 1879. *Begriffsschrift, eine der arithmetischen nachgebildete Formelsprache des reinen Denkens*. Halle: L. Nebert. Translated by S. Bauer-Mengelberg in van Heijenoort (ed.) 1967, 1–82.

1884. *Die Grundlagen der Arithmetik.* Breslau: Koebner. Reprinted and translated by J. Austin as *The Foundations of Arithmetic,* Evanston: Northwestern University Press, 1978.

1892. "Uber Sinn und Bedeutung," *Zeitschrift für Philosophie und philosophische Kritik* 100, 25–50. Translated by M. Black in P. Geach and M. Black (eds.) 1970, 56–78.

1892–95. "Comments on Sense and Reference," in Hermes et al. 1979, 118–125.

1893. *Grundgesetze der Arithmetik,* Vol. I. Jena: H. Pohle. Translated in part by R. Furth as *The Basic Laws of Arithmetic.* Berkeley: University of California Press, 1964.

1894. "Review of E. Husserl, *Philosophie der Arithmetik,*" *Zeitschrift für Philosophie und philosophische Kritik* 103, 313–332. Translated by E. Kluge, in *Mind* 81, 1972, 321–337.

1903. *Grundgesetze der Arithmetik,* Vol. II. Jena: H. Pohle. Translated in part in P. Geach and M. Black (eds.) 1970, 173–244.

1918. "Der Gedanke: Eine logische Untersuchung," *Beiträge zur Philosophie des deutschen Idealismus* 1, 58–77. Translated by A. Quinton and M. Quinton in *Mind* 65, 1956, 289–311.

1924–25a. "A New Attempt at a Foundation for Arithmetic," in Hermes et al. 1979, 278–281.

1924–25b. "Numbers and Arithmetic," in Hermes et al. 1979, 275–277.

1924–25c. "Sources of Knowledge of Mathematics and the Mathematical Natural Sciences," in Hermes et al. 1979, 267–274.

Friedman, M., 1999. *Reconsidering Logical Positivism.* Cambridge: Cambridge University Press.

2000. *A Parting of the Ways: Carnap, Cassirer, and Heidegger.* La Salle, Ill.: Open Court.

Geach, P., and Black, M. (eds.), 1970. *Translations from the Philosophical Writings of Gottlob Frege.* Oxford: Blackwell.

George, A., 1988. "The Conveyability of Intuitionism: An Essay on Mathematical Cognition," *Journal of Philosophical Logic* 17, 133–156.

1993. "How Not to Refute Realism," *Journal of Philosophy* 90, 53–72.

Gödel, K., 1929. "On the Completeness of the Calculus of Logic," in Feferman et al. (eds.) 1986, 60–101.

1931. "On Formally Undecidable Propositions of *Principia Mathematica* and Related Systems I," in Feferman et al. (eds.) 1986, 145–195.

1933e. "On Intuitionistic Arithmetic and Number Theory," in Feferman et al. (eds.) 1986, 286–295.

1933f. "An Interpretation of the Intuitionistic Propositional Calculus," in Feferman et al. (eds.) 1986, 300–303.

*1933o. "The Present Situation in the Foundations of Mathematics," in Feferman et al. (eds.) 1995, 45–53.

1934. "On Undecidable Propositions of Formal Mathematical Systems," in Feferman et al. (eds.) 1986, 346–371.

*1938a. "Lecture at Zilsel's," in Feferman et al. (eds.) 1995, 87–113.

*193?. "Undecidable Diophantine Propositions," in Feferman et al. (eds.) 1995, 164–175.

1940. "The Consistency of the Axiom of Choice and of the Generalized Continuum Hypothesis with the Axioms of Set Theory," in Feferman et al. (eds.) 1990, 33–101.

*1941. "In What Sense Is Intuitionistic Logic Constructive?" in Feferman et al. (eds.) 1995, 189–2000.

1944. "Russell's Mathematical Logic," in Feferman et al. (eds.) 1990, 119–143.

1946. "Remarks Before the Princeton Bicentennial Conference on Problems in Mathematics," in Feferman et al. (eds.) 1990, 150–153.

1947. "What Is Cantor's Continuum Problem?" in Feferman et al. (eds.) 1990, 176–187.

1949. "An Example of a New Type of Cosmological Solutions of Einstein's Field Equations of Gravitation," in Feferman et al. (eds.) 1990, 190–198.

1949a. "A Remark About the Relationship Between Relativity Theory and Idealistic Philosophy," in Feferman et al. (eds.) 1990, 202–207.

*1951. "Some Basic Theorems on the Foundations of Mathematics and Their Implications," in Feferman et al. (eds.) 1995, 304–323.

*1953/59, III and V. "Is Mathematics Syntax of Language?" in Feferman et al. (eds.) 1995, 334–363.

1958. "On an Extension of Finitary Mathematics Which Has Not Yet Been Used," in Feferman et al. (eds.) 1990, 240–251.

*1961/? "The Modern Development of the Foundations of Mathematics in the Light of Philosophy," in Feferman et al. (eds.) 1995, 374–387.

1964. "What Is Cantor's Continuum Problem?" in Feferman et al. (eds.) 1990, 254–270. Revised version of Gödel 1947.

1972. "On an Extension of Finitary Mathematics Which Has Not Yet Been Used," in Feferman et al. (eds.) 1990, 271–280. Revised version of Gödel 1958.

1972a. "Some Remarks on the Undecidability Results," in Feferman et al. (eds.) 1990, 305–306.

Goldfarb, W., 1988. "Poincaré Against the Logicists", in P. Kitcher and W. Aspray (eds.), *History and Philosophy of Modern Mathematics*, Minneapolis: University of Minnesota Press, 61–81.

Greenberg, M. J., 1993. *Euclidean and Non-Euclidean Geometries*, 3rd ed. New York: W. H. Freeman.

Gurwitsch, A., 1974. *Phenomenology and the Theory of Science*. Evanston, Ill.: Northwestern University Press.

Hallett, M., 1984. *Cantorian Set Theory and Limitation of Size*. Oxford: Oxford University Press.

Hardy, L., and Embree, L. (eds.), 1993. *Phenomenology of Natural Science*. Dordrecht: Kluwer.

Hauser, K., forthcoming. "Gödel's Program Revisted," in *Husserl Studies*.

Heelan, P., 1983. *Space Perception and the Philosophy of Science*. Berkeley: University of California Press.

1989. "Husserl's Philosophy of Science," in J. N. Mohanty and W. McKenna (eds.), *Husserl's Phenomenology: A Textbook*. Lanham, Md.: Center for Advanced Research in Phenomenology/University Press of America.

Heidegger, M., 1962. *Being and Time.* Translated by J. Macquarrie and E. Robinson. New York: Harper & Row. Originally published in German in 1927.

Heinzmann, G., 1997. "Umfangslogik, Inhaltslogik, Theorematic Reasoning," in E. Agazzi and G. Darvas (eds.), *Philosophy of Mathematics Today.* Dordrecht: Kluwer, 353–361.

Hermes, H., Kambartel, F., and Kaulbach, F. (eds.), 1979. *Gottlob Frege: Posthumous Writings.* Translated by P. Long and R. White. Chicago: University of Chicago Press.

Heyting, A., 1931. "Die intuitionistische Grundlegung der Mathematik," *Erkenntnis* 2, 106–115. Translated by E. Putnam and G. Massey in Benacerraf and Putnam (eds.) 1983, 52–61.

——— 1962. "After Thirty Years," in E. Nagel, P. Suppes, and A. Tarski (eds.), *Logic, Methodology and Philosophy of Science.* Stanford, Calif.: Stanford University Press, 194–197.

——— 1971. *Intuitionism: An Introduction,* 3rd ed. Amsterdam: North-Holland.

Hilbert, D., 1926. "Über das Unendliche," *Mathematische Annalen* 95, 161–190. Translated by S. Bauer-Mengelberg in van Heijenoort (ed.) 1967, 367–392.

Hill, C., and Rosado-Haddock, G., 2000. *Husserl or Frege?* La Salle, Ill.: Open Court.

Husserl, E., 1891. *Philosophie der Arithmetik. Psychologische und logische Untersuchungen.* Halle: Pfeffer.

——— 1900. *Logische Untersuchungen,* Erster Theil: Prolegomena zur reinen Logik. Halle: M. Niemeyer. 2nd ed. 1913.

——— 1901. *Logische Untersuchungen,* Zweiter Theil: *Untersuchungen zur Phänomenologie und Theorie der Erkenntnis.* Halle: M. Niemeyer. 2nd ed. in two parts 1913/22.

——— 1911. "Philosophie als strenge Wissenschaft," *Logos I,* 289–341.

——— 1913. "Ideen zu einer reinen Phänomenologie und phänomenologischen Philosophie," *Jahrbuch für Philosophie und phänomenologische Forschung* 1, 1–323. Published simultaneously as Separatum by M. Niemeyer.

——— 1928. "Vorlesungen zur Phänomenologie des inneren Zeitbewusstseins," edited by M. Heidegger in *Jahrbuch für Philosophie und phänomenologische Forschung* 9, viii–x, 367–498. Published simultaneously by M. Niemeyer.

——— 1929. "Formale und transzendentale Logik. Versuch einer Kritik der logischen Vernunft," *Jahrbuch für Philosophie und phänomenologische Forschung* 10, v–viii, 1–298. Published simultaneously as Separatum.

——— 1931. *Méditations cartésiennes. Introduction à la phenoménologie.* Paris: Librairie Armand Colin.

——— 1936. "Die Krisis der europäischen Wissenschaften und die transzendentale Phänomenologie," *Philosophia* 1, 77–176.

——— 1939. *Erfahrung und Urteil: Untersuchungen zur Genealogie der Logik.* Edited by L. Landgrebe. Prague: Akademia Verlagsbuchhandlung.

——— 1950a. *Cartesianische Meditationen und Pariser Vorträge.* Husserliana I. 2nd ed. 1973. The Hague: Nijhoff.

——— 1950b. *Ideen zu einer reinen Phänomenologie und phänomenologischen Philosophie. Erstes Buch: Allgemeine Einführung in die reine Phänomenologie.* Husserliana III. Revised ed. in two parts 1976. The Hague: Nijhoff.

1952a. *Ideen zu einer reinen Phänomenologie und phänomenologischen Philosophie. Zweites Buch: Phänomenologische Untersuchungen zur Konstitution.* Husserliana IV. The Hague: Nijhoff.

1952b. *Ideen zu einer reinen Phänomenologie und phänomenologischen Philosophie. Drittes Buch: Die Phänomenologie und die Fundamente der Wissenschaften.* Husserliana V. The Hague: Nijhoff.

1954. *Die Krisis der europäischen Wissenschaften und die transzendentale Phänomenologie.* Husserliana VI. The Hague: Nijhoff.

1960. *Cartesian Meditations: An Introduction to Phenomenology.* Translated by D. Cairns. The Hague: Martinus Nijhoff.

1962. *Phänomenologische Psychologie. Vorlesungen Sommersemester 1925.* Husserliana IX. The Hague: Nijhoff.

1965. "Philosophy as Rigorous Science," Translated by Q. Lauer in *Phenomenology and the Crisis of Philosophy.* New York: Harper, 71–147.

1966. *Zur Phänomenologie des inneren Zeitbewusstseins (1893–1917).* Husserliana X. Revised ed. 1966. The Hague: Nijhoff.

1969. *Formal and Transcendental Logic.* Translated by D. Cairns. The Hague: Nijhoff.

1970a. *Philosophie der Arithmetik (1890–1901).* Husserliana XII. The Hague: Nijhoff.

1970b. *The Crisis of the European Sciences and Transcendental Phenomenology.* Translated by D. Carr. Evanston, Ill.: Northwestern University Press.

1973a. *Ding und Raum.* Husserliana XVI. The Hague: Nijhoff.

1973b. *Experience and Judgment.* Translated by J. Churchill and K. Ameriks. Evanston, Ill.: Northwestern University Press.

1973c. *Logical Investigations,* Vols. I and II. Translation of the 2nd ed. by J. N. Findlay. London: Routledge and Kegan Paul.

1974. *Formale und transzendentale Logik. Versuch einer Kritik der logischen Vernunft.* Husserliana XVII. The Hague: Nijhoff.

1975. *Logische Untersuchungen. Erster Band: Prolegomena zur reinen Logik.* Husserliana XVIII. The Hague: Nijhoff.

1977. *Phenomenological Psychology: Lectures, Summer Semester 1925.* Translated by J. Scanlon. The Hague: Nijhoff.

1979. *Aufsätze und Rezensionen (1890–1910).* Husserliana XXII. The Hague: Nijhoff.

1980. *Ideas Pertaining to a Pure Phenomenology and to a Phenomenological Philosophy: Third Book.* Translated by T. Klein and W. Pohl. Dordrecht: Kluwer.

1982. *Ideas Pertaining to a Pure Phenomenology and to a Phenomenological Philosophy: First Book.* Translated by F. Kersten. Dordrecht: Kluwer.

1983. *Studien zur Arithmetik und Geometrie (1886–1901).* Husserliana XXI. The Hague: Nijhoff.

1984a. *Logische Untersuchungen. Zweiter Band: Untersuchunger zur Phänomenologie und Theorie der Erkenntnis.* Husserliana XIX, 2 vols. The Hague: Nijhoff.

1984b. *Einleitung in die Logik und Erkenntnistheorie. Vorlesungen 1906/07.* Husserliana XXIV. The Hague: Nijhoff.

1989. *Ideas Pertaining to a Pure Phenomenology and to a Phenomenological Philosophy: Second Book.* Translated by R. Rojcewicz and A. Schuwer. Dordrecht: Kluwer.

1990. *On the Phenomenology of the Consciousness of Internal Time (1893–1917)*. Translated by J. Brough. Dordrecht: Kluwer.

1994. *Early Writings in the Philosophy of Mathematics and Logic*. Translated by D. Willard. Dordrecht: Kluwer.

1997. *Thing and Space: Lectures of 1907*. Translation by R. Rojcewicz. Dordrecht: Kluwer.

2003. *Philosophy of Arithmetic: Psychological and Logical Investigations*. Translated by D. Willard. Dordrecht: Kluwer.

Jackson, F., 1986. "What Mary Didn't Know," *Journal of Philosophy* 83, 291–295.

Kac, M., Rota, G., and Schwartz, J. T., 1986. *Discrete Thoughts: Essays on Mathematics, Science and Philosophy*. Boston: Birkhaüser.

Kant, I., 1973. *Critique of Pure Reason*. Translated by N. K. Smith. London: Macmillan. 1st ed. originally published in 1781; 2nd ed. originally published in 1787.

Kaufmann, F., 1930. *Das Unendliche in der Mathematik und seine Ausschaltung*. Wien: Deuticke. Translated by P. Foulkes as *The Infinite in Mathematics*. Dordrecht: Reidel, 1978.

Kim, J., 1981. "The Role of Perception in *A Priori* Knowledge," *Philosophical Studies* 40, 339–354.

Kleene, S., 1952. *Introduction to Metamathematics*. Amsterdam: North-Holland.

Klein, F., 1893. "Vergleichende Betrachtungen über neuere geometrische Forschungen," *Mathematische Annalen* 43, 63–100. This is a revised text of the "Erlanger Program" paper that was first published in 1872.

1948. *Elementary Mathematics from an Advanced Standpoint*. Part II. *Geometry*. Translated by E. Hedrick and C. Noble. New York: Dover.

Kockelmans, J., 1993. *Ideas for a Hermeneutical Phenomenology of the Natural Sciences*. Dordrecht: Kluwer.

Kockelmans, J., and Kisiel, T. (eds.), 1970. *Phenomenology and the Natural Sciences*. Evanston, Ill.: Northwestern University Press.

Kolmogorov, A., 1932. "Zur Deutung der Intuitionistischen Logik," *Mathematische Zeitschrift* 35, 58–65. Translated by P. Mancosu in P. Mancosu (ed.) 1998, 328–334.

Kreisel, G., 1967. "Informal Rigour and Completeness Proofs," in I. Lakatos (ed.), *Problems in the Philosophy of Mathematics*. Amsterdam: North-Holland, 138–86.

1971. "Observations on Popular Discussions of Foundations," in Scott (ed.) 1971, 189–198.

1980. "Kurt Gödel, 28 April 1906–14 January 1978," *Biographical Memoirs of Fellows of the Royal Society* 26, 148–224.

Leonardi, P., and Santambrogio, M. (eds.), 1995. *On Quine*. Cambridge: Cambridge University Press.

Lohmar, D., 1989. *Phänomenologie der Mathematik*. Dordrecht: Kluwer.

Lucas, J. R., 1961. "Minds, Machines and Gödel," *Philosophy* 36, 112–127.

MacLane, S., and Moerdijk, I., 1995. *Sheaves in Geometry and Logic: A First Introduction to Topos Theory*. Berlin: Springer Verlag.

Maddy, P., 1980. "Perception and Mathematical Intuition," *Philosophical Review* 89, 163–196.

1983. "Proper Classes," *Journal of Symbolic Logic* 48, 113–139.

1990. *Realism in Mathematics.* Oxford: Oxford University Press.

Mancosu, P., forthcoming. "'Das Abenteuer der Vernunft': O. Becker and D. Mahnke on the Phenomenological Foundations of the Exact Sciences." (ed.), 1998. *From Brouwer to Hilbert.* Oxford: Oxford University Press.

Mancosu, P., and Ryckman, T., 2002. "The Correspondence Between O. Becker and H. Weyl," *Philosophia Mathematica* 10, 130–202.

Forthcoming. "Geometry, Physics and Phenomenology: Four Letters of O. Becker to H. Weyl."

Martin-Löf, P., 1982. "Constructive Mathematics and Computer Programming," in H. Rose and J. Shepherdson (eds.), *Logic, Methodology and Philosophy of Science,* Vol. VI. Amsterdam: North-Holland, 73–118.

1983–84. "On the Meanings of the Logical Constants and the Justifications of the Logical Laws," in *Atti Degli Incontri di Logica Matematica,* Vol. 2. Siena: Università di Siena, 203–281.

1984. *Intuitionistic Type Theory,* Napoli: Bibliopolis.

1987. "Truth of a Proposition, Evidence of a Judgment, Validity of a Proof," *Synthese* 73, 407–420.

Merleau-Ponty, M., 1962. *Phenomenology of Perception.* Translated by C. Smith. London: Routledge & Kegan Paul. Original French ed. 1945.

Meserve, B., 1953. *Fundamental Concepts of Geometry.* New York: Dover.

Miller, I., 1984. *Husserl, Perception, and Temporal Awareness.* Cambridge, Mass.: MIT Press.

Miller, J. P., 1982. *Numbers in Presence and Absence.* Dordrecht: Nijhoff.

Mohanty, J. N., 1982. *Husserl and Frege.* Bloomington: Indiana University Press.

Natanson, M. (ed.), 1973. *Phenomenology and the Social Sciences,* 2 vols. Evanston, Ill.: Northwestern University Press.

Pagels, H., 1988. *The Dreams of Reason: The Computer and the Rise of the Sciences of Complexity.* New York: Simon & Schuster.

Paris, J., and Harrington, L., 1977. "A Mathematical Incompleteness in Peano Arithmetic," in Barwise (ed.) 1977, 1133–1142.

Parsons, C., 1965. "Frege's Theory of Number," in M. Black (ed.), *Philosophy in America.* London: Allen & Unwin. Reprinted in Parsons 1983a, 150–175.

1977. "What Is the Iterative Conception of Set?" in R. Butts and J. Hintikka (eds.), *Logic, Foundations of Mathematics, and Computability Theory.* Dordrecht: Reidel, 335–367. Reprinted in Parsons 1983a, 268–297.

1980. "Mathematical Intuition," *Proceedings of the Aristotelian Society,* 80, 145–168.

1983a. *Mathematics in Philosophy: Selected Essays.* Ithaca, N.Y.: Cornell University Press.

1983b. "Quine on the Philosophy of Mathematics" in Parsons 1983a, 176–205.

1986. "Intuition in Constructive Mathematics" in J. Butterfield (ed.), *Language, Mind and Logic.* Cambridge: Cambridge University Press, 211–229.

1990. "Introductory Note to 1944," in Feferman et al. (eds.) 1990, 102–118.

1992. "The Impredicativity of Induction," in M. Detlefsen (ed.), *Proof, Logic and Formalization.* London: Routledge, 139–161.

1995a. "Platonism and Mathematical Intuition in Kurt Gödel's Thought," *Bulletin of Symbolic Logic* 1, 44–74.

1995b. "Quine and Gödel on Analyticity," in Leonardi and Santambrogio (eds.) 1995, 297–313.

Penrose, R., 1989. *The Emperor's New Mind*. Oxford: Oxford University Press.

1994. *Shadows of the Mind: A Search for the Missing Science of Consciousness*. Oxford: Oxford University Press.

Poincaré, H., 1902. *La Science et l'hypothèse*. Paris: Ernest Flammarion. Translated by G. Holsted as *Science and Hypothesis*. New York: Dover, 1952.

1905. *La valeur de la science*. Paris: Ernest Flammarion. Translated by G. Halsted as *The Value of Science*. New York: Dover, 1958.

1908. *Science et méthode*. Paris: Ernest Flammarion. Translated by F. Maitland as *Science and Method*. New York: Dover, 1952.

1913. *Dernières pensées*. Paris: Ernest Flammarion. Translated by J. Bolduc as *Mathematics and Science: Last Essays*. New York: Dover, 1963.

Pollard, S., 1987. "Introduction to *The Continuum*," in Weyl 1987, xv–xxvi.

Prawitz, D., 1965. *Natural Deduction: A Proof Theoretical Study*. Stockholm: Almqvist and Wiksell.

1977. "Meaning and Proofs: On the Conflict Between Classical and Intuitionistic Logic," *Theoria* 4, 1–40.

1978. "Proof and the Meaning and Completeness of the Logical Constants," in J. Hintikka, I. Niiniluoto, and E. Saarinen et al. (eds.), *Essays on Mathematical and Philosophical Logic*. Dordrecht: Reidel, 25–40.

Quine, W. V., 1937. "New Foundations for Mathematical Logic," *American Mathematical Monthly* 44, 70–80. Reprinted with revisions in Quine 1953, 80–101.

1940. *Mathematical Logic*. New York: Norton. Revised 2nd ed., Cambridge, Mass.: Harvard University Press, 1951.

1951. "Two Dogmas of Empiricism," reprinted in Quine 1953, 20–46.

1953. *From a Logical Point of View*. Cambridge, Mass.: Harvard University Press.

1960. *Word and Object*. Cambridge, Mass.: MIT Press.

1966a. "Necessary Truth," in Quine 1966b, 68–76.

1966b. *The Ways of Paradox and Other Essays*. New York: Random House.

1969. *Set Theory and Its Logic*, 2nd ed. Cambridge, Mass.: Harvard University Press.

1970. *Philosophy of Logic*. Englewood Cliffs, N.J.: Prentice-Hall.

1974. *The Roots of Reference*. La Salle, Ill.: Open Court.

1984. "Review of Charles Parsons' *Mathematics in Philosophy*," *Journal of Philosophy* 81, 783–794.

1986. "Reply to Charles Parsons," in L. Hahn and P. A. Schilpp (eds.), *The Philosophy of W. V. Quine*. La Salle, Ill.: Open Court, 396–403.

1992. *Pursuit of Truth*, 2nd ed. Cambridge, Mass.: Harvard University Press.

1995. *From Stimulus to Science*. Cambridge, Mass.: Harvard University Press.

Resnik, M., 1979. "Frege as Idealist and Then Realist," *Inquiry* 22, 350–57.

1980. *Frege and the Philosophy of Mathematics*. Ithaca, N.Y.: Cornell University Press.

Riemann, B., 1959. *Über die Hypothesen, welche der Geometrie zugrunde leigen*. Darmstadt: Wissenschaftliche Buchgesellschaft. Reprint of the 1st ed., published in 1867 in *Abhandlungen der Kgl. Gesellschaft der Wissenschaften zu Göttingen*, Vol. 13.

Rosado-Haddock, G., 1987. "Husserl's Epistemology of Mathematics and the Foundation of Platonism in Mathematics," *Husserl Studies* 4, 81–102. Reprinted in Hill and Rosado-Haddock 2000, 221–239.

Saaty, T. L., and Weyl, F. J., 1969. *The Spirit and the Uses of the Mathematical Sciences.* New York: McGraw-Hill.

Schmit, R., 1981. *Husserls Philosophie der Mathematik. Platonistische und konstruktivistische Momente in Husserls Mathematikbegriff.* Bonn: Grundmann.

Scott, D. (ed.), 1971. *Axiomatic Set Theory,* Proceedings of Symposia in Pure Mathematics, Vol. 13, part 1. Providence, R. I.: American Mathematical Society.

Searle, J., 1980. "Minds, Brains and Programs," *Behavioral and Brain Sciences* 3, 417–457.

1983. *Intentionality.* Cambridge: Cambridge University Press.

Seebohm, T., Føllesdal, D., and Mohanty, J. N. (eds.), 1991. *Phenomenology and the Formal Sciences.* Dordrecht: Kluwer.

Sluga, H., 1980. *Gottlob Frege.* London: Routledge & Kegan Paul.

Smith, D., 1984. "Content and Context of Perception," *Synthese* 61, 61–87.

Smith, D., and McIntyre, R., 1982. *Husserl and Intentionality.* Dordrecht: Reidel.

Steiner, M., 1975. *Mathematical Knowledge.* Ithaca, N.Y.: Cornell University Press.

Ströker, E., 1979. *Lebenswelt und Wissenschaft in der Philosophie Edmund Husserls.* Frankfurt am Main: Vittorio Klostermann.

1987a. *The Husserlian Foundations of Science.* Lanham, Md.: Center for Advanced Research in Phenomenology/University Press of America.

1987b. *Investigations in the Philosophy of Space.* Translated by A. Mickunas. Athens, Ohio: Ohio University Press.

1988. "Husserl and the Philosophy of Science," *Journal of the British Society for Phenomenology* 19, 221–234.

Sundholm, G., 1983. "Constructions, Proofs and the Meanings of the Logical Constants," *Journal of Philosophical Logic* 12, 151–172.

1986. "Proof Theory and Meaning," in D. Gabbay and F. Guenthner (eds.), *Handbook of Philosophical Logic,* Vol. III. Dordrecht: Kluwer/Reidel, 471–506.

Tait, W. W., 1981. "Finitism," *Journal of Philosophy* 78, 524–546.

Tieszen, R., 1984. "Mathematical Intuition and Husserl's Phenomenology," *Noûs* 18, 395–421.

1989. *Mathematical Intuition: Phenomenology and Mathematical Knowledge.* Dordrecht: Kluwer.

1990. "Frege and Husserl on Number," *Ratio* 3, 150–164.

1991. "Review of D. Lohmar, *Phänomenologie der Mathematik,*" *Husserl Studies* 7, 199–205,

1994a. "What Is the Philosophical Basis of Intuitionistic Mathematics?", in D. Prawitz, B. Skyrms, and D. Westerståhl (eds.), *Logic, Methodology and Philosophy of Science IX.* Amsterdam: Elsevier, 579–594.

1994b. "Mathematical Realism and Gödel's Incompleteness Theorems," *Philosophia Mathematica* 2, 177–201.

1997a: "Review of *Kurt Gödel: Unpublished Philosophical Essays,*" Edited by F. Rodriguez-Consuegra, *Annals of Science* 54, 99–101.

1997b. "Science Within Reason: Is There a Crisis of the Modern Sciences?" in M. Otte and M. Panza (eds.), *Analysis and Synthesis in Mathematics.* Dordrecht: Kluwer, 243–259.

2004. "Husserl's Logic," in D. Gabbay and J. Woods (eds.), The *Handbook of the History of Logic: The Rise of Modern Logic: From Leibniz to Frege.* Amsterdam: Elsevier, 207–321.

Tonietti, T., 1988. "Four Letters of E. Husserl to H. Weyl and their Context," in Deppert (ed.) 1988, 343–384.

Torretti, R., 1978. *Philosophy of Geometry from Riemann to Poincaré.* Dordrecht: Reidel.

Tragesser, R., 1977. *Phenomenology and Logic.* Ithaca, N.Y.: Cornell University Press.

1984. *Husserl and Realism in Logic and Mathematics.* Cambridge: Cambridge University Press.

Troelstra, A., 1969. "Principles of Intuitionism," *Lecture Notes in Mathematics* 95. Berlin: Springer Verlag.

1982, "Logic in the Writings of Brouwer and Heyting," in *Atti del convegno internazionale di Storia della Logica. San Gimignano, dicembre 4–8, 1982.* Bologna: CLUEB, 193–210.

1990. "On the Early History of Intuitionistic Logic," in P. Pethov (ed.), in *Proceedings of the Heyting '88 Summer School and Conference on Mathematical Logic, September 13–23, 1988.* New York: Plenum, 3–17.

Troelstra, A., and van Dalen, D., 1988. *Constructivism in Mathematics,* 2 vols. Amsterdam: North-Holland.

van Atten, M., 1999. "Phenomenology of Choice Sequences," Proefschrift, Universiteit Utrecht.

van Atten, M., and Kennedy, J., 2003. "On the Philosophical Development of Kurt Gödel," *Bulletin of Symbolic Logic* 9, 425–476.

van Atten, M., van Dalen, D., and Tieszen, R., 2002. "Brouwer and Weyl: The Phenomenology and Mathematics of the Intuitive Continuum," *Philosophia Mathematica* 10, 203–226.

van Dalen, D. 1973. "Lectures on Intuitionism," *Lecture Notes in Mathematics* 337. Berlin: Springer Verlag.

1984. "Four Letters from Edmund Husserl to Hermann Weyl," *Husserl Studies* 1, 1–12.

1991. "Brouwer's Dogma of Languageless Mathematics and Its Role in His Writings," in E. Heijerman and H. Schmit (eds.), *Significs, Mathematics and Semiotics.* Amsterdam: Benjamins, 33–43.

1995. "Hermann Weyl's Intuitionistic Mathematics," *Bulletin of Symbolic Logic* 1, 145–169.

1999. *Mystic, Geometer and Intuitionist: The Life of L. E. J. Brouwer,* Vol. 1. Oxford: Oxford University Press.

van Heijenoort, J. (ed.), 1967. *From Frege to Gödel.* Cambridge, Mass.: Harvard University Press.

Wang, H., 1974. *From Mathematics to Philosophy.* New York: Humanities Press.

1978. "Kurt Gödel's Intellectual Development," *Mathematical Intelligencer* I, 182–184.

1981. "Some Facts About Kurt Gödel," *Journal of Symbolic Logic* 46, 653–659.

1986. *Beyond Analytic Philosophy: Doing Justice to What We Know.* Cambridge, Mass.: MIT Press.

1987. *Reflections on Kurt Gödel.* Cambridge, Mass.: MIT Press.

1996. *A Logical Journey: From Gödel to Philosophy.* Cambridge, Mass.: MIT Press.

Weyl, H., 1910. "Über die Definitionen der mathematischen Grundbegriffe," *Mathematische-naturwissenschaftliche Blätter* 7, 93–95, 109–113. Reprinted in Weyl 1968, I, 298–304.

1918a. *Das Kontinnum.* Leipzig: Veit.

1918b. *Raum, Zeit, Materie.* Berlin: Springer Verlag.

1919. "Der circulus vitiosus in der heutigen Begründung der Analysis," *Jahrebericht der Deutschen Mathematiker-Vereinigung* 28, 85–92. Reprinted in Weyl 1968, I, 43–50. English translation in Weyl 1987.

1921. "Über die neue Grundlagenkrise der Mathematik," *Mathematische Zeitschrift* 10, 39–79. Reprinted in Weyl 1968, II, 143–80. Translated by B. Müller in Mancosu (ed.) 1998, 86–118.

1925. "Die heutige Erkenntnislage in der Mathematik," *Symposion* 1, 1–23. Reprinted in Weyl 1968, II, 511–542. Translated by B. Müller in Mancosu (ed.) 1998, 123–142.

1926. *Philosophie der Mathematik und Naturwissenschaft.* München: Leibniz Verlag. Expanded and translated as *Philosophy of Mathematics and Natural Science.* Princeton, N.J.: Princeton University Press.

1928. "Diskussionsbemerkungen zu dem zweiten Hilbertschen Vortrag über die Grundlagen der Mathematik," *Abhandlungen aus dem mathematischen Seminar der Hamburgischen Universität* 6, 86–88. Translated by S. Bauer-Mengelberg and D. Føllesdal in van Heijenoort (ed.) 1967, 480–84.

1949. *Philosophy of Mathematics and Natural Science.* Princeton, N.J.: Princeton University Press.

1955. "Erkenntnis und Besinnung (Ein Lebensrückblick)," *Studia Philosophica* 15, 17–38. Reprinted in Weyl 1968, IV, 631–49. Translated in Saaty and Weyl 1969, 281–301.

1968. *Gesammelte Abhandlungen*, I–IV. Edited by K. Chandrasekharan. Berlin: Springer Verlag.

1985. *Space, Time, Matter.* New York: Dover.

1987. *The Continuum.* Translated by S. Pollard and T. Bole. Kirksville, Mo.: Thomas Jefferson Press.

Willard, D., 1981. "Introduction to 'On the Concept of Number'," in F. Elliston and P. McCormick (eds.), *Husserl: Shorter Works.* Notre Dame, Ind.: University of Notre Dame Press.

1984. *Logic and the Objectivity of Knowledge.* Athens, Ohio: Ohio University Press.

Wright, C., 1982. "Strict Finitism," *Synthese* 51, 203–282.

Zermelo, E., 1908. "Neuer Beweis für die Möglichkeit einer Wohlordnung," *Mathematische Annalen* 65, 107–128. Translated by S. Bauer-Mengelberg as "A New Proof of the Possibility of a Well-Ordering," in van Heijenoort (ed.) 1967, 183–198.

1930. "Über Grenzzahlen und Mengenbereiche," *Fundamenta Mathematicae* 16, 29–47.

Index

absolutism, 44, 332
abstraction, 34, 36–37, 78, 84, 325, 327, 328
  formal and material, 28
  see also founding/founded distinction
absurdity
  formal a priori, 27
  material a priori, 27
algorithmic methods, 32
alienation, 42
analysis, paradox of, 320
analytic a priori
  judgments, 28
  see also analyticity; analytic/synthetic distinction
Analytic and Continental philosophy, 1–2, 44–45, 66
analyticity, 185–190
  rational intuition and, 188–190
  see also Frege, G.; Gödel, K.; Poincaré, H.; Quine, W. V. O.
analytic/synthetic distinction, 318
  in Husserl, 27–28
antinomies, 131
  see also paradoxes
antireductionism
  in Husserl, 33
  see also reductionism
apophantic analytics, 28–29
Aristotelian realism, 127
Aristotle, 127
arithmetic
  in Husserl, 2–3, 22, 32
  Husserlian transcendental view of, 319, 325–336

  see also Frege, G.; natural numbers; Peano arithmetic (PA); Poincaré, H.; primitive recursive arithmetic (PRA); Weyl, H.
artificial intelligence, 288
associationism, 210
authentic/inauthentic distinction, 3, 42
axiom systems
  definite formal, 4, 11–12, 29

Becker, O., 8, 9, 14, 62, 83, 126, 237, 247, 254, 260, 268
Bell, D., 329
Benacerraf, P., 58, 64, 172
Bernays, P., 108, 153, 245
Beth models, 289, 293
BHK interpretation, 229, 232, 238, 242, 245
biologism, 23
Bishop, E., 228, 288
Bolzano, B., 24, 51, 154, 318
Boolos, G., 187
Brentano, F., 1, 22
Brouwer, L. E. J., 7, 8, 118, 227, 228, 231, 234, 235, 248, 249, 253, 254, 266, 278, 283, 296
  see also intuitionism

calculation
  in science, 36, 40, 41–42
Cantor, G., 9, 86, 126, 285
Carnap, R., 1, 9, 70, 178, 179, 182
Carnap's program, 136–138, 182
Cartesian dualism, 224
Cayley, A., 77